EUGENE G. ROCHOW
Harvard University

MODERN DESCRIPTIVE CHEMISTRY

SAUNDERS GOLDEN SUNBURST SERIES

W. B. SAUNDERS COMPANY Philadelphia / London / Toronto

W. B. Saunders Company: West Washington Square
Philadelphia, PA 19105

1 St. Anne's Road
Eastbourne, East Sussex BN21 3UN, England

1 Goldthorne Avenue
Toronto, Ontario M8Z 5T9, Canada

Cover is a photograph of the Sun, courtesy NASA

Modern Descriptive Chemistry ISBN 0-7216-7628-6

 Made in the United States of America. Press of W. B. Saunders Company. Library of Congress catalog card number 76-19610.

Last digit is the print number: 9 8 7 6 5 4 3

PREFACE

The current practice of presenting introductory college chemistry almost solely in terms of theoretical principles is being questioned in many of our universities. Too much theory by itself, it seems, has a hollow ring. Without having had experience in other areas of chemistry, students find it difficult to imagine how these abstract principles can help them meet the very real problems of a troubled world. Their professors, in turn, are finding that physicochemical principles do not stay in a student's memory, no matter how forcibly they may be underscored by multitudes of numerical problems, if they are not related to the real world by illustrations from descriptive and practical chemistry. Unfortunately, there are too few books on modern descriptive chemistry at the right level to help.

This little volume is intended to fill that void. It presents from the author's personal experience some interesting and important aspects of the chemical elements and their compounds which can serve to connect chemical principles with reality, in terms of our industrial culture and its current problems.

In order to accomplish its purpose, *Modern Descriptive Chemistry* should be used concurrently or sequentially with a proven textbook, such as Masterton and Slowinski's *Chemical Principles.* The successive chapters are intended as collateral reading to go with the textbook, starting quite early in the course, if desired, even before the student gains proficiency in writing chemical equations and working out their stoichiometry. Such early use makes it impossible for this book to be rigorous in its treatment of inorganic chemistry or to take up all the elements impartially. These matters are best left to a subsequent course in systematic inorganic chemistry, to which this could serve as an introduction.

The presentation begins with an introductory survey of the chemical elements and how they fit into the periodic system. It then discusses hydrogen and helium as a unique period of two elements, the second derived from the first in an all-important way. The lightest representative elements are then considered as a progression from base-formers to acid-formers, with appropriate attention given to important reactions and compounds. Chapter 4 takes up the second period of eight elements in the same way. The congeners of these representative *s* and *p* elements are then considered by groups. The remaining five chapters are devoted to the first, second, and third transition series, the lanthanide and actinide elements, and the gases of the zero group. Obviously some of the chapters could be omitted, but the instructor should consider very carefully the relative importances of the affected elements before making deletions. A glossary of unfamiliar terms is included in each chapter, followed by some thought questions and selected problems.

It is apparent that this book presents the author's own views and choices, but he is glad to acknowledge the invaluable help of thousands of students in

Chemistry 1 at Harvard, and of hundreds of practicing chemists elsewhere, all of whom have contributed to its content. He is grateful in particular to the General Electric Company and its Chemical Division for a dozen years of exposure to the economic and industrial realities of chemistry, including matters of safety and practicability which never come out in academic training. He appreciates also the 22 years of advice and stimulation he received from his Harvard colleagues, especially an early orientation by George Kistiakowsky. "It is far more important to be interesting than to be thorough or erudite," said George, "for if we have the interest of a beginning student, we can easily lead him to read more on his own or to take further courses that will be rigorous and complete. If we do not gain his interest, we simply give chemistry and this department a bad name." The direction taken by this book reflects that view.

A first draft of these chapters was reviewed by Professor William L. Masterton of the University of Connecticut, by Professor Robert J. Hanrahan of the University of Florida, by Professor Jon M. Bellama of the University of Maryland, by Professor W. R. Robinson of Purdue University, and by hundreds of students at the University of Florida at Gainesville, all of whom made valuable suggestions for revision. The author is grateful for all criticisms, and he appreciates the constant encouragement of John J. Vondeling, who as Chemistry and Physics Editor of W. B. Saunders Company has seen the project through to completion.

Captiva, Florida
March 1976

EUGENE G. ROCHOW

CONTENTS

LIST OF TABLES

INTRODUCTION: THE CHEMICAL ELEMENTS AND THEIR RELATIONSHIPS

1

This little book is about the chemical elements and the compounds they form with each other. It is also concerned with the energy lost or gained when such combinations of elements occur. The collective name for all such descriptions of substances and events is, of course, descriptive chemistry. If the description leaves out most of the compounds of carbon, considering that these are the special province of organic chemistry, then we call it descriptive inorganic chemistry, and that is what we shall consider here.

This book cannot take the place of a textbook of chemical principles, for it does not explain such theoretical matters as atomic structure, the nature of chemical bonding, the gas laws, thermochemistry, kinetics, or electrochemistry. These subjects necessarily are left to text and lectures while we explore the elements themselves, their similarities and differences, and the way they behave toward each other and toward common substances like water and oxygen. We shall seek (and find) relationships that organize the study and make our task easier.

We cannot take up all the elements impartially, for there are nearly a hundred of them in our surroundings, and it would take five times the space in this book just to explore their main interactions.* Instead we shall consider only those elements which are prominent in our material world and which have particular economic, biological, or political importance. This means, for example, that we shall consider carefully the element calcium, which lends

*Such a comprehensive exploration is offered in most colleges as a course in Systematic Inorganic Chemistry, which comes after the fundamental chemical and physical principles have been mastered.

strength to our bones, our teeth, and the walls of our buildings, but we shall ignore its neighbor scandium, which contributes nothing to our nutrition, clothing, or shelter. Similarly we shall consider tungsten but not its neighbor rhenium, and so on.

It is expected that these chapters will be assigned as collateral reading, either interspersed with the discussion of chemical principles as these are taken up in your textbook and lectures, or following them. For this reason the level of presentation herein must be kept low initially, and raised gradually as your acquaintance with theoretical principles widens. If your school background is particularly good, please do not be offended at the early simplicity. Further along you will find some concepts which are not encountered either in school or in ordinary college chemistry, but which arise naturally in the expansion of your view of the material universe according to the purpose of this book.

The style of these chapters is purposely narrative, to be read for pleasure and general education. It is important to look for trends and for points of general interest, and not to try to memorize all the facts and numbers. If you come across terms which are unfamiliar, look them up in the glossary which follows each chapter.

The first few questions at the end of the chapter will give you some idea of what is expected on an examination. These are followed by some "think" questions which require extended application of what you have learned. Lastly, the "think" questions are followed by some pertinent numerical problems to supplement those in your textbook.

HOW MANY ELEMENTS, AND WHY?

Before considering the chemical elements individually, we should like to know how large the task is and how we might go about it. People now accept the fact that air is mostly a mixture of the elements nitrogen and oxygen, that we ourselves are composed of carbon, hydrogen, oxygen, nitrogen, and many other elements, and that the earth is made up of silicon, oxygen, aluminum, iron, and many other elements in complicated combination. They also accept the fact that the various elements just named exist as *atoms*,* which are the smallest subdivisions of an element that still have the properties of that element and no other. People take comfort in the recent knowledge that the moon and the planets are made of the same elements as the earth, and that even the most distant stars contain no elements other than those already known to us on earth. In all, there are some 89 elements which we can find, and concentrate, and weigh; there are about 14 more which we can recognize in minute amounts by their properties, even though they are unstable and very rare. Of the 89 or so that can be seen and weighed, about half are important to us in the structure of the earth and in our food, clothing, shelter, and commerce.

Why should there not be thousands, or perhaps even millions, of chemical elements? Considering the enormity of the universe, and its complexity, how is it that everything is composed of a mere hundred or so elements? Are there internal details in atoms that serve to limit the possibilities? These are questions to be answered in the rest of this section. To find the answers, we shall have to start by going back about a hundred years.

---

*Students with inquiring minds are not required to take atoms (or any other concept of chemistry) purely on faith. They may read about the proofs that atoms exist by turning to H. M. Leicester, *The Historical Background of Chemistry*, New York: Dover (1971), or to the quantitative experiments outlined by L. K. Nash in *Stoichiometry*, Reading, Mass.: Addison-Wesley (1966).

John Dalton (1766–1844), a self-taught English school teacher, moved to Manchester in 1793 and devoted the rest of his life to scientific investigations. His presentation of atomic theory in the early part of the 19th Century served as the basis from which modern chemical theories have grown.

THE DAYS OF SOLID ATOMS. In 1808, when John Dalton proposed the atomic theory that is still with us, he considered that atoms were small, hard, indivisible spheres with some kind of hooks on them, and that the hooks engaged each other when elements combined to form compounds. Some elements had one hook, others two, some three, and so on. Here at once is a scheme of classifying the elements according to the number of hooks (or their valences, as we now would say). Dalton insisted that the atoms of one element are all alike, but differ from those of other elements. Furthermore, he said, the atoms of a particular element have distinctive properties, chief among them being a characteristic *atomic weight.*

During the nineteenth century, relative atomic weights were measured by chemical means, using hydrogen and then oxygen as standards. When about 50 atomic weights were known, some regularities began to appear. In 1829 J. W. Döbereiner pointed out his famous triads, groups of three elements with very similar chemical properties and a regular progression of atomic weights (Table 1–1). This was all very well, but no basic reason for the progression was forthcoming. Furthermore, many other groups of three similar elements did *not* show the same regularity of atomic weights. Consider the obviously similar metals copper, silver, and gold: the atomic weights* are: Cu, 63.6; Ag, 107.9; and Au, 197.2. The difference between the first two weights is 44.3, but the difference between the second two is 89.3. The regularity is missing. The same can be said for the trio zinc, cadmium, and mercury, for the trio nickel, palladium, and platinum, and so on.

In 1864 John A. R. Newlands pointed out that when the elements were arranged in order of increasing atomic weights, with hydrogen left out, each element resembled the eighth one after it. Thus sodium resembled lithium, and potassium in turn resembled sodium; magnesium resembled beryllium, and calcium in turn resembled magnesium.† Newlands called this the Law of

*Here and in Table 1–1 old atomic weights of the period (corrected to oxygen = 16.00) are used.

†A glance at the Periodic Table at the front of this book will show how this worked out. Ignore Group 0; it was not known then.

TABLE 1–1 THE DOBEREINER TRIADS

Element	*Atomic Weight*	*Difference*
Lithium, Li	7.0	
		16.0
Sodium, Na	23.0	
		16.1
Potassium, K	39.1	
Calcium, Ca	40.1	
		47.5
Strontium, Sr	87.6	
		49.8
Barium, Ba	137.4	
Sulfur, S	32.1	
		47.1
Selenium, Se	79.2	
		48.3
Tellurium, Te	127.5	
Chlorine, Cl	35.5	
		44.5
Bromine, Br	80.0	
		47.0
Iodine, I	127.0	

Octaves. It is said that people ridiculed the idea because it seemed to be based on the octaves of music, which was considered unscientific, but that is no reason at all. The musical scale itself is based on the theory of numbers, which is the basis of all science. The real reason why the Law of Octaves got nowhere is that it applies only to the elements up to calcium, and then fails. For example, consider the three closely similar elements called nitrogen, phosphorus, and arsenic: phosphorus is indeed the eighth element after nitrogen, but arsenic is *not* the eighth element after phosphorus. The very dissimilar metal chromium is. So, after the first score of elements, the "law" obviously does not hold any more, and therefore it is of no use in predicting anything.

It remained for Dmitri Ivanovich Mendeleev to bring order among the growing numbers of chemical elements in a way that holds true to this day. He de-

TABELLE II

REIHEN	GRUPPE I. — R^2O	GRUPPE II. — RO	GRUPPE III. — R^2O^3	GRUPPE IV. RH^4 RO^2	GRUPPE V. RH^3 R^2O^5	GRUPPE VI. RH^2 RO^3	GRUPPE VII. RH R^2O^7	GRUPPE VIII. — RO^4
1	H=1							
2	Li=7	Be=9,4	B=11	C=12	N=14	O=16	F=19	
3	Na=23	Mg=24	Al=27,3	Si=28	P=31	S=32	Cl=35,5	
4	K=39	Ca=40	—=44	Ti=48	V=51	Cr=52	Mn=55	Fe=56, Co=59, Ni=59, Cu=63.
5	(Cu=63)	Zn=65	—=68	—=72	As=75	Se=78	Br=80	
6	Rb=85	Sr=87	?Yt=88	Zr=90	Nb=94	Mo=96	—=100	Ru=104, Rh=104, Pd=106, Ag=108.
7	(Ag=108)	Cd=112	In=113	Sn=118	Sb=122	Te=125	J=127	— — — —
8	Cs=133	Ba=137	?Di=138	?Ce=140	—	—	—	
9	(—)	—	—	—	—	—	—	
10	—	—	?Er=178	?La=180	Ta=182	W=184	—	Os=195, Ir=197, Pt=198, Au=199.
11	(Au=199)	Hg=200	Tl=204	Pb=207	Bi=208	—	—	— — — —
12	—	—	—	Th=231	—	U=240	—	

Figure 1–1. Mendeleev's early periodic table, published in 1872. From *Annalen der Chemie und Pharmacie, VIII,* Supplementary Volume for 1872, p. 151.)

serves the credit for the success of the Periodic System because he alone stood by it through a quarter-century of controversy, defending its inconsistencies and turning them into tools for predicting new discoveries.* The blank spaces he left in his table in 1872 (Fig. 1–1) in order to make it come out even did indeed become the rightful places of new elements he predicted. The atomic weights he insisted must be wrong because they placed some elements in the wrong spaces in his table were almost all found to be wrong, indeed. At the time of his death in 1907 only three small inconsistencies remained: an apparent reversal of atomic weights for argon and potassium, another reversal for cobalt and nickel, and one for tellurium and iodine. It is a pity that Mendeleev did not live nine years more to see this little confusion cleared up by Moseley's brilliant demonstration that *atomic number* controls the ordering of the elements in the Periodic System, not the closely associated atomic weight.†

Just what is the Periodic Law, and why was it such a success? Mendeleev stated it this way: When the chemical elements are arranged in the increasing order of their atomic weights, they show a periodicity of chemical and physical properties. The reason, he stated, is that the properties of an element are a function of its atomic weight. That was a very radical idea for its time! What was there about the *weight* of an atom that enabled weight to control chemical properties? And if the properties are periodic, then something about the weight must also be periodic. Mendeleev could not answer all these new questions, but his Periodic Table led the way to the elucidation of the electronic structures of atoms of the elements, and those structures explain the properties very satisfactorily. The

*Lothar Meyer discovered the Periodic Law simultaneously and independently, but did not pursue it, defend it, or use it in the way Mendeleev did. Nevertheless, Mendeleev was very generous in sharing the praise with Meyer, and insisted on his being on the platform with him when the Davy Medal of the Royal Society was awarded.

†If you are interested in the three "inconsistencies," and understand the meaning of *isotopes* as explained in your textbook, then the "inconsistencies" are seen to be accidents which arise from the averaging of atomic weights of an element's isotopes according to their natural abundance. For example, it just so happens that potassium exists as 93.08% ^{39}K and 6.91% ^{41}K, so the weighted average atomic weight comes out to be just slightly higher than the atomic weight of argon.

structures also turn out to be periodic, as shown by many independent experiments. So the discussion of atomic structure in your textbook stems from the organization of the chemical elements according to the Periodic System, and this we owe to Mendeleev.

Suppose our ideas of atomic structure were to change radically as a result of future experiments. Would the new results invalidate the Periodic System, or relegate it to the history books? Not likely. The System has survived all the drastic changes in atomic theory that have occurred in the past century, and is likely to survive all future changes, for one good and simple reason: Mendeleev made no assumptions whatever about the structure of atoms, and the System depends on none. Therein lies its durability. It has been said that atoms could be made of green cheese, and the Periodic System would still be valid.

THE UPPER LIMIT OF ATOMIC WEIGHT. We return now to the question of how many chemical elements there are, or can be. Granted that there are only spaces for so many elements in the Periodic Table, couldn't it be expanded to accommodate still heavier elements? The answer to this takes us out of the world of solid atoms and into the world of subatomic particles and the nuclear atom. There seems to be a natural upper limit to the size and weight of atoms, for the atoms of all elements above lead and bismuth (see the Periodic Table in the front of this book) fall apart spontaneously. They simply disintegrate, at a rate which we cannot influence by temperature or pressure or any other means. Furthermore, the falling-apart process does not involve the separation of a nucleus from its electrons, or anything like that, but begins *in* the nucleus instead. It is the nuclei that become unstable in the heavy elements, and they disintegrate by shooting out radiations and rearranging themselves into new nuclei surrounded by a different number of electrons. That is, the heaviest elements transmute themselves spontaneously and at an inevitable rate to lighter elements, usually in small regular steps but occasionally by spontaneous fission into two middle-weight elements.

To understand the reason for this we must consider the nuclei of atoms. We cannot consider how many elements there are, or how much of each one, without bringing in something about nuclear structure, for it is the structure of its nucleus that determines whether an atom is stable or unstable, rare or abundant. It is even the nuclear structure that determines chemical behavior, for the nuclear structure dictates the number of electrons and hence their arrangement. So how can we proceed?

As you know, nuclei are not hard, massive, little spheres any more than atoms are; they are composed of nuclear particles, especially protons, neutrons, and mesons. Nuclei are not random jumbles of these particles, but instead have a definite nuclear structure.* The structure of a nucleus determines its stability or instability, and hence the chances of its being formed in the first place (in the stars) and its chances of survival through billions of years.

We begin by inquiring about the relative numbers of neutrons and protons in nuclei of the various elements, with the aim of learning which nuclear compositions and structures are most stable among the common elements, and what the trends are. Some information can be gained simply by comparing atomic weights with atomic numbers, but very soon we need to look at the separate *isotopes* of each element. Fortunately, the subject of isotopes is explained early in most textbooks: atoms that have the same atomic number (the number of positive charges on the nucleus) do not necessarily have the same atomic weight, because some of the nuclei may contain more neutrons than others do. The re-

*See, for example, R. T. Overton, *Basic Concepts of Nuclear Chemistry*, New York: Reinhold (1963), or G. Choppin, *Nuclei and Radioactivity*, New York: W. A. Benjamin Co. (1964).

sulting different atoms of the same element, differing only in nuclear mass, are called isotopes. There is nothing in the Periodic Table about isotopes because they were unknown when it was proposed; to learn something about them we must turn to a table of isotopes* or to an isotope chart.† What we find there can be summarized in a few conclusions which can be verified by anyone:

1. Of all the possible combinations of neutrons and protons, only a surprisingly few are stable. Consider an isotope chart, which is just a plot of the number of protons (0 to 103) against the number of neutrons (0 to 158). On the chart there are $103 \times 158 = 16{,}274$ spaces for possible combinations, and only 280 of these are found to be occupied by stable isotopes.

2. Among the light elements, the stable isotopes usually contain equal numbers of neutrons and protons. There are some exceptions, but this generalization holds all the way up to calcium, 96.97% of which is $^{40}_{20}Ca$ (this symbol means that the mass number is 40 and the atomic number is 20; hence there are $40 - 20 = 20$ neutrons).

3. Above calcium all the stable isotopes contain more neutrons than protons. The neutron surplus increases steadily, so that the principal isotope of iron has 26 protons and 30 neutrons, iodine has 53 protons and 74 neutrons, and gold has 79 protons and 118 neutrons.

4. Above bismuth, which has 83 protons and 126 neutrons, *no* isotope of any element is stable. This means that no amount of extra neutron glue is sufficient to hold the very heavy nuclei together; all come apart at characteristic rates. Moreover, the rate of disintegration increases as the nuclei become still heavier: for uranium (99.27% $^{238}_{92}U$), half disintegrates in 10^9 years, but for the heaviest element in the list (lawrencium, $^{257}_{103}Lw$) half disintegrates in 8 seconds.

These observations define a trend that establishes a clear cutoff point: at an atomic weight of 209, at which point there is a 52% excess of neutrons over protons, the possibilities for stable nuclear bonding are exhausted. Above that weight, no matter what the proton-neutron ratio, nuclear rearrangement takes place spontaneously. This sets an upper limit, and we need not worry about thousands of elements.

Of course a few elements heavier than bismuth, such as radium, thorium, uranium, and plutonium, are extremely important and useful despite their instability, and they will receive proportionate attention in later chapters. Above uranium all the elements have to be made artificially by high-energy processes, at rapidly rising cost and difficulty.

HOW MUCH OF EACH ELEMENT, AND WHY?

Finding ourselves limited to 89 elements we can gather in practical amounts, of which about half are of biological or utilitarian importance, we should like to know whether they all occur in equal weights, or equal numbers of atoms. To put it another way, were they all made impartially in the stars, and have they come down to us that way? If not, what are the proportions in star-material and in earth-material? What are the probable reasons for these proportions? Is there any sense or regularity?

STABILITY VS. ABUNDANCE. A close look at any isotope chart or table shows that there are quite a few elements *lighter* than bismuth which are radioactive

---

*Such a table may be found in the *Handbook of Chemistry and Physics*, published by the Chemical Rubber Co., Cleveland, Ohio.

†This chart is furnished free on request by the Educational Relations Dept., General Electric Co., Schenectady, N.Y. 12305.

in their natural state. In all cases, however, the unstable isotope or isotopes which impart that radioactivity constitute only a tiny proportion of that element as we find it in nature. The reason is simple: the earth is several billion years old, and any unstable species may be expected to have made itself rare by now. For example, natural potassium is radioactive because it still contains 0.012% of the unstable isotope of mass 40 ($^{40}_{19}K$, with 19 protons and 21 neutrons). Since potassium is a major constituent of plants and animals as well as rocks, the amount of radiation produced by it is appreciable even though so little of the active isotope remains. Similarly vanadium, which is an essential trace element in plants, is radioactive because of the 0.24% of $^{50}_{23}V$ it contains. Rubidium, indium, tellurium, and a dozen other middleweight elements also are radioactive, but in each case the activity is due to a very small surviving proportion of an unstable isotope.

Projecting backward, the earth's crust must have been far more radioactive billions of years ago, before so much of the active material decayed. Going back still further, before the earth was formed, the implacable decay processes must have gone on for more billions of years. What we have left, for the most part, are *the most stable nuclear structures*, the ones which have survived the passage of eons upon eons. We should like to know what features of nuclear structure enable them to survive, so that we may understand any underlying regularity.

ABUNDANCES OF ELEMENTS IN THE STARS. The light we receive from a star was radiated by very hot ("excited") atoms in the white-hot gas on the surface of the star, and the energy in that light beam came from the recapture of one or more electrons by the ions in that gas. The particular amount of energy evolved

TABLE 1–2 COSMIC ABUNDANCES OF THE ELEMENTS

Atomic Number, Z	*Symbol*	*Atoms per 10,000 atoms of Si*
1	H	3.5×10^8
2	He	3.5×10^7
6	C	8×10^4
7	N	16×10^4
8	O	21×10^4
9	F	90
10	Ne	10,000
11	Na	462
12	Mg	8,870
13	Al	882
14	Si	10,000
15	P	130
16	S	3,500
17	Cl	190
18	Ar	130
19	K	69
20	Ca	670
21	Sc	0.18
22	Ti	26
23	V	2.5
24	Cr	95
25	Mn	77
26	Fe	18,300
27	Co	99
28	Ni	1,340
29	Cu	4.6
30	Zn	1.6

by each recapture is definite and distinctive; it is characteristic only of the element that produced it. This means that starlight is a collection of signals from individual elements and, if we process the signals by separating the light into its components by means of a spectrograph,* we can find out which elements are emitting the light. Furthermore, from relative intensities we can determine what proportion of each element is present. Plate 1 shows the characteristic colors of light emitted by several common elements. Each colored line is an image of the slit which admits light to the spectrograph, and so there is a line for each color (each frequency) emitted by that element. By comparing the intensities of the lines, and averaging the results from a great many stars, we arrive at the cosmic abundances of the elements shown in Table 1–2.

We find the same elements as on earth, but with a vast preponderance of hydrogen and helium. The stars seem to be mostly hydrogen, present in the form of protons and electrons because of the intense heat. There also are neutrons, and from these three kinds of elementary particles helium is continually being formed with the evolution of energy. The significance and the mechanism of this transformation on an enormous scale will be taken up in Chapter 2, which is devoted to a discussion of hydrogen and helium. Large amounts of carbon, nitrogen, oxygen, neon, magnesium, silicon, and iron also are formed, together with smaller amounts of the other elements listed.

A closer look at the relative abundances of all the elements after hydrogen in Table 1–2 shows that nine of the ten most abundant have *even atomic numbers*. Among all 29 elements the abundances alternate high and low as the atomic numbers alternate even and odd. Furthermore, there seems to be something special about neon, silicon, and iron that makes them more plentiful than the rest of the elements which come after the first four. This "something special" is filled energy levels of neutrons and protons in the nuclear structures of these elements.

Such filled nuclear shells confer nuclear stability in much the same way as filled shells of electrons confer chemical stability and unreactivity on the inert gases. Filled nuclear shells are often the end result of nuclear reaction, just as filled electron shells† or "completed octets" are so often the end result of chemical reactions. The reason is the same: those configurations correspond to minimum residual potential energy. The filled atomic shells contain 2, 8, 18, or 32 electrons, corresponding to the lengths of the periods in the Periodic System, but filled nuclear shells contain 2, 8, 20, 28, 50, or 82 nucleons (neutrons or protons). These "magic numbers" correspond to helium, oxygen, calcium, and nickel in Table 1–2. There are also filled subshells at 6, 14, 16, and 26 nucleons, corresponding to carbon, silicon, sulfur, and iron.

ABUNDANCES OF ELEMENTS ON EARTH. Having analyzed the stars by spectroscopy, we can do something similar with the rocks and minerals on earth: we can take samples, heat them to a very high temperature in an electric arc in the laboratory, and examine the light they emit by putting it through a spectrograph and photographing the colored lines. By comparing the lines in the spectrum with those obtained from pure elements in known amounts, we determine which elements and what proportion of each are present. Other methods of analysis are used, too, and when we average all the results from all the speci-

---

*This can be a glass-prism spectrograph, or more likely a grating spectrograph (see any textbook of physics), attached to a telescope. The light is so weak that the spectrum must be photographed.

†For "shells" read "completely filled orbitals for that energy level," if you have advanced to the college-level vocabulary.

TABLE 1–3 TERRESTRIAL ABUNDANCES OF THE ELEMENTS (in parts per million)*

Atomic No.	Symbol	Ppm	Atomic No.	Symbol	Ppm
1	H	8,700	40	Zr	220
2	He	0.003	41	Nb	24
3	Li	65	42	Mo	8
4	Be	6	46	Pd	0.010
5	B	3	47	Ag	0.10
6	C	800	48	Cd	0.15
7	N	300	49	In	0.1
8	O	495,000	50	Sn	0
9	F	270	51	Sb	1
11	Na	26,000	53	I	0.3
12	Mg	19,000	55	Cs	7
13	Al	75,000	56	Ba	250
14	Si	257,000	57	La	18.3
15	P	1,200	58	Ce	46.1
16	S	600	59	Pr	5.53
17	Cl	1,900	60	Nd	23.6
18	Ar	400	62	Sm	6.47
19	K	24,000	63	Eu	1.06
20	Ca	34,000	64	Gd	6.36
21	Sc	5	65	Tb	0.91
22	Ti	5,800	66	Dy	4.47
23	V	150	67	Ho	1.15
24	Cr	200	68	Er	2.47
25	Mn	1,000	69	Tm	0.20
26	Fe	47,000	70	Yb	2.66
27	Co	23	71	Lu	0.75
28	Ni	80	72	Hf	4.5
29	Cu	70	73	Ta	2.1
30	Zn	132	74	W	34
31	Ga	15	78	Pt	0.005
32	Ge	7	79	Au	0.005
33	As	5	80	Hg	0.30
34	Se	0.09	81	Tl	1.8
35	Br	1.62	82	Pb	16
37	Rb	310	83	Bi	0.2
38	Sr	300	90	Th	11.5
39	Y	28.1	92	U	4

*Elements present to the extent of less than 0.003 ppm are omitted.

mens we can obtain from the mountains and fields and geology museums, we arrive at the terrestrial abundances given in Table 1–3. These figures apply only to the parts of the earth's crust which we can sample and be sure about; they do not take into account the material of the core (believed to be metallic iron and nickel) or the material of the fluid mantle that surrounds the core (believed to be iron, calcium, and magnesium silicates). No authentic sample of either has ever been obtained.

Table 1–3 seems long and complicated, even though 30 elements have been left out to simplify it (all elements present in amounts less than 0.003 parts per million, ppm, have been omitted). However, some aspects fairly jump out at you. Silicon and oxygen together make up 752,000 parts per million, or 75.2% of the earth's crust. The first ten elements, in order of their abundance, are oxygen, silicon, aluminum, iron, calcium, sodium, potassium, magnesium, hydrogen, and titanium. Together these elements make up 99.2% of the solid earth as we know it. All are light elements, with atomic numbers from 1 to 26. The average atomic weight (a weighted average taking into account the proportions) is 22.38. Furthermore, all the heaviest elements listed in Table 1–3, from tungsten through uranium, together constitute less than one thousandth of one per cent of the crust of the earth.

Considering all the evidence, we may list our conclusions:

1. The light elements are far more abundant than the middleweight and heavy ones, indicating that there was a genesis of elements in the stars that began with protons, neutrons, and electrons. When the process of generation reached those elements with exceptionally stable nuclei, corresponding to filled energy levels as indicated above, it usually stopped right there. This left us mostly oxygen, silicon, aluminum, iron, calcium, sodium, potassium, magnesium, hydrogen, and titanium, composing over 99% of the earth's crust in the form of solid compounds of these light elements.

2. Of these ten most abundant elements, seven have even atomic numbers and three do not. Reference to an isotope chart shows that the three with odd atomic numbers (Al, Na, and K) all have principal isotopes with an even number of neutrons. In fact, 100% of natural sodium is $^{23}_{11}Na$, with 12 neutrons, 100% of natural aluminum is $^{27}_{13}Al$, with 14 neutrons, and 93% of natural potassium is $^{39}_{19}K$, with 20 neutrons. These numbers of neutrons correspond to the previously-designated filled energy levels for nucleons. So the principal abundances fit within the theory of nuclear structure which calls for paired neutrons and protons within a system of energy levels, and are not haphazard or arbitrary.

3. Hydrogen and helium are notably low on the terrestrial list, compared with their cosmic abundances. Helium is too light to be held by the earth's gravitational field, and was lost. The same is true of gaseous elementary hydrogen; only combined hydrogen, present as water or hydrated minerals or as petroleum and coal, remains with us. Since helium forms no compounds, it left us long ago, except for small amounts trapped in natural gas and radioactive minerals.

4. Of all the remaining elements, which constitute 0.8% of the earth's crust, those of even atomic number are more plentiful than those of odd atomic number. The general abundance declines with rising atomic weight, but among the heavy elements notable exceptions occur at tungsten, lead, and thorium. These are all even-even elements (even numbers of protons and neutrons) particularly favored by the scheme of energy levels in nuclei.

5. With the exception of Jupiter, Saturn, Uranus, and Neptune, which are made up mostly of hydrogen and helium, the rest of the solar system appears to be made of much the same stuff as the earth. The only extraterrestrial body we can be absolutely sure of is the moon; the Apollo astronauts brought back samples from many areas of the moon, and their compositions are not much different from that of the earth's crust, except that there is more titanium and iron.

These conclusions, and the facts from which they were drawn, are very significant in two ways: they contribute immensely to our understanding of the physical world around us and the rest of the universe beyond that, and they necessarily influence our utilization of natural resources to satisfy human needs. Whether it be choice of a structural metal, or a new alloying ingredient, or development of new building material, or a new battery for electric cars, the first considerations must be the abundance and availability of the chemical elements involved. Application of just three simple rules (light elements are more plentiful than heavier ones, even-numbered elements are more plentiful than odd-numbered ones, and certain elements are favored by the "magic numbers" corresponding to filled nuclear shells) will enable anyone to make sensible choices. Conversely, schemes such as the sweeping substitution of boron hydrides for petroleum fuels, or utilization of a cadmium-iodine battery for automotive propulsion, betray a lamentable ignorance of elemental chemistry.

RELATIONSHIPS OF THE ELEMENTS

We have seen that the chemical elements may be ordered according to atomic number, or atomic weight, or abundance, or similarities of behavior, or any of several other ways, but the only completely comprehensive scheme that embraces all the elements in all their relationships is the Periodic System. Hence the rest of this book will be organized according to the various sequences in the periodic table. At the same time, it should be emphasized here and now that the different kinds of matter do *not* represent a continuum, with the properties of one element grading imperceptibly into those of the next. Far from it. The elements are all chemical individuals, with distinctive properties that set them apart from all others. Each element has a characteristic set of responses that amounts to a chemical personality, so to speak. Each will repay study by revealing just how active or passive, friendly or unfriendly, and helpful or hostile it really is. Only by learning about them in this way can an educated person be on familiar and comfortable terms with the physical world around us.

ARRANGEMENT OF THE PERIODIC TABLE. The table as we shall use it arranges the elements according to successive periods of varying length. The periods are the horizontal sequences, and there are seven of them, numbered at the left as shown in Figure 1–2 (which is the same as the large Periodic Table on the inside front cover except that it is not cluttered with atomic weights). The first period contains only two elements, hydrogen and helium; the second and third periods each contain eight elements, the fourth and fifth periods each contain 18 elements, and the sixth period contains 32 elements. Presumably the seventh period

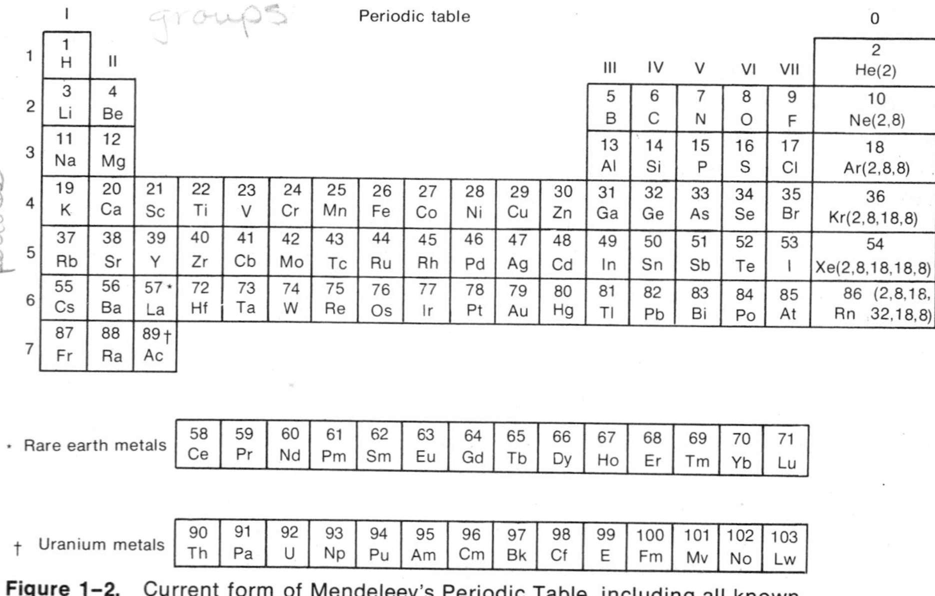

Periodic table

	I	II											III	IV	V	VI	VII	0
1	1 H																	2 He(2)
2	3 Li	4 Be											5 B	6 C	7 N	8 O	9 F	10 Ne(2,8)
3	11 Na	12 Mg											13 Al	14 Si	15 P	16 S	17 Cl	18 Ar(2,8,8)
4	19 K	20 Ca	21 Sc	22 Ti	23 V	24 Cr	25 Mn	26 Fe	27 Co	28 Ni	29 Cu	30 Zn	31 Ga	32 Ge	33 As	34 Se	35 Br	36 Kr(2,8,18,8)
5	37 Rb	38 Sr	39 Y	40 Zr	41 Cb	42 Mo	43 Tc	44 Ru	45 Rh	46 Pd	47 Ag	48 Cd	49 In	50 Sn	51 Sb	52 Te	53 I	54 Xe(2,8,18,18,8)
6	55 Cs	56 Ba	57• La	72 Hf	73 Ta	74 W	75 Re	76 Os	77 Ir	78 Pt	79 Au	80 Hg	81 Tl	82 Pb	83 Bi	84 Po	85 At	86 Rn (2,8,18,32,18,8)
7	87 Fr	88 Ra	89† Ac															

• Rare earth metals	58 Ce	59 Pr	60 Nd	61 Pm	62 Sm	63 Eu	64 Gd	65 Tb	66 Dy	67 Ho	68 Er	69 Tm	70 Yb	71 Lu
† Uranium metals	90 Th	91 Pa	92 U	93 Np	94 Pu	95 Am	96 Cm	97 Bk	98 Cf	99 E	100 Fm	101 Mv	102 No	103 Lw

Figure 1–2. Current form of Mendeleev's Periodic Table, including all known chemical elements. The Roman numerals across the top label the *groups,* and the Arabic numerals at far left label the *periods.*

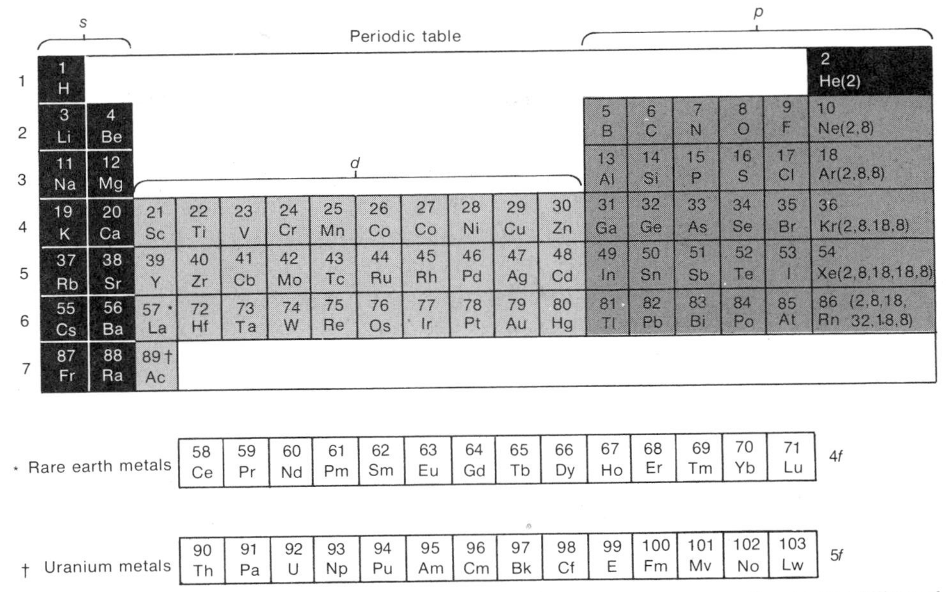

Figure 1–3. The Periodic Table, showing areas corresponding to the filling of *s, p, d,* and *f* orbitals. (From E. G. Rochow, G. Fleck, and T. H. Blackburn, *Chemistry: Molecules that Matter,* Holt, Rinehart & Winston, New York, 1974, p. 43. By permission.)

will also contain 32 elements, if it is ever completed by synthesizing the remaining missing elements.

The numbers 2, 8, 18, and 32 may seem to form no regular sequence, or sound that way to ears accustomed to the decimal system, but actually the sequence 2, 8, 18, and 32 is very regular. It is the sequence $2n^2$, where n is simply a cardinal number, 1, 2, 3, or 4.

The vertical columns of Figure 1–2 are called *groups*, and are numbered from one to seven in Roman numerals at the top. The zero group at the right comprises the monatomic gases which were unknown in 1872 but were added when they were discovered at the turn of the century (see Chapter 10). They could just as well be placed at the extreme left of the table.

Although the Periodic System does not embrace or depend upon any particular theory of atomic structure, it certainly must agree with any such theory, and vice versa. In terms of the modern view of atomic structure, various portions of the periodic table may be blocked out in a way that corresponds to filling of the *s, p, d* and *f* orbitals* within any one energy level. Figure 1–3 shows how different sections of the table may so be designated. The first period and Groups I and II contain elements which have electrons only in *s* orbitals, and which consequently have no *p, d,* or *f* electrons available for chemical bonding. Their properties are distinctive, as we shall see in the succeeding chapters. The block in different shading on the right, comprising Groups III, IV, V, VI, VII, and 0, contains elements which have one or more electrons in their *p* orbitals, and hence can supply *p* electrons for chemical bonding of a distinctive and directional character. The middle block of elements, in still another shading, contains those elements in which *d* orbitals are being filled with electrons. These are called *transition elements* because they come between the *s* and *p* blocks. All are capable of supplying one or more *d* electrons for the characteristic type of chemical bonding which we associate especially with coordination compounds

*See glossary at end of chapter.

(see Chapters 6 and 7). The last two blocks of elements, shown in white at the bottom of Figure 1–3, contain those elements in which *f* orbitals are being filled. These elements fit into Period 6 and Period 7 in the places shown by asterisks; the only reason why they are not inserted there is that to do so would give an unwieldy table 32 spaces long. The *f* elements behave in their own special way (see Chapters 8 and 9), and are all very similar in chemical properties.

METALS VS. NONMETALS. The Periodic Table can also be divided in a way that distinguishes the metallic elements from the nonmetallic ones. Figure 1–4 shows the regions where all the elements are metals (note iron, nickel, chromium, aluminum, copper, silver, gold, etc.), where all are nonmetals (note oxygen, chlorine, sulfur, iodine, etc.), and where they have an in-between character and are called *metalloids* (the elements used in transistors come in this category). It is seen that about three fourths of all the elements are metals, and only 16 elements are distinctly nonmetallic in character. Some of the eight elements designated as metalloids may be disputed, but the existence of a buffer zone of metalloids between the metals and nonmetals is conceded by all.

The metals are seen to be concentrated in the left, center, and bottom of the Periodic Table, and to extend all the way over to Group V in Period 7. This is a consequence of the nature of the metallic bond between atoms, a type of bonding that favors elements with few electrons in comparison to the number of empty orbitals able to receive them. Chemically, metals tend to *lose* electrons as they combine with nonmetals, and their oxides often are alkaline. All of the Group I and Group II elements are metals, and so are all the transition elements, including the 4*f* and 5*f* sequences shown at the bottom of Figure 1–3. Nonmetals tend to *gain* electrons during chemical combination, and their oxides are acid-formers. All the nonmetals are concentrated in the upper right corner of the table, indicating that their orbitals are almost filled with electrons to begin with. All nonmetals except the zero-group gases actively seek electrons to fill the remaining empty places. Much more about this will develop in the succeeding chapters.

A QUICK LOOK AT GROUP I. All the useful elements will be considered individually in later chapters, but at present it would be well to see just how close the family relationships are in a few groups. Let us choose three groups: Group I at the left of the table, Group VII at the right, and Group IV in between. Groups I and VII contain the first and last of Döbereiner's triads listed in Table 1–1, but not all of the elements in Group IV were known in Döbereiner's time, so they were not included in that table.

Periodic table

1	1 H																	2 He
2	3 Li	4 Be											5 B	6 C	7 N	8 O	9 F	10 Ne
3	11 Na	12 Mg											13 Al	14 Si	15 P	16 S	17 Cl	18 At
4	19 K	20 Ca	21 Sc	22 Ti	23 V	24 Cr	25 Mn	26 Fe	27 Co	28 Ni	29 Cu	30 Zn	31 Ga	32 Ge	33 As	34 Se	35 Br	36 Kr
5	37 Rb	38 Sr	39 Y	40 Zr	41 Nb	42 Mo	43 Tc	44 Ru	45 Rh	46 Pd	47 Ag	48 Cd	49 In	50 Sn	51 Sb	52 Te	53 I	54 Xe
6	55 Cs	56 Ba	57* La	72 Hf	73 Ta	74 W	75 Re	76 Os	77 Ir	78 Pt	79 Au	80 Hg	81 Tl	82 Pb	83 Bi	84 Po	85 At	86 Rn
7	87 Fr	88 Ra	89† Ac															

Figure 1–4. Abbreviated form of the Periodic Table showing location of metals (lightly shaded squares), metalloids (heavily shaded squares), and nonmetals (unshaded squares). (From E. G. Rochow, G. Fleck, and T. H. Blackburn, *op. cit.*, p. 44. By permission.)

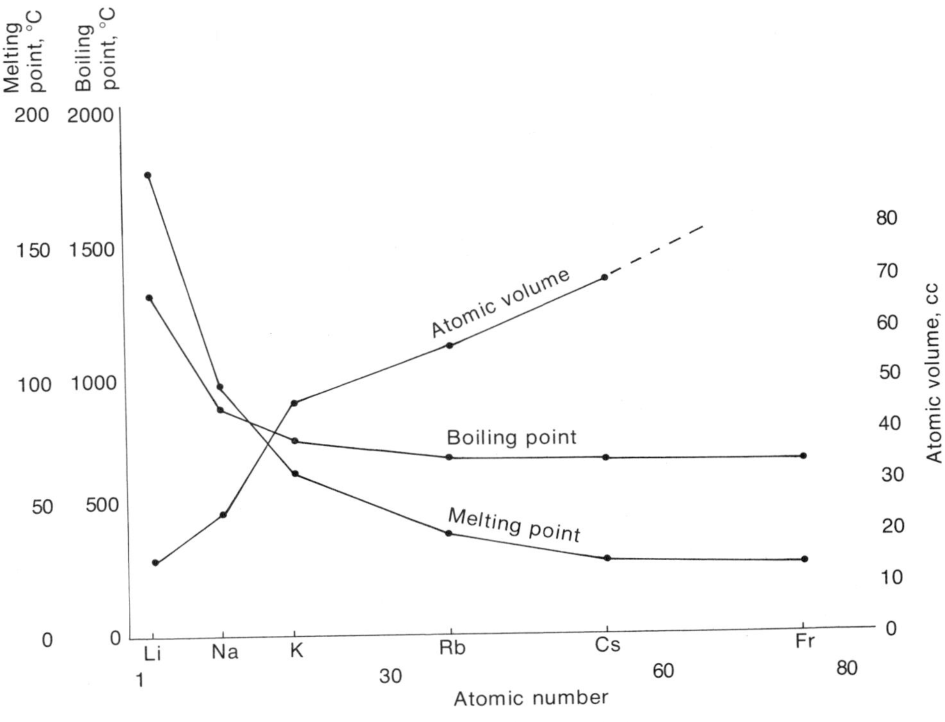

Figure 1–5. Properties of Group I elements.

Figure 1–5 shows how some physical properties of the elements in Group I vary. The melting points of the elements are seen to decline rapidly at first, then more slowly; all are surprisingly low for metals, and four of the five are below 100° C. The scale for boiling points is ten times that for melting points, but still the boiling points of the Group I metals are far below those of most common metals such as iron (bp 2750° C) and copper (bp 2567° C).

The Group I elements are electron-suppliers, and are very reactive for this reason. They react rapidly with air, violently with water, and even attack glass. In each case they give up electrons which "reduce" the substance attacked: water is reduced and one of its elements, hydrogen, is liberated; glass is reduced and one of its elements, silicon, is liberated. The more energetically a reaction partner seizes the released electrons, the greater the total energy liberated in the reaction. The Group I elements, combining with electron-hungry elements like oxygen and chlorine, give up some of the largest amounts of energy, per mole and per gram, in all of chemistry.

The only curve in Figure 1–5 that trends upward is that for atomic volume, which is the volume in cubic centimeters occupied by 1 mole (or gram atomic weight) of the element in the massive solid state. Since the atoms of lithium are smaller than those of sodium or potassium, and are more tightly bound, we expect that the volume occupied by 1 mole will be less. Compared with the atomic volumes of other elements, though, the atomic volumes of all the Group I elements are very large. They stand at the peaks of the curve for atomic volumes of all the elements versus their atomic numbers (Fig. 1–6). The extreme ups and downs of atomic volume reflect the periodic variations of properties among all

Dmitri Mendeleev: (1834–1907) Born in Siberia, Mendeleev rose to Professor of Chemistry at St. Petersburg (now Leningrad) and then to director of the Russian Bureau of Weights and Measures. Although a prolific writer, a versatile chemist and inventor, and a popular teacher, the fame of this brilliant scientist rests on his discovery of the periodic law.

the elements, with the peaks becoming more widely separated as the periods become longer and longer.

A QUICK LOOK AT GROUP VII. The same three properties depicted in Figure 1–5 for the elements of Group I are shown in Figure 1–7 for the elements of Group VII, a group of nonmetals usually referred to as the halogen (salt-forming) elements. The range of the scales has had to be changed because the first two elements of Group VII (fluorine and chlorine) are gases at room temperature, the third (bromine) is a liquid, the fourth is an easily vaporized solid (iodine), and the fifth member (astatine) is a high-melting solid. The scales for the melting and boiling points are the same.

Notice that the atomic volumes of these halogen elements in the solid state are only one third as large as those of the Group I metals. The halogen elements hold on to their electrons more tightly, resulting in atomic structures that are very compact, while the Group I metals have a single far-flung electron that uses up space.

The halogen elements hang on to their electrons strongly, so much more energy is required to knock one loose. It is much more likely that a halogen element will *gain* an electron than lose one, as in the equation below (where E represents a Group VII element):

$$e^- + E \rightarrow E^-$$

In this respect the halogens act like oxygen in taking electrons away from other elements, especially from metals. Hence the halogen elements are spoken of as oxidizing agents, and the process is called oxidation. Their favorite reactions are with the reducing agents mentioned in the previous section, reactions in which a great deal of energy is liberated and very stable compounds are formed:

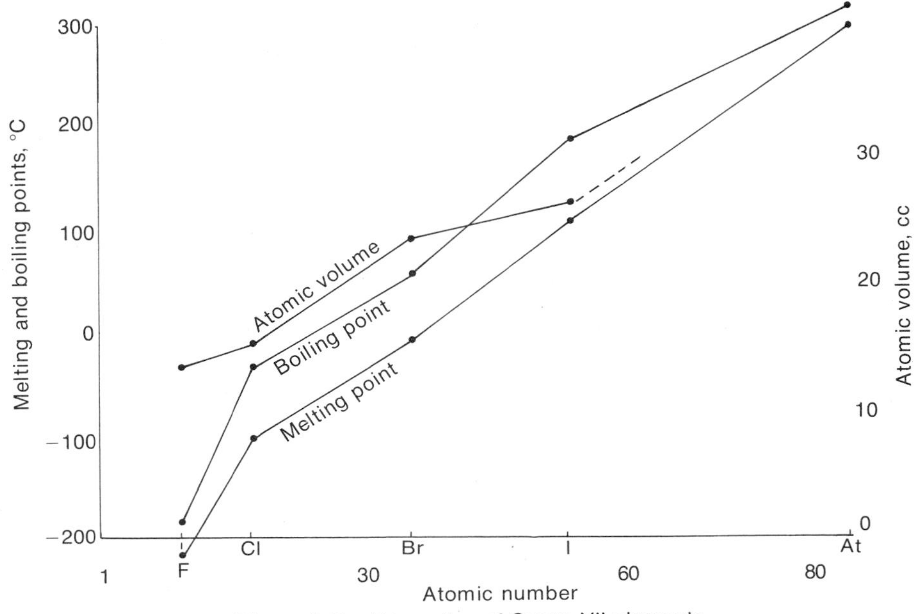

Figure 1–7. Properties of Group VII elements.

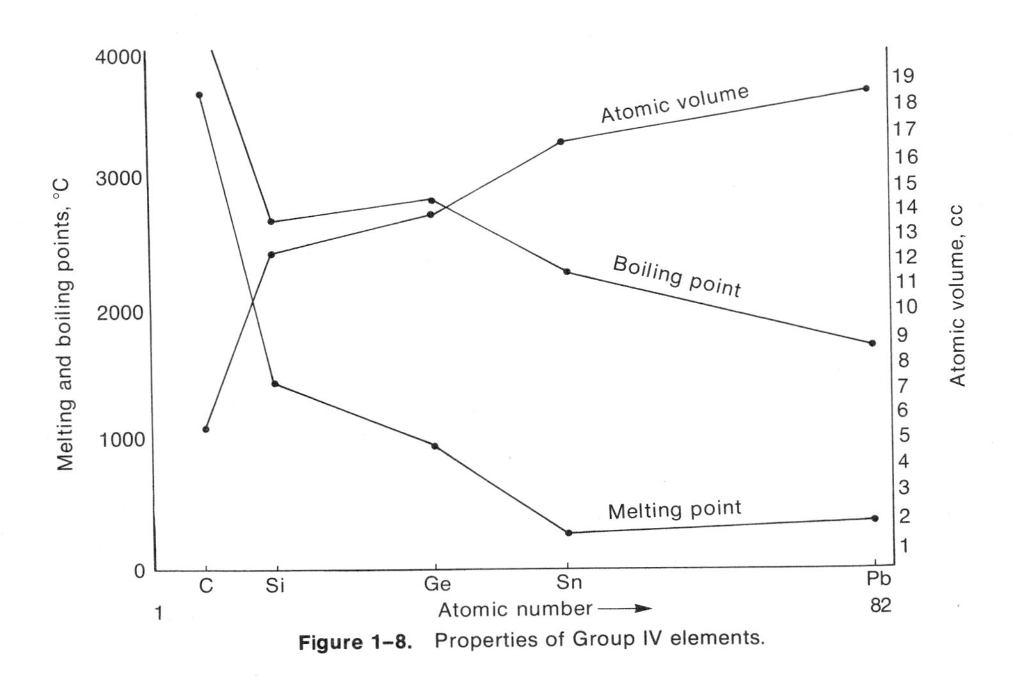

Figure 1–8. Properties of Group IV elements.

$$Na(s) \rightarrow Na^+ + e^-$$

and

$$e^- + Cl(g)^* \rightarrow Cl^-$$

so

$$2Na(s) + Cl_2(g)^* \rightarrow 2NaCl + 196 \text{ kcal}$$

One other point is noteworthy. In Figure 1–5 the melting points and boiling points of the Group I elements all trend sharply *downward*, whereas in Figure 1–7 the same properties all trend sharply *upward*. This indicates that in Group I the bonding between atoms to form a metallic solid is strongest in the light elements, and becomes weaker as the atoms become larger. Among the halogen elements the situation is reversed. There is no metallic bonding because there are no readily available electrons to participate in that kind of bonding. Instead, atoms of these elements combine to form diatomic molecules (F_2, Cl_2, Br_2, I_2). There is only a very weak attraction between neighboring molecules of this sort. This attraction increases as the atoms become larger, and eventually it becomes great enough to allow the heaviest molecules to coalesce into a solid. So in Group VII we have the entire range from a very low-boiling gas (fluorine) to a black solid (iodine) within the space of four elements.

A QUICK LOOK AT GROUP IV. We have looked at the group relationships at the two extremes of the Periodic Table, and now to balance things we should look at the group that lies midway between them. For the time being we shall ignore the transition elements that expand the left side of the table, and consider the system as though it consisted only of periods of eight elements. Group IV then does indeed lie in the middle, and a closer look at it should tell us whether the group relationships are any different in that region of the table.

Figure 1–8 shows the same three properties for the elements of Group IV plotted against atomic number, but again using different scales to conform to the small atomic volumes and the extreme range of temperatures encountered. The Group IV elements are all compact; their atomic volumes are by far the smallest we have found in this brief survey. It is apparent that the elements shrink at first, as we progress from Group I toward the right, and then expand again.

The elements of this group resemble the metals of Group I, rather than the halogen elements, in having decreasing melting points and boiling points. The curves drop steeply from extremely high values for carbon to moderate (and rather similar) values for tin and lead. Notice that germanium has a higher boiling point (and also a higher ionization energy and heat of vaporization) than either of its neighbors in Group IV. Such a reversal, which we did not encounter in either Figure 1–5 or Figure 1–7, indicates that something different happens halfway down Group IV, and the something turns out to be the effect of the first transition series on the further development of this period.

While we are at it, let us take a closer look at the melting point of tin; it is lower than that of its heavier neighbor, lead. The difference may not seem like much on the highly compressed scale made necessary by the 4000° C melting and boiling points of carbon, but on the melting-point scales of Figures 1–5 and 1–7 the difference would seem huge and the ups and downs would cover half the page. The difference is actually nearly 100° C; the melting point of tin is 232° C, while that of lead is 327.4° C. Suppose, then, that tin were unknown and

---

*(g) stands for a substance which is a gas under the conditions of the reaction. If it were liquid the symbol would be (l). If it were solid, the symbol would be (s), as in the previous equation. Cl_2 is the ordinary form of chlorine at room temperature.

we were trying to predict its melting point from those of its neighbors, germanium and lead, by interpolation. The calculation is

$$\text{expected mp} = \frac{\text{mp of Ge} + \text{mp of Pb}}{2} = \frac{937.4^\circ\text{C} + 327.4^\circ\text{C}}{2} = 632.4^\circ\text{C}$$

This is almost three times the actual melting point; we are off by 400° C, or 173%!

Obviously there is much more going on in the relationships between elements than can be described by simple averaging of numerical values within a given group. What would happen if we tried to estimate the melting point of tin by averaging the melting points of its neighbors to the *right* and the *left*, in Groups III and V, instead of using its neighbors above and below? The neighbor to the left is indium, and the neighbor to the right is antimony, so we have

$$\text{expected mp Sn} = \frac{\text{mp In} + \text{mp Sb}}{2} = \frac{156^\circ\text{ C} + 630.7^\circ\text{ C}}{2} = 393^\circ\text{ C}$$

This is much closer to the measured value of 232° C than we came before; we are off by only 69% this time. It still isn't much of a prediction, but the result teaches us that *relationships within a period are at least as important as those within a group*, and we must study the chemical elements *both* by periods and by groups if we hope to arrive at a reasonable understanding of them. Furthermore, we must have an appropriate respect for the individuality of each element as it goes through changes of form and structure in response to outside influences. In this frame of mind the reading of the remaining chapters will lead to more understanding and greater benefit.

SUMMARY

In the physical world there are some 89 chemical elements we can recognize, measure, and weigh. About half of the 89 have geological, biological, or commercial importance, and these are the subject of this book. The elements do not come to us in equal proportions, but in extremely variable abundances. It appears that all were formed billions of years ago from protons, neutrons, and electrons by nuclear reactions that took place at temperatures of millions of degrees within the sun and other stars, and that only those with the most stable nuclear structures have survived. Such favored structures follow three simple rules: light elements are generally more plentiful than heavy ones, those with even numbers of protons or neutrons (or both) are vastly more prevalent than the others, and nuclei with filled energy levels according to a quantum-mechanical pattern are especially prominent. The ten most abundant elements, with their percentages in the earth's crust, are oxygen (49.5%), silicon (25.7%), aluminum (7.5%), iron (4.7%), calcium (3.4%), sodium (2.6%), potassium (2.4%), magnesium (1.9%), hydrogen (0.87%), and titanium (0.58%). Together these amount to 99.2% of the world around us.

All the elements fit within a scheme proposed by Mendeleev over a century ago and represented by the modern Periodic Table. There are no longer any blank spaces left in the table. All the elements heavier than lead and bismuth are radioactive because of instability of their nuclear structures. The table is arranged in horizontal sequences (periods) of 2, 8, 8, 18, 18, 32, and 17 elements, respectively, the last period being incomplete and consisting principally of elements synthesized by self-limiting processes of high-energy physics. The numbers 2, 8, 18, and 32 follow the numerical sequence $2n^2$, where n is a digit from one to four. This arrangement of periods corresponds to a vertical ordering in groups numbered from I to VII, plus a zero group and a block of 30 transition elements (ten in each of three periods). The groups and the ten trios of transition elements consitute the so-called families of elements. The Periodic Table may also be divided vertically into regions corresponding to the filling of *s, p, d,* and *f* atomic orbitals, and it may be divided diagonally into a major region of metals, a minor region of nonmetals, and a band of metalloid elements in between.

The chapter concluded with a series of three graphs on which some physical properties of the elements in Groups I, IV, and VII were plotted against atomic number. The first impression may be one of a bewildering array of criss-crossing crooked lines, but closer examination shows that melting and boiling points (but not atomic volume) trend *downward* among the Group I metals but *upward* among the Group VII elements. Brief explanations for these trends and for the large periodic variations of atomic volume were given, principally in terms of the bonding that holds solids together.

The corresponding properties for Group IV were then considered, and some reversals of trend were noticed. Attempts to predict the melting point of tin solely from the melting points of its Group IV neighbors proved to give a result which was off by 200° C to 400° C. This, plus the crookedness of all the lines on the graphs, showed that the properties of the elements within any one group cannot really be said to vary smoothly and gradually; the relations of elements within the Periodic Table are much more complex than that. Consideration of the progressive changes within periods as well as groups is necessary, and to this we must add an allowance for individuality among the elements arising from many subtle details of atomic structure. With these aspects in mind, individual study of the most interesting and the most important elements can begin.

GLOSSARY

anisotopic: having no isotopes in the natural form.

atomic number: the number of protons in the nucleus of an atom, and hence the number of electrons in the neutral atom.

atomic weight: the mass of an atom on a relative scale wherein the mass of the $^{12}_{6}C$ isotope is taken as 12.00000.

electron: an elementary particle with unit negative charge and a mass of 5.49×10^{-4} on the atomic mass scale.

gram atomic weight: the weight in grams of 6.023×10^{23} atoms of a particular element, and hence a sample of that element having a weight in grams numerically equal to its atomic weight.

heat of vaporization: the heat required to vaporize 1 mole of an element.

ion: a charged atom; an atom containing a number of electrons different from its atomic number, and hence bearing a positive or negative electrical charge.

ionization energy: the amount of energy required to separate one electron from each of the atoms in 1 mole of that element.

isotopes: atoms with the same atomic number, but differing atomic weights, due to different numbers of neutrons in the nucleus.

metalloid: an element such as silicon or germanium, having properties intermediate between those of metals and nonmetals.
mole: a sample of pure substance having a mass in grams numerically equal to the molecular weight of that substance; 6.023×10^{23} molecules of a pure substance; 6.023×10^{23} atoms of an element; in slang, 6.023×10^{23} of any item.
molecular weight: the sum of atomic weights of all the atoms in a particular molecule.
monatomic: having molecules consisting of a single atom, such as molecules of helium, neon, or argon.
neutron: an elementary particle with a mass of 1.008665 on the atomic mass scale and with zero electric charge.
orbital: in atomic structure, a domain in space characterized by a definite potential energy level. Both the shape of the space and the magnitude of its energy level are determined by the values of the quantum numbers assigned by the theory of atomic structure.
proton: an elementary particle with a mass of 1.007825 on the atomic mass scale and unit positive charge.
transition metal: an element characterized by a partly filled *d* subshell; an element lying between Groups II and III in Periods 4, 5, or 6 of the Periodic Table.

EXAMINATION QUESTIONS

1. Trace briefly the historical evolution of the periodic system during the nineteenth century.

2. Draw an outline of the Periodic Table, and without necessarily identifying all the elements indicate therein the positions of

   a. the metallic elements.
   b. the nonmetals.
   c. the metalloids.
   d. the transition metals.
   e. Groups I, II, III, IV, V, VI, and VII.
   f. Periods 1, 2, 3, 4, 5, 6, and 7.

3. How many elements can you think of which are liquid on a hot summer's day (that is, having melting points of 30° C or lower, and ordinarily are liquids)? There are four, two of which have appeared in this chapter.

4. Why is helium so plentiful in the sun's atmosphere? Why is there so little on earth?

5. What are the two most abundant elements on earth? Given that the two combine to form very stable solid compounds, approximately how much of the earth's crust is made up of these two elements?

6. Gold has an atomic number of 79 and comes three spaces before lead in the periodic table. Give two good reasons why it is so rare.

7. Thorium, uranium, and plutonium (atomic numbers 90, 92, and 94) are the elements which are important in nuclear power plants. Why are the neighboring elements of atomic numbers 89, 91, 93, and 95 not used? Give two good reasons why you think they would not be practical.

8. Why do the melting points, boiling points, and heats of vaporization of the Group I alkali-metal elements trend downward as we descend in the group, whereas the same properties of the Group VII halogen elements trend upward?

9. Would you expect the physical properties of the Group IV elements to resemble those of the Group I elements more closely than those of the Group VII elements? Why? In what way do the elements of Group IV differ most from those of Groups I and VII?

10. Write the names and symbols of 20 elements already known to you before you began this course. For bonus points, how many more can you write now?

"THINK" QUESTIONS

A. Using the Table of Isotopes in the *Handbook of Chemistry and Physics*, or any equivalent source, show arithmetically that the three unresolved inconsistencies in the order of elements in Mendeleev's periodic table (K before Ar, Co before Ni, Te before I) are indeed due to accidental distribution of isotopes in the natural forms of those elements.

B. Platinum is a very useful metal, with an extremely high ability to further chemical reactions such as the oxidation of residual fuel in an automobile's catalytic converter. It is also a very expensive metal, selling for $160 or more per ounce. Could the valuable properties of platinum be matched by a judicious mixture of neighboring elements in the periodic table? Why?

C. There are some exceptions to the rules for nuclear stability mentioned in this chapter, particularly among the light elements. Using Table 1–3, can you pick them out? How many are there?

D. If all our iron ore should become used up (a definite possibility!), what are the chances of replacing iron (and its derivative, steel) with a structural metal chosen from the elements *heavier* than iron? What candidates would you choose from that category to replace steel? What candidates would you choose from the *lighter* elements?

E. The elements in the 4*f* series, at the bottom of the Periodic Table, are called the lanthanide rare-earth elements. Are they really rare? Compare their rarity with that of chromium, silver, and iodine.

PROBLEMS

1. The abundance of nickel is given in Table 1–3 as 80 parts per million (ppm). What percentage of the earth's crust is this?

2. Natural bromine consists of two isotopes, 50.69% $^{79}_{35}Br$ and 49.31% $^{81}_{35}Br$. Calculate the atomic weight of bromine.

3. Iodine is anisotopic, consisting of 100% $^{127}_{53}I$. What is the neutron excess in its nucleus? Compare this with the neutron excess in gold (atomic no. 79) which also is anisotopic and has an atomic weight of 197.

4. The melting points of the Group III elements are: boron, 2300° C; aluminum, 660° C; gallium, 29.8° C; indium, 157° C; and thallium, 304° C. The neighbors to the left and right of gallium are zinc (mp 420° C) and germanium (mp 937° C). Calculate the melting point of gallium by interpolation, first by obtaining the average melting point of its neighbors above and below, and then of its neighbors to the left and right. Which calculated value is nearer the observed value?

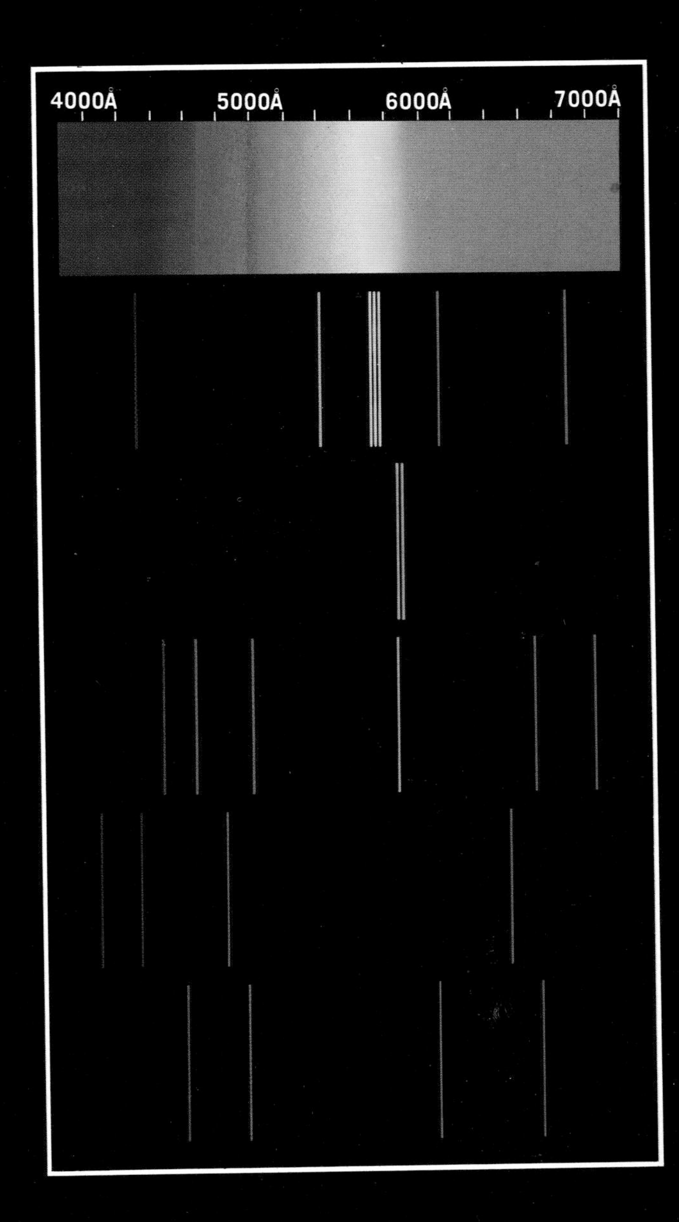

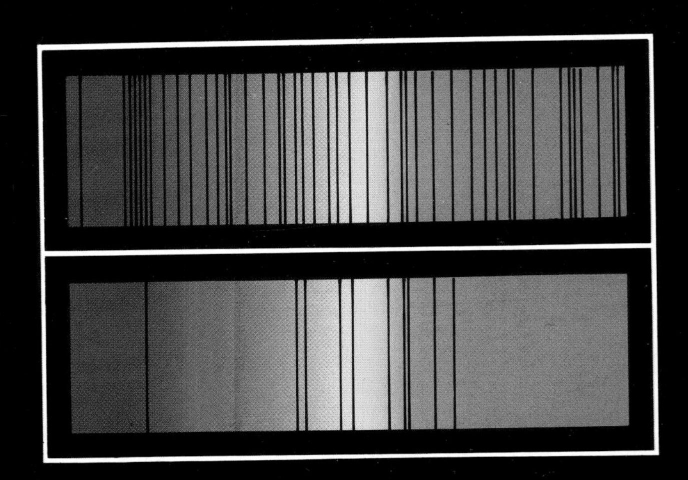

PLATE 1

Above, Emission spectra of some common elements. The continuous spectrum is shown at the top. *Below,* The solar spectrum is compared to an absorption spectrum of iron (at bottom). The correspondence of lines in these spectra indicates the presence of iron in the sun's atmosphere. (From J. S. Faughn and K. F. Kuhn, *Physics for People Who Think They Don't Like Physics,* Philadelphia, W. B. Saunders Co., 1976, Plate 1.)

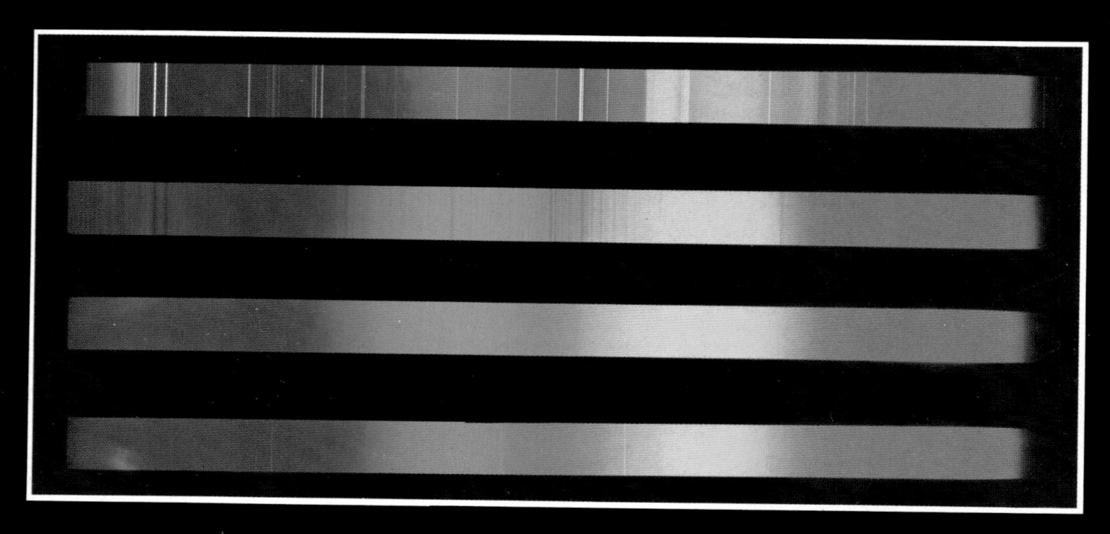

PLATE 2 From top to bottom, the emission spectrum of ionized calcium, the Fraunhofer spectrum of the sun, a continuous spectrum, and the emission spectrum of a fluorescent lamp. At the left of the ionized calcium spectrum are the H and K lines, which are the strongest absorption lines in the visible part of the solar spectrum. (Photographed in Bausch and Lomb, Incorporated, Research Laboratory.)

5. Germanium tetrachloride, $GeCl_4$, melts at −49.5° C; lead tetrachloride, $PbCl_4$, melts at −15.0° C. Calculate the expected melting point of the tetrachloride of tin, which lies between germanium and lead in Group IV. Compare your result with the observed value of −33° C for $SnCl_4$, and comment upon it.

2 HYDROGEN, HELIUM, AND WHAT MAKES THE SUN SHINE

Hydrogen is the lightest element, having the simplest atomic structure. It is also a common and indispensable element; its most prevalent compound, H_2O, permeates everything, covers four fifths of the earth's surface, and constitutes 70% of our bodies. Helium, atomic number 2, is much less well-known

An eruption from the sun, photographed in the ultraviolet by the Naval Research Laboratory experiment aboard the manned orbiting Skylab. The image on the right is in the 304Å radiation of ionized helium, which is radiated by gas at a temperature of about 50,000 K. The image on the left shows radiation from multiply ionized iron, typical of coronal regions, which have temperatures of millions of degrees Kelvin. (Courtesy of Naval Research Laboratory/NASA.)

but is interesting and useful. Most important of all, helium is derived from hydrogen by nuclear reactions that take place in the intensely hot interior of the sun, providing the enormous quantities of energy which the sun radiates. In turn, the sunshine keeps us warm, provides our fuel, and grows our food.

At the very high temperatures which prevail within the sun and other stars (about 100 million degrees C), atoms are stripped of their electrons. The nuclei and electrons seethe in a white-hot "plasma," pulled together by immense gravitational forces and thrown apart by their high-energy collisions. Every once in a great while (about once in every billion collisions) a nuclear reaction takes place. Since the plasma is made up mostly of elementary particles such as protons, electrons, and neutrons, plus some alpha particles (helium-4 nuclei), nuclear reactions can take place with the formation of light elements by the process of *fusion.* In such reactions a minor part of the combined mass of the particles is converted to radiated energy and the rest of the mass is held together by a nuclear binding energy equivalent to the amount of emitted energy. This energy, E, is related to the loss of mass, m, by the well-known Einstein equation

$$E = mc^2$$

where c is the velocity of light, 3×10^{10} cm per second. It follows that the more energy emitted, the more stable will be the reaction product. In fact, the only way to destroy the product nucleus is to inject into it more energy than was lost during its formation.

The major product of fusion from the initial reaction of protons is helium, but successive accretions of protons and neutrons produce heavier elements up to nickel and iron. The production of heavy elements, such as gold and platinum and uranium, seems to require special conditions of intense neutron flux for a very short time, and hence most of the heavy elements we enjoy on earth are extremely rare in the rest of the universe. Just why and how they accumulated on the earth remain to be explained.

HELIUM-4 FROM HYDROGEN

Forty years of study of the sun's composition, temperature, and rate of emission of energy, supplemented by extensive laboratory experimentation, have led to the conclusion that there are two principal mechanisms whereby hydrogen is converted to helium in the sun. The first of these is a simple accretion reaction, in which protons are bound successively, and the second is a cyclic reaction in which carbon-12 is the recycled material. Both can be illustrated by simple nuclear equations.

PROTON-PROTON FUSION. In the accretion reaction, two protons first collide and fuse to form the hydrogen-2 isotope:

$${}^{1}_{1}H + {}^{1}_{1}H \rightarrow {}^{2}_{1}H + e^{+} \text{ (a positron)} \qquad \text{(a)}$$

The positron which is emitted at the same time is best described as a positive counterpart to the familiar electron, with a mass of 0.00055 and a unit positive charge. The ${}^{2}_{1}H$ nucleus is stable and is the source of the 0.015% of "heavy hydrogen" in our natural hydrogen on earth. Under the conditions inside the sun, however, such a nucleus may collide with and hold another proton, resulting in the formation of a helium-3 nucleus:

$${}^{2}_{1}H + {}^{1}_{1}H \rightarrow {}^{3}_{2}He \qquad \text{(b)}$$

This product, unstable and dissatisfied with its neutron-proton ratio, is likely to collide and react with a similar nucleus to produce the extremely stable helium-4 nucleus (an alpha particle). Two protons are released thereby, and these can again return to the initial reaction:

$${}^{3}_{2}He + {}^{3}_{2}He \rightarrow {}^{4}_{2}He + 2{}^{1}_{1}H$$

Doubling Equations (a) and (b) to get the reactants for (c), the overall reaction is

$$4{}^{1}_{1}H \rightarrow {}^{4}_{2}He + 2e^{+}$$

From the mass of the reactants (the four protons used up) minus the mass of the products (^4_2He and two positrons), we can find the quantity of mass lost during the production of helium from hydrogen by this mechanism, and hence determine the energy liberated:

(1) mass of four protons = 4 × 1.007825 = 4.031300
(2) mass of helium-4 nucleus = 4.002603
(3) mass of two positrons = 2 × 0.000550 = 0.001100
(4) sum of (2) and (3) = 4.003703

So,

$$\begin{aligned} \text{mass of reactants} &= 4.031300 \\ \text{mass of products} &= \underline{4.003703} \\ \text{difference} &= 0.027597 \end{aligned}$$

From the Einstein equation, 1 atomic mass unit (amu) is equivalent to 2.1472 $\times 10^{10}$ kcal per mole of material converted; thus, the energy liberated during the formation of 1 mole of helium (4.0026 grams of it) is

$$0.027597 \times 2.1472 \times 10^{10} \text{ kcal/mole} = 5.92563 \times 10^{8} \text{ kcal/mole}$$

This is 592 *million* kilocalories for each mole of helium produced. No wonder the sun is so hot! This heat output is about ten million times what we would get by burning 4 grams of natural gas. The energy-producing capability of nuclear reactions is just that much greater than that of conventional chemical reactions such as the burning of coal or oil or gas.

THE CARBON CYCLE. The other mechanism which operates to make helium from hydrogen in the sun and the stars is a cycle of reactions involving carbon, nitrogen, and oxygen. First a very fast-moving proton strikes a carbon-12 nucleus and fuses with it:

$$^1_1H + ^{12}_6C \rightarrow ^{13}_7N \qquad (1)$$

The nitrogen-13 isotope is very unstable and decays rapidly to carbon-13, emitting a positron:

$$^{13}_7N \rightarrow ^{13}_6C + e^+ \qquad (2)$$

Carbon-13 then captures a second and third proton:

$$^{13}_6C + ^1_1H \rightarrow ^{14}_7N \qquad (3)$$

$$^{14}_7N + ^1_1H \rightarrow ^{15}_8O \qquad (4)$$

From laboratory experiments oxygen-15 is known to be unstable; half of it decays every 122 seconds, emitting a positron:

$$^{15}_8O \rightarrow ^{15}_7N + e^+ \qquad (5)$$

Lastly, capture of a fourth proton regenerates the original carbon-12 and liberates helium-4:

$$^{15}_7N + ^1_1H \rightarrow ^{12}_6C + ^4_2He \qquad (6)$$

If Equations (1) through (6) are added up, cancelling all the terms that appear on both sides, the net result is

$$4\,{}^1_1H \rightarrow {}^4_2He + 2e^+$$

which is just what we had before from the previous mechanism (Equations 2 to 4). It follows that the energy output, 5.92563×10^8 kcal per mole of helium produced, will be just the same.

Suppose we were to use four atoms of hydrogen as fuel in a rocket engine, as it was used in the later stages of the Apollo vehicles. The hydrogen (as H_2 molecules) would undergo combustion with oxygen in the rocket engine according to the chemical equation

$$2H_2 + O_2 \rightarrow 2H_2O$$

The heat produced by this reaction, using molar quantities, is only 115.6 kcal. The nuclear reaction produces *five million times* as much energy as the chemical reaction, from the same amount of hydrogen!

There are other nuclear reactions going on in the sun and the other stars, of course. The reader will find it interesting to calculate the nuclear energy emitted during the formation of 1 mole of oxygen, or silicon, or any other stable, abundant type of atom found on earth.

With such energetic processes going on, we can understand the vast outpouring of heat and light from our sun. The temperature at the center of the sun is estimated at 100,000,000° C, and its surface temperature is 5510° C. Each *second* it pours forth 92,256 *billion billion* kilocalories of radiation of all kinds. To do so, it loses 4,264,000 *tons* of matter each second!

Obviously the sun is getting lighter and smaller all the time, but since it is large enough at present to contain a million earths, a long time will go by before the sun's output diminishes perceptibly. If it went on shining at the present rate, right down to a sudden blink-off, the sun would last 14,860 million million years. Long before that it will have changed its character and become a red giant-type star. At present the sun is an average-sized star of the most common type, and is already several billion years old. Its present composition is approximately 80% hydrogen, 19% helium, and 1% other elements, with most of the helium and heavier elements concentrated at the center. When most of the hydrogen in the central region is used up, the helium will contract gravitationally and the hydrogen-rich exterior will expand and cool, giving it a reddish appearance. Other nuclear reactions will then take over, and the sun will be on a different path of stellar evolution.

The earth is a tiny ball 91.41 million miles away from the sun, so it receives only about one two-billionths of the sun's output. Nevertheless, each day the earth receives on every 1.5 square miles of its surface an average amount of energy equivalent to an atomic bomb of the Hiroshima type. In three days we receive the equivalent of *all* our coal and oil reserves. Much of this is reflected and most of the rest goes into gentle heat, simply warming the surface and evaporating water. A small part is converted to chemical energy through the growth of plants, of course, and eventually some of the plant and animal life may produce coal and petroleum. However, we simply do not know how to store sunlight effectively, or even how to use it efficiently. Even our best and most expensive methods for converting sunlight to electrical energy have an efficiency of only 10 or 12 per cent, and the apparatus is very fragile and costly. If we could convert *all* of it to electricity, the solar energy falling on 15 sq ft of a rooftop would supply all the power needed to run the appliances in the house (on a cloudless day at the average latitude of the U.S.). How to convert it, and what to do on cloudy days and at night, present huge unsolved problems. There is a tremendous challenge to inventiveness here, particularly as our conventional energy resources dwindle.

In the meantime, considerable sums are being spent searching for ways to carry out nuclear fusion reactions of hydrogen on earth safely and continuously. The advantages of such a process would be enormous: there would be a cheap and abundant source of fuel in the form of hydrogen from water; this would be available to all nations; the chief products would be safe and nontoxic (in sharp contrast to the highly poisonous radioactive fission products of present nuclear power plants); and no recovery of starting materials would be necessary. The development of earthbound fusion reactors has proved to be an elusive goal, however, and the constant presence of that huge fusion reactor 91 million miles away reminds us just how puny is man and how crude his technology. Science as a purposeful endeavor has really just begun; our greatest problems await its further development for their solution.

FATE OF THE HYDROGEN LEFT ON EARTH

Very likely the earth's gravitational field was (and is) too weak to hold the light molecules of gaseous hydrogen—hence, whatever hydrogen there may have been when the earth was assembled from many pieces has been lost. Hydrogen *does* still exist in the atmospheres of the larger and heavier planets Jupiter, Saturn, and Neptune. On earth, the liberation of water during the geologic heating of metamorphic minerals has made possible the liberation of ammonia from nitrides, and methane from carbides (see Chapter 3), so that the atmosphere of the earth before the appearance of life probably consisted of water vapor, methane, and ammonia. Experiments begun by S. L. Miller in 1953 have shown that electric discharges (like lightning) and ultraviolet light (like that in sunlight), acting on such a reducing atmosphere, produced amino acids, hydroxy acids, urea, and simple sugars. Within the warm dilute "soup" composed of such organic substances dissolved in water, self-duplicating organisms could develop and probably did. These eventually evolved into the simple plant life of the oceans and the less extensive but more highly developed plant life of the land. The action of sunlight then caused the plants to liberate oxygen by photosynthesis, and gradually an oxidizing atmosphere developed in which the ammonia was oxidized to nitrogen and water, and the methane to water and carbon dioxide (which was consumed by the plants). So the hydrogen which remains on earth is tied up in the stable forms of water, ice, and hydrated minerals, with a very small proportion in the less stable forms of living protoplasm, wood, coal, oil, and natural gas.

We can recover free gaseous hydrogen from water by electrolysis or by the action of potent reducing agents, and we can get it more easily by splitting hydrocarbons. At present the chief source in our industrial economy is petroleum, for during refining operations large quantities of hydrogen are given off as by-products. Examples of such refining reactions, for those who are interested in the how and why, or want to know what makes the nation's wheels turn and its crops successful, are:

1. Cracking of the crude oil paraffins, in which large hydrocarbon molecules like $C_{11}H_{24}$ are split into C_2, C_3, C_4, and C_5 unsaturated hydrocarbons, liberating the surplus hydrogen:

$$C_{11}H_{24} \xrightarrow[\text{catalyst}]{450^\circ\,C} C_5H_{10} + C_4H_8 + C_2H_4 + H_2$$

2. Alkylation, in which small hydrocarbon molecules from the cracking reaction are used to replace hydrogen in a larger hydrocarbon, liberating the replaced hydrogen:

$$C_4H_8 + 2C_2H_6 \xrightarrow[\text{catalyst}]{HBF_4} C_8H_{18} \text{ (octane)} + H_2$$

3. Hydroforming or cyclization, in which linear hydrocarbons are converted to cyclic ones:

$$\underset{\text{n-hexane*}}{C_6H_{14}} \xrightarrow[\text{High temperature}]{Cr_2O_3 + Mo_2O_5} \underset{\text{benzene*}}{C_6H_6} + \underset{\text{hydrogen}}{4H_2}$$

Part of the by-product hydrogen goes into other refining operations, such as desulfuring petroleum products, and the rest is used to make ammonia for fertilizer:

$$N_2 \text{ (from air)} + 3H_2 \xrightarrow[\text{450° C, 500 atm}]{\text{Iron catalyst}} 2NH_3(g)$$

Most of the ammonia is converted to ammonium nitrate, to the tune of 11 million tons per year, to grow the grain the world requires. Other uses for hydrogen are as a gaseous fuel (in fuel cells and industrial operations), as a liquid fuel (in rockets), as a reducing agent in metallurgy, and as a reducing atmosphere for the fabrication of metals. Lithium and sodium hydrides, which are used for bleaching paper and recovering silver from photographic film, are made from hydrogen.

HYDRIDES. Compounds of hydrogen with one other element are called hydrides, and are of three general types:

1. The salt-like hydrides, which contain the H^- ion. In most of its chemical reactions hydrogen acts as a reducing agent and gives up electrons. However, there are about a dozen elements which give up electrons more freely than hydrogen, and when one of these combines with hydrogen it supplies a second electron to the hydrogen atom, thereby filling its 1*s* orbital. The resulting H^- ion is held electrostatically in a crystal lattice, just the way Cl^- ions are held in NaCl. For example, sodium (in Group I) combines with hydrogen at a temperature of 300° C to form sodium hydride, NaH, a white crystalline salt:

$$2Na(l) + H_2(g) \rightarrow 2Na^+(c) + 2H^-(c)$$

(Here the symbol (c) indicates that the ions are packed together in a crystal lattice.)

Only the metals of Groups I and II form salt-like binary hydrides and, of these, LiH is the only one that melts. The others decompose at temperatures near 800° C before melting. All such hydrides are strong reducing agents, and liberate hydrogen whenever they meet H^+ ions, even those in water:

$$KH(s) + H_2O(l) \rightarrow K^+(aq) + H_2(g) + OH^-(aq)$$

*Hydrocarbons such as the two indicated here are named by a system that designates their molecular structure in a specific manner. The system will be revealed when you study organic chemistry, and is not essential here. The names need not be memorized, but the splitting out of hydrogen is important.

When dissolved in molten LiCl and electrolyzed, the salt-like hydrides evolve hydrogen at the *positive* electrode:

$$2H^- \rightarrow H_2 + 2e^-$$

In addition to these simple salts of the H^- ion, there are some ternary hydrides which contain "complex" hydride ions of the type XH_4^-. For example, $LiBH_4$ and $NaBH_4$ are white solids containing the BH_4^- ion, and $LiAlH_4$ contains the AlH_4^- ion. All these are potent reducing agents, and evolve hydrogen when they encounter acids. They and the binary hydrides LiH, NaH, and CaH_2 are commercial products with many uses.

2. The covalent hydrides, in which hydrogen shares electrons with a nonmetal. This is a large body of familiar liquid and gaseous hydrides, such as H_2O, H_2S, HCl, HF, NH_3, and so on, plus many less familiar volatile hydrides such as B_2H_6 (see Chapter 3) and SiH_4 (see Chapter 4). Covalent hydrides of the metalloids (such as B_2H_6 and SiH_4) act as reducing agents and will release hydrogen when they meet water, but covalent hydrides of the strongly electronegative elements (such as HCl and HF) release H^+ ions when they dissolve in water. Ammonia, NH_3, releases OH^- ions and is basic when it dissolves in water (see Chapter 3). Covalent hydrides are formed by all of the elements in Groups IV, V, VI, and VII.

3. The metallic or interstitial hydrides, in which hydrogen is not bound as an ion or as a covalent partner, but rather is held within the crystal lattice of a metal. All the transition metals and the rare-earth elements absorb hydrogen and hold it within their structures in one way or another, but with variable tenacity and behavior. In most instances the products are not compounds in the usual sense, but depend upon the number of spaces of appropriate size where H atoms can be held between the atoms of metal in the lattice. Sometimes the resulting ratio of hydrogen to metal turns out to approach a small whole number, as in UH_3. In all cases the product is metallic in appearance and electrical conductivity, but the presence of hydrogen embrittles the metal, and in high enough ratio will weaken it or even cause it to fall to powder. For this reason dissolved hydrogen is undesirable in steel and in titanium alloys. The metallic hydrides have no uses.

WATER AND ITS PROPERTIES. Water is vital to living organisms and indispensable to industry. It shapes the face of our planet, covers most of the earth, and permeates its rocks and its atmosphere. It is so important to all aspects of chemistry, from equilibrium theory to nuclear reactors, that it is considered or discussed in almost every chapter of your textbook. We need only look briefly into three aspects: water as an acid, "light water" versus "heavy water," and water as a solvent.

Pure water is usually thought to be the quintessence of mildness, neutrality, and harmlessness, but actually it can be a vigorous chemical reagent. In the ordinary system of acids and bases, pure water is indeed neutral because it furnishes very few H^+ or OH^- ions, and these only in equal numbers:

$$\text{pure } H_2O(l) \rightarrow H^+(aq;\ 10^{-7} \text{ moles per liter}) + OH^-(aq;\ 10^{-7} \text{ moles per liter})$$

Nevertheless, water can act as an acid, and frequently does. We need only think of one of the fundamental properties of aqueous acids, the reaction with metals to liberate hydrogen, to see how this is so. Dilute sulfuric acid reacts with zinc in the laboratory to produce hydrogen and leave zinc sulfate behind as dissolved zinc and sulfate ions:

$$Zn(s) + H_2SO_4(aq) \rightarrow H_2(g) + Zn^{+2}(aq) + SO_4^{-2}(aq)$$

The reaction of more active metals, such as those of Group I, liberates hydrogen from water alone in an exactly analogous way:

$$2Na(s) + 2H_2O(l) \rightarrow H_2(g) + 2Na^+(aq) + 2OH^-(aq)$$

In this respect water is acting as an acid, and a very effective one.

With common metals such as iron we ordinarily do not notice such acid attack by water because we do not see any hydrogen; we think only of slow rusting of the iron. At elevated temperatures, however, water becomes increasingly destructive as an acid. At 350° C it corrodes ordinary steel rapidly, and at 500° C steam under high pressure attacks even stainless steel, converting its iron, nickel, and chromium to oxides and liberating hydrogen:

$$3Fe(s) + 4H_2O(g) \rightarrow Fe_3O_4(s) + 4H_2(g)$$

$$Ni(s) + H_2O(g) \rightarrow NiO(s) + H_2(g)$$

$$2Cr(s) + 3H_2O(g) \rightarrow Cr_2O_3(s) + 3H_2(g)$$

This acidic attack on alloy steel sets an upper limit on the operating temperatures of steam power stations for generating electricity, and so limits their efficiency sharply. The result is higher cost for electric power and more waste heat going into rivers and lakes, two aspects which naturally cause much public concern. Any student of chemistry can explain why this has to be so as long as we rely upon heat-power engineering to produce electricity. The maximum efficiency of any heat engine in converting heat energy to useful electrical energy is limited by the second law of thermodynamics to

$$\text{maximum efficiency} = \frac{Ta - Tr}{Ta}$$

where Ta is the temperature at which the engine absorbs heat (the temperature of the incoming steam), and Tr is the temperature at which the engine rejects waste heat (the temperature of the cooling water in the condensers at the end of the turbines). A century ago steam engines operated at an efficiency of about 10%, and people were satisfied because the supply of fuel was considered inexhaustible and nobody objected much to the waste heat going into the surrounding air and water. Today, in the midst of a critical fuel shortage, power stations still operate at less than 50% efficiency, and the major part of the heat from the fuel is wasted. Worse than that, the waste heat is an embarrassment. A major reason why efficiencies cannot be pushed higher is that corrosion of the turbine machinery by superheated steam becomes too severe. Alloys with greater mechanical strength at high temperatures have been developed, but chemical reactivity of the metals with steam still limits the design.

Water is the standard substance on which the scales for measuring many physical properties are based. Thus, water has a density of 1 g/cm³, a specific heat of 1 cal/g° C, a freezing point of 0° C, a boiling point of 100° C, and so on. A more explicit listing of properties is given in Table 2–1. The simplicity of the numbers may lead the reader to think that water is a simple compound, but ac-

TABLE 2–1 PROPERTIES OF H_2O AND D_2O

	H_2O	D_2O
Melting point	0.0° C	3.8° C
Boiling point	100.0° C	101.4° C
Density at 4° C	1.000 g/cm^3	1.108 g/cm^3
Temperature of maximum density	4° C	11.2° C
Critical temperature	374.2° C	371.5° C
Critical pressure	218.5 atm	218.6 atm

tually its internal structure is complex and its properties are very unusual. For example, the specific heat of iron is 0.106 – it takes only one tenth as much heat to raise the temperature of a pound of iron 1 degree C as it does to raise the temperature of a pound of water 1 degree C. Similarly, the heats of fusion and vaporization of water are excessive in comparison with those of other compounds of similar molecular weight. Yet water is the substance we most often want to heat or boil, and we need to be reminded of the large quantities of fuel required to do so.

HEAVY WATER. As a matter of principle, the five isotopes of iron are all called iron because they *are* iron and they make up all iron; the ten isotopes of tin are all called tin because they *are* tin, and so on. However, an exception is made with hydrogen because its isotopes are so exceptional. Nowhere else do we find one isotope of an element that weighs twice or three times as much as another, and nowhere else does the isotopic composition of an element affect the physical properties of its compounds so much. So, by common consent, the isotopes of hydrogen have special names; ${}^1_1H$ is called protium, ${}^2_1H$ is called deuterium, and ${}^3_1H$ is called tritium, with the symbols H, D, and T, respectively. Tritium is radioactive, with a half-life of 12 years; it does not occur in ordinary hydrogen in appreciable amounts, and is of interest only in nuclear weapons. Deuterium is with us all the time, however. It differs enough from ${}^1_1H$ to be worth the comparison of physical properties given in Table 2–2. Pure D_2 is separated from D_2O by electrolysis, and is used in research as a stable isotope marker or tag to follow the path of hydrogen in chemical or biological processes. Com-

TABLE 2–2 PROPERTIES OF THE GASES H_2 AND D_2

	H_2	D_2
Boiling point	−252.8° C	−249.6° C
Critical temperature	−239.9° C	−234.8° C
Critical pressure	12.8 atm	16.4 atm
Density, g/liter	0.0899	0.1798
Heat of fusion	28 cal per mole	47 cal per mole
Heat of vaporization	216 cal per mole	293 cal per mole

pounds of interest can be synthesized from D_2, and the pathway of hydrogen is followed by mass spectrographic analysis of samples taken along the way.

The compound D_2O, deuterium oxide, is popularly called heavy water to distinguish it from H_2O, light water. Ordinary water in the lakes, rivers, and oceans contains .015% of D_2O, the presence of which does not bother us one bit because its chemical and physical properties are so much like those of H_2O that we don't notice it. Pure D_2O is desired at times, though, for two good reasons: it is needed to obtain deuterium for research purposes, as above, and it is preferred for nuclear reactors because it absorbs neutrons only one six-hundredth as much as H_2O. Despite its very small concentration in ordinary water, the separation of D_2O from H_2O can be accomplished commercially by at least four different methods. Electrolysis was the first method used; D_2 is lost only one eighth as fast as H_2 during electrolysis, and so D_2O tends to concentrate in the electrolytic cell. During the 1940's this separation was carried out in central Norway because cheap hydroelectric power was available there. Today D_2O is recovered from ordinary water by distillation (see Table 2–1 for the difference in boiling points of H_2O and D_2O). This is accomplished in Canada by complicated heat-conserving fractional distillation towers, and the pure D_2O is used in the CANDU*design of Canadian nuclear power stations. The great advantage of D_2O over H_2O as coolant and moderator in nuclear reactors is that the neutron absorption is low enough to permit the use of natural uranium, despite its low concentration (1 part in 138) of the fissionable isotope ^{235}U. American nuclear power reactors all use ordinary water as moderator and coolant, requiring a very expensive isotopic separation for enrichment of the uranium to about 3% ^{235}U.

WATER AS A SOLVENT AND SUSPENDING MEDIUM. Much of the importance of water lies in its ability to dissolve so many substances and to transport them to the sites of reactions. Think of blood, a watery tissue, carrying salts, sugars, and protein-derived nutrients to the bones and muscles of the body, and carrying dissolved carbon dioxide back to the lungs. Think also of the sea, carrying the dissolved gases, salts, and nutrients for a vast ecology embracing so much plant and animal life. The sea also transports the dissolved ions and gases for building layers of carbonate and phosphate rock, as well as coral reefs and the shells of seabottom creatures.

It is perhaps equally important that what water cannot dissolve it can usually carry in suspension. The blood carries red blood cells which in turn carry oxygenated hemoglobin; blood also transports white cells, platelets, and other organized tissues suspended in the plasma. Oils and fats, which do not dissolve

*CANDU is an acronym for the CANadian Deuterium oxide–Unenriched uranium design of power reactor. See Chapter 9.

in the watery body fluids, are carried in suspension as emulsified droplets. These remain suspended because water has a density sufficiently close to that of the insoluble fat to delay gravitational settling. (Constant motion of the liquid helps keep the insoluble matter suspended, of course.) The oceans also carry suspended matter—tiny plants remain suspended on the ever-moving surface and become the basic foodstuff of the sea.

All this activity involves basic principles of solution and equilibrium which are treated adequately in all modern textbooks. In order to lend substance to all that theoretical discussion, however, it would be well to consider at least one practical aspect. We may well choose the largest and most interesting one, the wealth that lies in the sea. We shall by-pass the plants and animals of the sea because those are the province of biology; we shall consider only the material dissolved in seawater, and ask about it as a possible storehouse of raw material. What is there, and how much? Can anything of value be recovered from that vast reservoir?

THE WEALTH OF THE SEA. Our task is simplified at the start by a surprising uniformity of composition. All the major oceans of the world are interconnected, and the various ocean currents serve to mix the water quite well. While some shallow areas near the mouths of large rivers may contain little dissolved material, the major oceans all contain about 3.5% of dissolved inorganic solids. If we evaporate the water from samples taken from the Atlantic, Pacific, Arctic, and Indian oceans, we find that the dry residues all contain 30.5 to 30.9% sodium, 0.9 to 1.2% potassium, about 1.2% calcium, 3.7 to 3.8% magnesium, 55.3 to 55.4% chloride ion, 7.69 to 7.79% sulfate ion, and 0.13 to 0.19% bromide ion. Such residues ("sea salt") are mostly sodium chloride, with minor amounts of magnesium chloride (which is deliquescent and makes the salt stay damp) and of sulfates of all four metals (which give the salt its bitter taste). For thousands of years crude salt has been recovered from seawater by evaporation of the water, wherever the climate permits it, and of course the salt can be purified by recrystallization. In other regions of the world the slow evaporation of prehistoric seas has deposited large amounts of rock salt, some already fractionated by successive resolution and recrystallization. Under the plains of northern Germany there are some deposits which are 99.9% pure sodium chloride, and at other locations the deposits are so rich in potassium chloride that they are mined and treated to extract almost pure potassium salt for use in fertilizer.

Consideration of the area and the average depth of the oceans shows that there are 320,000,000 *cubic miles* of seawater, and each cubic mile contains 4 billion tons of water. Analysis of a sample from the Atlantic shows that each cubic mile contains 128 million tons of sodium chloride, 18 million tons of magnesium chloride, 8 million tons of magnesium sulfate, and 4 million tons of potassium chloride and sulfate. It also contains 350,000 tons of bromine, 45 tons of silver, 9 pounds of gold, and even 5 grams of radium.

At present magnesium is recovered from seawater and used as a light, plentiful, and relatively inexpensive structural metal, as will be described in Chapter 4. Bromine also is recovered commercially from seawater, as described in Chapter 5. Other valuable materials may in time be obtained from the sea and from the ocean bottoms; intensive investigation is being directed toward that end.

Huge quantities of potassium chloride are needed for raising crops, and with farming becoming more intensive each year as the world food shortage worsens, it may soon be necessary to devise a commercial process for extracting the potassium ions from seawater. Sea plants already do this, of course, and in some parts of the world kelp and other seaweeds are collected, dried, and burned to recover the potassium for agricultural use.

As for the precious metals in seawater, with silver at $5 per troy ounce, the amount in 1 cubic mile is worth $5,400,000—a considerable incentive to find the means for getting it out. At the time of this writing, the gold in that cubic mile is worth only $17,820, and it seems likely that the cost of pumping and treating a cubic mile of seawater would exceed that sum. Someone may even one day find plants that will do the work of collecting the gold for us.

We should not ignore the manganese dioxide and phosphate sediment which are precipitated from the water by the inexorable laws of chemical equilibrium, and which lie on the bottom waiting to be gathered. Even the water itself is valuable, in places where fresh water is not available and where the necessary desalination processes can be carried out. The problem of drinking water is now so acute in some places that consideration is being given to towing icebergs south to a place where the ice can be melted to provide fresh water. Enormous quantities of water are immobilized in the polar ice caps, and it is estimated that, if all this ice were to melt, the level of the oceans would rise 193 feet and almost all the coastal cities would be inundated. At present the glaciers and polar ice caps are indeed receding, but very slowly.

A HYDROGEN-FUELED ECONOMY IN THE FUTURE?

Increasing scarcity and higher prices for petroleum and natural gas have pointed out the need for a substitute fuel, and the most logical substitute is hydrogen. Having once attained a copious source of energy, let us say by controlled fusion or efficient conversion of sunlight, it would then be possible to electrolyze water and obtain gaseous hydrogen and oxygen. The oxygen would be used in making iron and steel, and would have scores of other uses. The hydrogen would be transmitted all over the country through existing natural gas pipelines, making it available for heating homes and public buildings, for industrial fuel, for cooking, and for all other uses to which natural gas and bottled gas now are put. Hydrogen has a low density (only 0.09 g/liter at 0° C and 1 atm), so large volumes would need to be pumped, but the viscosity of the gas is low (making it easier to pump) and it gives 2.5 times as much heat per gram or per pound as methane, the chief constituent of natural gas.

Using hydrogen as a fuel would eliminate almost all air pollution, since pure hydrogen burns cleanly to pure water. Being derived from water by electrolysis, it contains no sulfur to produce obnoxious oxides of sulfur upon combustion. Neither would it give any carbon monoxide, nor any dust or fumes, if burned in clean air. The nitrogen in the air does raise the possibility of production of residual oxides of nitrogen (which result whenever nitrogen and oxygen are raised to a high temperature), but slow cooling of the combustion gases ordinarily allows these unstable oxides to decompose. The result would be much cleaner air within cities and industrial areas.

Fuel for automobiles, trucks, and busses poses a particular problem in a hydrogen-fueled economy. Their engines would have to be modified, and all such vehicles would have to carry unwieldy gas bags, or heavy cylinders of compressed gas. The advocates of hydrogen fuel* point out that liquid hydrogen is now commonly transported by tank truck and used as a rocket fuel, so that the technology and the safety precautions are already worked out for large con-

*See J. Bockris, "A Hydrogen Economy," *Science*, **176**:1323, 1972. It has been pointed out since then by David Whitten that sunlight can decompose water directly into H_2 and O_2 in the presence of a ruthenium catalyst.

sumers (say locomotives, large trucks, and heavy tractors, which could use up a tankful in a few hours). Since liquid hydrogen boils at 20.2° K (degrees absolute; −252.8° C), the evaporation loss from a small tank located in the family car would be prohibitive, and the explosion hazard of accumulated boiled-off gas would be worse. Inventors would have to come up with another fuel for the small automobile, or else a different method of propulsion.

HELIUM ON EARTH AND THE OTHER PLANETS

The "solar wind," a stream of protons, alpha particles, and other positively-charged particles from the sun, delivers some helium to all the planets, but the concentration in the earth's atmosphere is very low. Quite likely it is continuously lost from the atmosphere because the earth's gravitational field is too weak to hold such light molecules (molecular weight = 4.003). A measure of the equilibrium between acquisition of helium from all sources and loss from the atmosphere into space is given by the Mariner 10 report from Mercury on March 24, 1975: Mercury (40% of Earth's diameter) has a very thin atmosphere composed primarily of helium, at a surface pressure of only two millionths of our normal atmospheric pressure. Presumably this is all the planet can hold.

Helium is being produced continuously in the rocks of the planets by the radioactive decay of uranium and thorium. Earth's crust contains 4 parts per million of uranium and 12 parts per million of thorium, both of which decay by the emission of successive alpha particles and end up as stable isotopes of lead. The alpha particles are slowed down by the surrounding rock and pick up electrons from it, leaving atoms of helium trapped within the solid matrix. By powdering a sample of igneous rock and then heating it within a previously evacuated chamber, the helium can be collected and measured. Since uranium decays at a rate that destroys half of it every 4.51 billion years, we can measure the amount of uranium still left and the helium that was trapped, and from these calculate the age of the rock. Similarly, thorium-containing minerals can be analyzed and the trapped helium in them collected, and since half the thorium is known to disintegrate every 13.9 billion years, the age of the minerals can be calculated. The results show that the rocks and minerals were solid and remained in place about 1.8 billion years. The figure must be regarded as a lower limit, because helium does diffuse through rock and so all of it does not remain trapped. Other methods for estimating the age of the earth's solid crust (methods which depend on the ratios of stable and radioactive isotopes, without reference to gases) give results of 3.0 to 4.5 billion years.

Some of the helium which escaped from rocks containing uranium and thorium has collected in very old pockets of natural gas underground, and it can be isolated by compressing and cooling until it becomes liquefied and then distilling off the helium as the lowest-boiling constituent. Alternatively, absorption methods can be used. The helium so collected is expensive, but it is a valuable gas for inflating blimps and weather balloons, and it provides a blanket of unreactive gas for the arc welding of aluminum.

Molecules of helium have so little attraction for each other that it is the lowest-boiling substance known. It boils at −268.9° C (4.2° absolute, or ° K), and if it is allowed to boil off at low pressure (0.01 mm) a temperature only 0.7° above absolute zero has been obtained. Between 4.2° K and 2.18° K the liquid behaves normally, but below 2.18° K an abnormal liquid called helium II is observed. This exhibits stepwise changes of specific heat, compressibility, surface tension, and thermal conductivity as the temperature is lowered, due to the

TABLE 2–3 PROPERTIES OF HELIUM

Melting point	−272.1° C
Boiling point	−268.9° C
Density of gas at 0° C	0.1785 g/liter
Density of liquid at its boiling point	0.122 g/cm^3
Solubility in water at 1 atm pressure	9.26 ml/liter of water

removal of successive quanta of energy at so low a temperature. In fact, this behavior is the most striking proof we have of the validity of the quantum theory. Helium II has one other strange characteristic: it wets and spreads on every surface with which it comes in contact, and will climb right out of a suspended beaker and drip off the bottom.

Some physical properties of helium are listed in Table 2–3, for those who are curious. The U.S.A. is the only country that has a usable supply of helium, and it should be guarded as an irreplaceable natural resource (see Figure 2–1 for a nondestructive use that is familiar). Other uses which do not conserve the helium are: (1) as an inert-gas shield in arc welding; (2) as a diluent for oxygen to provide breathing gas for divers; (3) as a laboratory gas and carrier in gas chromatography; (4) in the liquid form, as a refrigerant at very low temperatures for experiments on superconductivity; and (5) as a moderator and coolant in experimental nuclear reactors. The last use depends on the extremely low neutron absorption of helium, beyond the limits of measurement (which denotes a completely stable nucleus, uninterested in more neutrons). The second use depends on the low solubility of helium in water (9.26 ml/liter vs. 23.3 ml/liter for nitrogen), which prevents excessive solution of the gas in the bloodstream at high underwater pressures and hence avoids the painful and dangerous nitrogen bubbles that would occur upon decompression if air were breathed. Researchers have lived for weeks in a submersible vessel filled with 20% oxygen and 80% helium, the only complaints being feeling cold (because helium transfers heat away from the body faster than nitrogen) and a squeaky high-pitched voice that makes recognition of the voice difficult. The raised pitch comes from more rapid vibration of light molecules in the larynx, an effect that can be verified by anyone simply by taking a breath of helium from a tank of the gas or a balloon filled with it, and then speaking right away.

Figure 2–1. Helium-filled Goodyear airships of the Columbia type. These "blimps" are 192 ft long, 59 ft high, and 50 ft wide and have a volume of 202,700 ft^3. The gross weight is 12,320 lb, and the maximum lift is 3281 lb (the design rule is that 1000 ft^3 of helium at STP will lift 60 lb). Six passengers and the pilot are accommodated in the cabin. These large airships use only two 6-cylinder engines of 210 hp each, and are very economical of fuel; one can operate for a week on the amount of fuel used by a big jet as it taxies from the ramp to the end of the runway.

SUMMARY

Hydrogen (at. no. 1) and helium (at. no. 2) comprise the first period of the Periodic System. They are both very light gases, the density of H_2 being only 0.09 g/liter and that of He 0.179 g/liter. Hydrogen is very reactive chemically; it burns in air, liberating 57.8 kcal per mole of H_2 consumed. Helium is completely inactive chemically. Both elements have extremely low boiling points, that of H_2 being only 20° C above absolute zero and that of He being only 4° C above absolute zero. Liquid hydrogen is used as a rocket fuel, and is proposed as a substitute for gasoline and diesel fuel for trucks and locomotives. Liquid helium is a convenient laboratory refrigerant. When it is cooled by rapid evaporation, liquid helium converts at 3° K to a strange new form that crawls out of beakers and shows quantum steps in its physical properties.

Helium is formed in the interior of the sun and stars by two nuclear processes that release 592 million kilocalories per mole of He formed. This is the source of the tremendous radiation that warms our planet and grows our food. The sun is an average star which has already "burned up" 20% of its nuclear fuel, but it should last for thousands of billions of years more. Some method for initiating and maintaining controlled nuclear fusion of hydrogen on earth is being sought avidly as an inexhaustible source of energy to replace fossil fuels and uranium.

Practically no hydrogen or helium is found in the earth's atmosphere. Helium exists in small concentrations in some pockets of natural gas, from which it can be obtained by chromatographic absorption or by distillation; it is used as an inert protective gas in welding active metals, and to fill balloons and airships. Hydrogen is obtained by electrolysis of water or by "cracking" hydrocarbons in refining petroleum; huge quantities of it are used to make ammonia for nitrogenous fertilizer and for desulfurizing petroleum products. If we had enough cheap electric power, we could replace natural gas with hydrogen derived from water electrolytically, thereby ending almost all air pollution from this source.

The ${}^2_1H$ isotope, deuterium, forms the oxide D_2O, which is called "heavy water." It has a density 10% greater than that of H_2O, melts at 3.8° C, and boils at 101.4° C. It is isolated from ordinary water by distillation, and is used as a moderator and coolant in Canadian nuclear reactors because its neutron absorption is only one six-hundredth that of H_2O, thus allowing the use of natural uranium as fuel without isotopic enrichment.

Water is not a universal solvent but it is an ever-present medium for dissolving an enormous variety of substances. What it cannot dissolve it can usually hold in suspension. The oceans were discussed as an example of what water can hold in solution. They constitute an immense storehouse of raw materials, from which we already derive drinking water, salt, magnesium as a structural metal, and bromine for chemical manufactures. There is also a possibility of obtaining potassium and phosphorus for fertilizer, manganese for alloy steel, and even silver.

GLOSSARY

aerosol: a suspension of liquid or solid particles in a gas. An aerosol differs from a spray in that the particles are often so small as to be invisible.

alpha particle: the nucleus of an atom of helium-4; an He^{+2} ion.

amino acid: an organic acid in which an amino group ($-NH_2$) is attached to one of the carbon atoms. In the amino acids which are the building blocks of all proteins, the amino group is attached to the carbon atom immediately adjacent to the acid group ($-COOH$).

critical pressure: the pressure exerted by a substance at its critical temperature.

critical temperature: the maximum temperature at which a liquid can exist, regardless of the pressure applied. Above the critical temperature thermal agitation is too intense to allow the molecules of gas to agglomerate as liquid.

emulsion: a suspension of droplets of one liquid in another, as of butterfat in milk or of rubber particles in latex. Usually the suspension is stabilized by a thin layer of a third phase ("emulsifying agent") which covers the suspended droplets and prevents their coalescence.

half-life: the time required for the decay of exactly one half of any quantity of a given radioactive species.

metamorphic mineral: a constituent of preexisting rock which has been altered by higher temperatures and greater pressures than those normally present at the earth's surface.

plasma, blood: the clear straw-colored liquid which is left after the red cells and white cells are removed by centrifugation. It consists of a solution of many salts, proteins, and sugars in water.

quantum (pl. quanta): a fundamental unit or discrete packet of energy, as called for in the quantum theory.

quantum theory: the concept (proposed by Max Planck in 1900) that energy, like matter, is atomistic and is radiated only in quanta defined by the relation

$$E = h\nu$$

where E is the amount of energy involved, ν is the frequency of its associated radiation, and h is a universal constant of nature with the value of 6.623×10^{-27} erg-seconds.

specific heat: the amount of heat required to raise the temperature of 1 gram of a particular substance 1 degree centigrade.

thermodynamics: the branch of physics which deals with the relation between heat and work, or heat and other forms of energy.

EXAMINATION QUESTIONS

1. Name the isotopes of hydrogen, and describe their occurrence and properties.

2. Why are the isotopes of hydrogen given specific names, whereas the isotopes of other elements are not?

3. Outline the two principal thermonuclear mechanisms whereby hydrogen is converted to helium within the sun and other stars.

4. Approximately how much more energy is released during the formation of 1 mole of helium from hydrogen by thermonuclear processes than is released from a comparable classical chemical reaction, say the oxidation of 2 moles of hydrogen to form water?

5. What are the properties of "heavy water," D_2O? How may it be isolated from ordinary water? Would you expect the isolation to be expensive? Why?

6. Describe at least two important uses for heavy water, and explain why they are important.

7. Is water a base? An acid? Why? How acidic or basic is water at room temperature?

8. By means of chemical equations, illustrate the preparation of hydrogen from water and a metal (a) at room temperature and (b) at red heat (say 600° C).

9. In what way is the acidity of water responsible for some of the high cost of electricity?

10. What is dissolved in seawater? (Name the principal ions present, and the approximate proportions of each.)

11. What valuable products are being isolated from sea water at present? Name at least four.

12. Give two important examples, drawn from living systems, of the ability of water to keep insoluble substances in suspension. Why does the insoluble matter stay suspended?

13. Give three current uses for helium, and describe how each depends on a particular property of helium.

14. Does any one nation have an advantage over all others in gaining access to the raw materials of seawater? Are there any nations at a disadvantage?

"THINK" QUESTIONS

A. Why do researchers in an undersea vessel filled with a mixture of helium and oxygen feel colder than they do in ordinary air at the same temperature? Refer to the kinetic theory of gases and come up with a quantitative explanation.

B. Consider the buoyancy of a blimp, as in Figure 2–1. The molecular weight of helium is 4.003, and the molecular weight of hydrogen is 2.016, so helium weighs twice as much as hydrogen per cubic foot. Why is it, then, that helium will lift 96% as much weight as the same volume of hydrogen, when used in a blimp?

C. Helium was first discovered in the spectrum of the sun's corona, and hence its name (from the Greek *helios*, meaning sun). Twenty-four years afterward, the same bright yellow line was found in the spectrum of the gas extracted from a uranium mineral, and so helium was discovered on earth. Why is the spectrum of an element unique, and why can it be used with such confidence to prove the presence of that element, even in a sample too small to weigh?

D. A mixture of hydrogen and air will ignite and explode, but only if the air contains more than 6.2% and less than 71.4% hydrogen. Why are there lower and upper limits for the inflammability of hydrogen in air? Why will not mixtures in all proportions ignite and explode?

E. A premixture of two volumes of hydrogen and one volume of oxygen will explode violently when ignited, but if the two gases are mixed only within the flame itself (by using a specially designed burner) the mixture will burn quietly and continuously. The maximum flame temperature is 2210° C.

a. Why is this temperature less than that at the surface of the sun?

b. Will the flame temperature be the same when air is substituted for pure oxygen? (Air is approximately one fifth oxygen and four fifths nitrogen.) Will it be higher? Lower? Why?

F. Look up the history of the Hindenburg, a large transatlantic airship that made regular crossings with passengers in the 1930's. What happened to it? What are the advantages and disadvantages of hydrogen over helium for use in rigid airships?

G. The gold content of seawater has been found to vary markedly from one place to another. Where would you expect the gold content to be highest? Where lowest? Has any serious attempt ever been made to extract gold from seawater economically? When, and why? What happened? (Hint: look up the life of Fritz Haber.)

H. Considering the composition of seawater given early in this chapter, do you think that the silver in it could be precipitated as the compound silver chloride, AgCl? (Look up the solubility of silver chloride in a chemical handbook, if the answer is not immediately apparent to you by logical reasoning.)

PROBLEMS

1. What volume of hydrogen at standard conditions would be liberated by the action of 10 g of sodium on water?

2. What volume of oxygen is needed to burn 10 liters of H_2 to H_2O, both volumes being considered at standard conditions?

3. The atmosphere of the planet Mercury consists of helium at a pressure of 0.000002 atm, or 0.00152 mm of mercury. How many molecules of helium are there in 1 liter of this atmosphere?

4. What is the pressure exerted by 20 g of helium contained in a laboratory apparatus with a volume of 10 liters?

5. Consider the following equation, which is not yet balanced:

$$Fe_2O_3 + H_2 \rightarrow Fe + H_2O$$

What volume of hydrogen at 27° C and 340 mm pressure is needed to obtain 0.25 mole (or gram atomic weight) of metallic iron?

6. A student collected 300 ml of H_2 over water at 740 mm pressure and 27° C. Calculate its volume at standard conditions.

7. A volume of hydrogen that occupied 300 ml at 194° Fahrenheit was cooled at constant pressure to 50° Fahrenheit. Calculate the final volume.

8. Calculate the ratio of the rate of diffusion of H_2 to the rate of diffusion of CO_2.

9. Using the expression $PV = 1/3\ nmv^2$, where P is pressure, V is volume, n is the number of moles of gas, m is the molecular weight, and v the velocity of the molecules, derive a relation between velocity and temperature, and then calculate the velocity of H_2 molecules at 0° C.

10. At what temperature will H_2 have the same rate of diffusion that He has at 20° C?

11. What volume of hydrogen (at standard conditions) would be needed to remove all the sulfur from 1000 kg of crude oil as H_2S? Assume that the oil contained 4.5% sulfur.

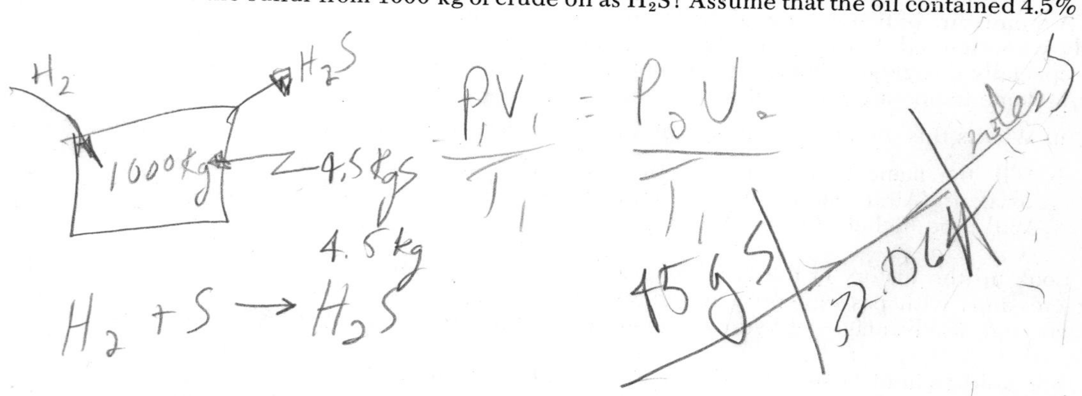

LITHIUM THROUGH FLUORINE: THE LIGHT ELEMENTS

3

We have seen that all the chemical elements are made of the same stuff, but all are not created equal. Now that we expect wide variation of occurrence and diversity of behavior, it is time to take a closer look at the seven chemically active elements of the second period, lithium through fluorine. Chemical change and the energy exchange accompanying it are the lifeblood of chemistry, so from here on these must be our chief concern. We shall be guided by two considerations: our study of chemical behavior must either satisfy some deep-seated intellectual curiosity about the material world around us, or it must respond to some important practical need in our culture or our economy. These are not two entirely separate considerations, for we often find that facts uncovered purely out of curiosity become very useful later on and, conversely, information gathered in response to some urgent practical problem often contributes essential building blocks to our intellectual conception of things. No knowledge is ever wasted.

WHAT THESE ELEMENTS HAVE IN COMMON

From your study of atomic structure, you will recognize that the seven elements with atomic numbers 3 through 9, lithium through fluorine, correspond to progressive filling of the 2*s* and 2*p* orbitals with electrons, giving rise to the whole range of chemical behavior open to light elements. The first element in the sequence, lithium, loses an electron readily and forms a unipositive ion, Li^+, so it is a reducing agent, and a base-former. The next three elements (Be, B, C) are not ionic in their behavior, but like to share their electrons. That is, they release more chemical energy and reach a more stable condition by sharing electrons than by losing or gaining them outright. Physically, they range from a metal through a metalloid to a nonmetal. The last three elements (N, O, F) may share electrons (and frequently do), but are also rabid electron-snatchers as they try to form negative ions, and so are electronegative elements, oxidizing agents, and acid-formers.

The chemical behavior of all seven elements is somewhat limited in that it does not encompass the magnetic or optical behavior we associate with *d*-orbital interactions in the transition metals. In other words, the elements of the second period usually form colorless, nonmagnetic compounds with each other, and with any other elements that employ only *s*- and *p*-type orbitals in their chemical bonding.* There are a few exceptions to the rule, notably among the oxides of nitrogen, but they are rare. This situation comes about because *d* orbitals are at too high an energy level to be available to second-period elements, and hence are almost never called into play.

Something else limits the chemical behavior of the present sequence of elements: they never combine with more than four partners. Regardless of which of the seven elements we pick, it never forms more than four covalent bonds. It may form less, of course, depending on the number of electrons available for shared-pair bonding, but its maximum covalency is limited to four. Heavier elements have a maximum covalency of six, or even eight, and their behavior is correspondingly more complex.

As would be expected, all the elements in the present sequence have small atoms and form relatively small ions. This small size favors covalent bonding for a sound physical reason: the smaller an atom, the closer its peripheral electrons are to the central positive charge, and the more tightly they are held. Hence, ionization of a small atom requires more energy than ionization of a larger atom of similar atomic structure, whereas covalent bonding is made easier. Furthermore, small positive ions attract *pairs* of electrons which were not previously used for covalent bonding, and may hold onto such electron pairs to establish a donor-acceptor bond, often called a coordinate covalent bond. This is the mechanism involved in the energetic hydration of lithium ions:

$$Li^{+}(aq) + :OH_2(l) \rightarrow (Li:OH_2)^{+}(aq)$$

These considerations are formalized in generalizations called Fajans' Rules,† which state that the smaller a positive ion and the greater its positive charge, the more likely it is to enter into covalent bonding; conversely, the larger a negative ion and the greater its negative charge, the more likely it is to contribute a pair of electrons for coordinate covalent bonding.

The reader should not draw false conclusions from Fajans' Rules about the *energy* liberated during chemical reaction. Ionic bonding does not necessarily lead to the liberation of more energy, per se, than covalent bonding; either type of combination may be energetic or mild, depending on other factors. All seven of the elements in the current sequence liberate comparatively large amounts of energy when they combine, no matter how they combine.

---

*When we say nonmagnetic in this book we mean not attracted by a magnet or pulled into a magnetic field. All matter, of whatever sort, displays a very weak repulsion from a magnetic field which we shall ignore here.

†After Kasmir Fajans, who stated it all very clearly in 1923. For details see B. E. Douglas and D. H. McDaniel, *Concepts and Models in Inorganic Chemistry*, New York: Blaisdell Publishing Company (1965), pp. 89–91.

THE POINTS OF DIFFERENCE AMONG THESE ELEMENTS

SIZE. If we measure atomic radii in angstrom units (Å), each being 10^{-8} cm, we find an abrupt drop from lithium (1.52 Å) to beryllium (1.11 Å), followed by a slow and steady further decrease to 0.64 Å for fluorine. As for radii of the ions, Li^{+} is only half the size of its parent atom (0.60 Å), and the Be^{2+} ion is truly tiny (only 0.31 Å). These ions are much smaller than the neutral atoms because they have lost their outermost electrons, leaving only the nucleus and the pair of 1*s* electrons. At the other end of the sequence, as electronegative atoms gain electrons to form negative ions, the additional electrons go into peripheral positions and increase the effective radius. Moreover, there has been no corresponding increase in positive charge on the nucleus during ion formation, so with a surplus of electrons *all* electrons are held more loosely, and mutual repulsion spreads them out in a larger sphere. Thus, we find the sizes of the pertinent atoms and ions, respectively, to be F = 0.64 Å, but F^{-} = 1.36 Å; O = 0.66 Å, but O^{-2} = 1.40 Å; N = 0.70 Å, but N^{-3} (when it forms) = 1.71 Å. These sizes govern the resulting density of a compound, and of course influence the degree of hydration in solution and in crystals.

ELECTRONEGATIVITY. Although commonly used in textbooks to describe the general degree of electropositive or negative behavior, electronegativity originally was defined by Linus Pauling in 1939 as the force of attraction exerted on an electron by a given element in one of its compounds. Since this attraction will vary from one compound to another, the calculated values must vary. Nevertheless, some charts print values of electronegativity to three or four significant figures, as though each were a precise value like the atomic weight. Figure 3-1 gives electronegativities more appropriately to two significant figures, which

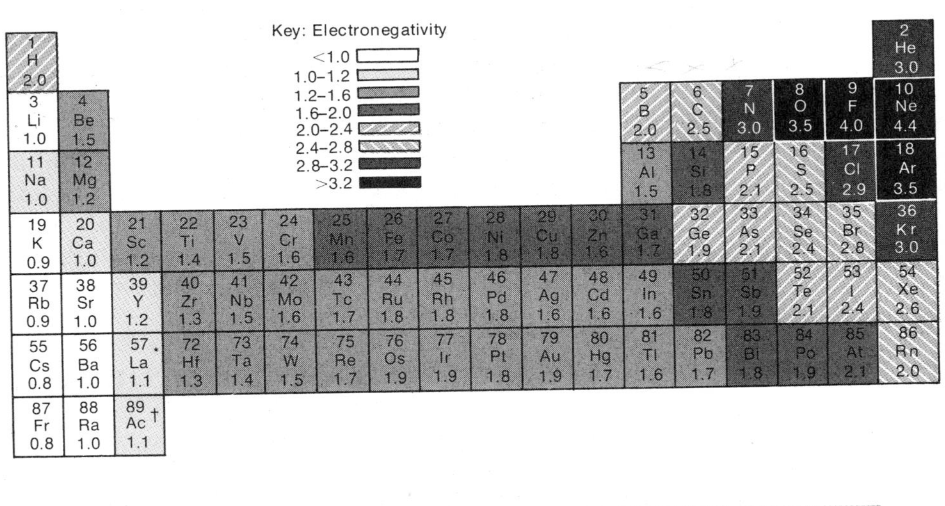

Figure 3–1. Electronegativity values of the elements. The shading of the blocks indicates the EN value of that element according to the key in the figure. (From E. G. Rochow, G. Fleck, and T. H. Blackburn, *Chemistry: Molecules that Matter*, Holt, Rinehart & Winston, New York, 1974, p. 187. By permission.)

are averages from several methods of calculation.* Notice that the figures for the present sequence are: Li, 1.0; Be, 1.5; B, 2.0; C, 2.5; N, 3.0; O, 3.5; and F, 4.0. This variation is the most extreme to be found among any sequence of *s* and *p* elements; the range of electronegativity from Li to F is 3.0 units, while for Na to Cl it is 1.9 units, for K to Br it is 1.9 units, for Rb to I it is 1.5 units, and for Cs to At it is only 1.3 units. Hence the small span of seven light elements in the second period encompasses more change in chemical character than anywhere else in the periodic table.

BEHAVIOR. Lithium is a silvery gray metal, light enough to float on water or oil, and harder than the other metals of Group I. It reacts with water with enthusiasm, evolving hydrogen and forming an alkaline solution of LiOH. Beryllium is a hard, light, strong metal which few people ever see. A thin foil of it will burn, with the evolution of a great deal of heat and light, to form an insoluble white oxide. Its oxide is very weakly basic; almost all its compounds other than the oxide are covalent. Its compounds are all white, as are those of lithium. Boron is a black, crystalline, nonmetallic element, extremely hard and high-melting. Its compounds are covalent; BF_3 and BCl_3 are gases, BBr_3 is a volatile liquid, and even boric acid and the glass-forming oxide B_2O_3 will sublime. The hydrides of boron have anomalous formulas and do not fit into the usual pattern of shared-pair covalent bonding (see p. 53); they are gases and liquids which usually ignite spontaneously in air and burn with a high heat of oxidation. Carbon is familiar as charcoal, graphite, and diamond, and of course as the structural basis of all organic matter and all living organisms. Its compounds number in the millions, almost all of them thermodynamically unstable in air but capable of surviving for centuries due to lack of a mechanism for their oxidation. The exceptions are the stable dioxide and the carbonates; in between lies the exceedingly poisonous monoxide CO, which forms coordination compounds with hemoglobin and with most transition elements. Nitrogen is an active element mostly present in an inactive gaseous form (triple-bonded N_2) in the atmosphere. It too is essential to all life, and enormous amounts of it must be supplied in suitably combined forms in order to grow the world's food supply. Oxygen is our great and common oxidizer, which allows combustion and respiration. Its oxidizing† power is exceeded only by that of fluorine, which combines vigorously with almost everything and is the most violently destructive gas we know. Paper, wood, iron, aluminum, and sulfur all ignite and burn in a stream of fluorine, and even platinum succumbs if heated. All covalent fluorides are gaseous or liquid or low-melting, and hence the name (from Latin *fluo*, to flow).

SPECIAL POINTS OF INTEREST ABOUT LITHIUM

Lithium is a rare element, compared with its cousins sodium and potassium, and it is widespread throughout the mineral world. Hence it is difficult to scrape together, and is expensive. The metal is obtained by electrolyzing melted LiCl, and is kept under mineral oil to prevent its reaction with O_2, H_2O, or CO_2 in the air:

---

*For a more detailed discussion of electronegativity and the methods for calculating it, see F. A. Cotton and G. Wilkinson, *Advanced Inorganic Chemistry*, 2nd ed., New York: Interscience Publishers (1966), pp. 720–747.

†As explained in Chapter 1, *oxidation* is a general term used in chemistry to denote a loss of electrons, as when Li is oxidized to Li^+. The *oxidizing agent* therefore strips electrons from another substance and holds them. Thus it is the fate of oxidizing agents to gain electrons (be reduced) in chemical reaction.

$$4Li(s) + O_2(g) \rightarrow 2Li_2O(s) \text{ (composed of } Li^+ \text{ and } O^{-2} \text{ ions)}$$

$$2Li(s) + CO_2(g) + H_2O(g) \rightarrow \text{ionic } Li_2CO_3(s) + H_2(g)$$

Lithium reacts violently with chlorine to form the white chloride

$$2Li(s) + Cl_2(g) = 2LiCl(s) \text{ (composed of } Li^+ \text{ and } Cl^- \text{ ions)}$$

and explosively with fluorine to form the white fluoride

$$2Li(s) + F_2(g) = 2LiF(s) \text{ (composed of } Li^+ \text{ and } F^- \text{ ions)}$$

The last two reactions can be portrayed more descriptively by pairs of equations:

$$Li \rightarrow Li^+ + e^- \text{ (oxidation)}$$

and

$$F + e^- \rightarrow F^- \text{ (reduction)}$$

or

$$Cl + e^- \rightarrow Cl^- \text{ (reduction)}$$

Lithium also reacts very vigorously with any aqueous acid to liberate hydrogen

$$2Li(s) + H_2SO_4 \text{ (dilute)} \rightarrow 2Li^+(aq) + SO_4^{-2}(aq) + H_2(g)$$

and reacts similarly with liquid water

$$2Li(s) + 2H_2O(l) \rightarrow 2Li^+(aq) + 2OH^-(aq) + H_2(g)$$

Metallic lithium reacts with a variety of organic compounds to form active organic derivatives which are exceedingly useful in organic synthesis, and therefore it is a common laboratory reagent. The melted metal is also used as a heat-transfer medium in experimental nuclear reactors.

Because of its highly electropositive character and the ease with which the Li^+ ion hydrates in water solution, lithium stands at the head of the electrochemical series of metals and has the highest standard oxidation potential. This means that if it is used in a battery as the negative electrode, lithium delivers a higher voltage than any other metal, *four times* as high as zinc (which is the electrode metal used in the common dry-cell battery). There are some experimental batteries that use lithium electrodes, but the lithium starts to react as soon as it meets the watery medium of the battery (the electrolyte), and so the assembly has short shelf-life. Some day when there is a more intense need for electric automobiles, research on the practical aspects of batteries may be stimulated to the point of producing a lithium battery that stands by, without reacting, until it is needed. The benefit is not just a matter of higher voltage; a lithium battery would require only one fourth as much weight of metal per ampere-hour of energy delivered as a zinc dry cell. Perhaps one day we shall drive into a service station and trade in the battery's used replaceable lithium electrode for a new one, instead of buying a tankful of gasoline. Metallic lithium could then be recovered from the spent electrodes at a central processing plant.

Lithium occurs with sodium and potassium in many minerals, especially mica. Since it is so much like sodium and potassium, you may wonder how it can be separated from them. We have seen that lithium compounds are more covalent than their sodium and potassium analogs, and we find that lithium chloride is sufficiently covalent to be soluble in alcohol whereas sodium and potassium chlorides are not. A mixture of all three chlorides can be recovered from hydrochloric acid treatment of the source mineral, and lithium chloride can then be extracted from the dry residue by repeated treatment with hot alcohol.

Lithium chloride itself, melting at 613° C, is used in liquid form as a solvent and reaction medium for electrolytic processes. For example, partial electrolysis

$$Li^+ + e^- \rightarrow Li(l)$$

leaves droplets of metallic lithium suspended in the chloride, and these react with hydrogen to form lithium hydride

$$2Li(l) + H_2(g) \rightarrow 2LiH(s)$$

a strong reducing agent that can be used immediately to make hydrides of boron and silicon.

Lithium hydride is an interesting compound because it is a white salt consisting of Li^+ and H^- ions, where each atom of lithium has lost its electron and each hydrogen atom has gained an electron. Whenever and wherever the H^- ion meets an H^+ ion, in water or acids or elsewhere, immediate reaction occurs to form H_2:

$$H^- + H^+ \rightarrow H_2(g)$$

Hence LiH is a portable source of hydrogen that requires no high-pressure cylinders.

The innocent-looking white powder of the composition LiH has a further significance which will be appreciated from three nuclear equations:

$$^7Li + {}^1H \rightarrow 2\ {}^4He + 399 \times 10^6 \text{ kcal/mole}$$

$$^6Li + {}^2H \rightarrow 2\ {}^4He + 517 \times 10^6 \text{ kcal/mole}$$

and

$$^6Li + {}^3H \rightarrow 2\ {}^4He + {}^1H + 378 \times 10^6 \text{ kcal/mole}$$

Here the two isotopes of lithium and the three isotopes of hydrogen undergo fusion reactions at a temperature of several million degrees, liberating an enormous amount of energy. The latter two equations are the basis of the French hydrogen bomb exploded on August 24, 1968. Evidently solid LiH can be used as a compact source of the lithium and hydrogen isotopes for this fusion, and the necessary temperature to set it off is reached by detonating a fission bomb within the mass. Knowing that this can be done, the more important question to be answered is how to start and then control a slow steady fusion of minute particles of lithium hydride as a continuous source of energy for domestic electric power. At present no one knows.

Lithium carbonate is an insoluble white powder which resembles magnesium carbonate and is very different from the very soluble sodium and potassium carbonates. Similarly, it is lithium and magnesium (not lithium and sodium) that occur together in many silicate minerals. In many other ways, also, lithium resembles magnesium rather than the other alkali metals of Group I. The reason for this is found in the sizes of the ions: Li^+ has a radius of 0.60 Å and Mg^{+2} a radius of 0.65 Å, but Na^+ has a radius of 0.95 Å and is too large to replace Li^+ or Mg^{+2} in a crystal structure. Here we have the first example of a *diagonal relation-*

ship between elements as they appear in the periodic table. We shall soon discover more examples of this type of relationship, and the explanation in terms of atomic structure will become obvious.

Soap may seem the least likely place where lithium might be encountered, but lithium soap is actually a valuable commodity. A soap is just an alkali-metal salt of an organic acid which has a long chain of carbon atoms, as in sodium stearate, $CH_3(CH_2)_{16}COONa$. Potassium soaps are soft and low-melting, and are used in liquid detergents. Sodium soaps have higher melting points and are the basis for "hard" soap, the familiar bar soap. Lithium soaps have still higher melting points. A lubricating grease is a combination of lubricating oil and some thickening agent which keeps the oil in place, and the usual thickening agent is sodium soap. At the high temperatures encountered in modern cars and trucks, however, the sodium soap melts and the grease runs out. This undesirable state of affairs is prevented by using higher-melting lithium soap as the thickening agent, so the grease stays put. If there is a question about which kind of grease is being used, it is easy to test. Lithium can be detected in small amounts by the bright red color it imparts to a Bunsen flame, and by the two red lines in its spectrum (see Plate 1).

BERYLLIUM, THE RARE ONE

One of the mysteries surrounding the genesis of the elements can be stated this way: Whatever happened to the nuclei with four protons and four neutrons? No one knows. We have seen how the even-even rule works, and we recognize the familiar and abundant elements corresponding to one, three, four, five, six, seven, and eight alpha particles (He, C, O, Ne, Mg, Si, and S). But there is no common or abundant atom of mass 8, corresponding to two alpha particles. Beryllium is element number 4, but there is no natural stable isotope ^{8}Be. Even when ^{8}Be is made by cyclotron techniques, it decays with a half-life of only 10^{-16} seconds. Natural beryllium is all in the form of ^{9}Be, and there isn't very much of that; beryllium constitutes only 0.0005% of the lithosphere (the rocky crust of the earth). So rare an element does not deserve much space here, but beryllium performs several important functions which should be mentioned.

The source of beryllium is the gemstone beryl, $Be_3Al_2Si_6O_{18}$, which occurs in various colors due to traces of various impurities. When it is a light blue-green it is called aquamarine, and when it is deep green it is called emerald. Of course emeralds are not used for the production of metallic beryllium; the very imperfect crystals of colorless or brown beryl are used. The metal is obtained by electrolyzing a mixture of $BeCl_2$ and BeF_2 at 800° C. It is used in copper alloys to strengthen and stiffen the copper for use in electrical switches and machinery. Pure beryllium finds its largest use in nuclear reactors, where its transparency to neutrons, its great strength, and its low density recommend it for structural parts. It is transparent to x-rays, and is used for windows in x-ray tubes. It is as strong as steel, but only one fourth as heavy (d = 1.85 g/cm³ vs. 7.85 g/cm³ for Fe), and it melts at 1285° C. One drawback is that the metal and its compounds are poisonous; Be^{+2} has such a small size and such a high charge that it attracts electrons from the nitrogen atoms in proteins, and so destroys the structure and function of body tissues. Beryllium in any form is especially toxic if it gets into cuts or is inhaled.

Beryllium oxide, BeO, is distinguished by a very high melting point (2570° C) and a general chemical inertness. It is a white powder which can be pressed and then sintered in an electric furnace to make crucibles and insulators for high-temperature use. It has another small but important use—a small amount of BeO mixed with a few micrograms of radium becomes a laboratory source of neutrons, because alpha particles from the radium react with ^{9}Be to form carbon and eject neutrons:

$$^{9}_{4}Be + ^{4}_{2}He \rightarrow ^{12}_{6}C + ^{1}_{0}n$$

The other compounds of beryllium are almost all 4-covalent, and they resemble the compounds of the Group III element *aluminum* more than those of magnesium and calcium, which are beryllium's relatives in Group II. Similarly, boron will soon be seen to resemble silicon, in Group IV, more closely than it resembles aluminum, in its own Group III. These examples point out a prominent diagonal relationship between the representative elements.

A glance at the Periodic Table shows that in all such diagonal relationships the diagonal always points from upper left to lower right, for two good reasons: the diagonally-placed ions are either the same size, or else increased size compensates for increased charge on the ion. In effect, the increased radius dictates an increased surface, because the surface of a sphere is $4\pi r^2$, where r is the radius. This necessarily spreads the electric charge over a larger surface, and the charge per unit area is correspondingly decreased. Hence the charge per unit area is likely to be more nearly the same for two ions in diagonal relationship than for two ions in the same group.

BORON, THE HARD AND QUEER ONE

Like beryllium, boron is a very rare element; it makes up only 0.0003% of the earth's crust. However, it has the advantage of being concentrated in a single water-soluble mineral, borax, which is the hydrated sodium salt of tetraboric acid, $Na_2B_4O_7(H_2O)_{10}$. This is found in the alkaline residue from dried-up ancient lakes in our Southwest desert, from which it can be recovered by extracting the residue with hot water and then crystallizing the borax from a saturated solution by cooling. Large quantities are used as a mild alkali and cleaning agent, usually in the form of laundry detergent mixtures. It also is used as a flux in welding operations because it melts at a low temperature, dissolves metal oxides, and leaves a clean surface.

Boron compounds are derived from borax by way of boric acid and its anhydride, boric oxide. Treatment of recrystallized borax with sulfuric acid gives boric acid, H_3BO_3, which is very sparingly soluble in cold water:

$$Na_2B_4O_7(H_2O)_{10}(s) + H_2SO_4(l) \rightarrow Na_2SO_4(aq) + 5H_2O + 4H_3BO_3(s)$$

Boric acid is used in water solution as a mild antiseptic. When the solid acid is melted and heated strongly, it gives up water and leaves a clear colorless melt of boric oxide, B_2O_3:

$$3H_3BO_3(s) \rightarrow B_2O_3(s) + 3H_2O(g)$$

If the melt is cooled quickly it retains all the properties of a glass. Hence B_2O_3 is one of the few glass-forming oxides, and it is used in conjunction with the only other glass-former of practical importance, SiO_2, to make borosilicate glasses such as the popular Pyrex ovenware and laboratory glassware. Here boric oxide replaces most of the sodium and calcium oxides which are used in window and bottle glass, without raising the softening point. The result is a lower thermal coefficient of expansion than in a soda-lime glass, and so the borosilicate glass is less likely to crack when subjected to sudden changes in temperature.

BORON HALIDES. Boron trifluoride is a highly corrosive gas (bp 100° C) made by treating a mixture of boric oxide and calcium fluoride with concentrated sulfuric acid:

$$CaF_2(s) + H_2SO_4(l) \rightarrow CaSO_4(s) + 2HF(g)$$

$$B_2O_3(s) + 6HF(g) \rightarrow 2BF_3(g) + 3H_2O(l)$$

The water is retained by the sulfuric acid, and the BF_3 is collected and compressed. Boron trifluoride is used as an acid catalyst in the refining of petroleum and in other aspects of industrial organic chemistry. The term "acid catalyst" does not refer to the usual water system of acids and bases, but to the more general concept of acids advanced by G. N. Lewis in 1923. A Lewis acid is defined as any molecule or ion that can accept a pair of electrons, and a Lewis base is any molecule or ion that can donate such an electron pair. Boron trifluoride is a potent Lewis acid because the three fluorine atoms draw the three pairs of bonding electrons toward themselves, leaving the boron atom strongly electron-deficient. Any donor atom, ion, or molecule can then supply the necessary pair of electrons to establish a coordinate covalent bond. The oxygen atom of ether (ordinary hospital ether, Et_2O, where Et represents the ethyl group C_2H_5) does this in the most familiar coordination compound of BF_3:

```
 Et             Et
 |    F         | F
:O: + B:F  →  :O:B:F
 |    F         | F
 Et             Et
```

This etherate is the usual form in which BF_3 is used, because it is so convenient—it is a liquid which boils at 125° C and freezes at −60° C, and thus can be poured and measured as a liquid, making unnecessary the steel cylinder, valves, and tubing required by gaseous BF_3 itself.

In use as a catalyst, BF_3 etherate loses its ether whenever a better electron donor (stronger base) is present to displace the ether. For example, the high electron concentration around the double bond between carbon atoms in an unsaturated petroleum hydrocarbon will do this, liberating the ether and leaving the BF_3 attached to the hydrocarbon:

$$Et_2O{:}BF_3 + C{::}C \rightarrow Et_2O + C{:}C{:}BF_3$$

(For clarity, the rest of the hydrocarbon molecule is not shown.) Once attached this way, the electron-withdrawing action of the BF_3 group strains the nearby carbon-carbon and carbon-hydrogen bonds so that further reactions take place, liberating the BF_3 to take part in another reaction. It is this repetitive activity on the part of the BF_3, without permanent change to itself, that qualifies it as a true catalyst.

The other boron halides are also volatile: BCl_3 boils below room temperature (at 12.5° C), BBr_3 boils at 90° C, and BI_3 is a low-melting solid (mp 43° C, bp 210° C). All four halides fume in moist air as they hydrolyze to boric acid and the corresponding hydrohalogen acid. If X represents any one of the halogen elements, then

$$2BX_3 + 6H_2O \rightarrow 2H_3BO_3 + 6HX + \text{heat}$$

BORON AND ITS CARBIDE AND NITRIDE. The vapor of BI_3, diluted with hydrogen and passed slowly at low pressure over a hot tungsten wire, is dissociated thermally to deposit hard, black, inert crystals of elementary boron. Another process yields slender fibers of boron (called "whiskers") which are extremely strong and refractory (mp 2300° C), and are used to strengthen special alloys and polymers for use in aircraft jet engines in much the same way that fiberglass is used to strengthen plastics.

Crystalline boron, boron nitride, and boron carbide are all extremely hard, high-melting solids which resemble diamond in their properties and can substitute for diamond in many industrial operations. Boron nitride, BN, can well be called pseudo-diamond because it has the same crystal structure as real diamond and very nearly the same hardness. Diamond is a dense close-packed crystalline form of carbon in which each atom is held tightly in the crystal lattice by four strong covalent bonds to neighboring atoms. Boron has one less electron per atom than carbon, and nitrogen has one more electron per atom, so boron and nitrogen can team together to form BN with the same electron complement as diamond, and very nearly the same interatomic dimensions and physical properties. Hence BN crystals are made by high-temperature technology to be used in cutting tools and as a high-quality abrasive. Boron nitride has one advantage over diamond itself: it does not oxidize as readily at high temperatures. A diamond disappears slowly when heated in air to 800° C, but at the same temperature a crystal of boron nitride stays there and is still useful.

Boron carbide has a hardness of 9.3 on the Mohs scale (diamond = 10) and it has the very strange crystal structure shown in Figure 3–2. Close-packed clusters of twelve boron atoms each alternate with short chains of three carbon atoms each, so the composition is $B_{12}C_3$ and the empirical formula is B_4C. The B_{12} clusters have the form of a dodecahedron, one of the five regular polyhedra of solid geometry. Each boron atom therein is bonded to its neighbors by soon-

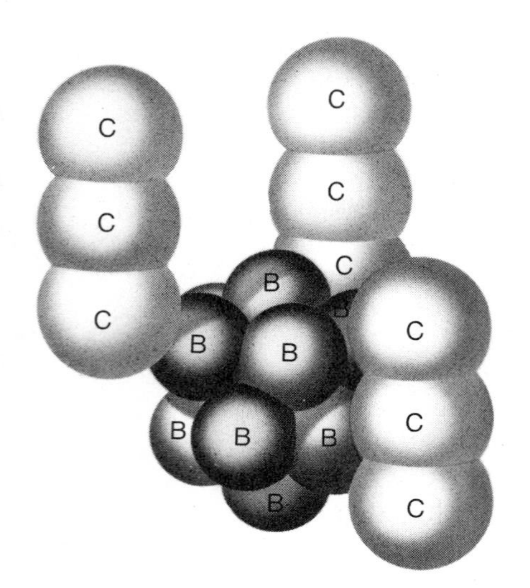

Figure 3–2. Crystal structure of boron carbide, $B_{12}C_3$.

TABLE 3–1 PRINCIPAL HYDRIDES OF BORON

Formula	*Name*	*Mp, °C*	*Bp, °C*	*Reaction With Air*
B_2H_6	diborane	−165.6	−92.5	spontan. inflammable
B_4H_{10}	tetraborane	−120.	18	not spontan. infl.
B_5H_9	pentaborane-9	− 46.6	48	spontan. inflammable
B_5H_{11}	pentaborane-11	−123	63	spontan. inflammable
B_6H_{10}	hexaborane-10	62.3	?	stable

to-be-discussed three-center covalent bonds, in which *three* atoms share a pair of electrons. Boron carbide is produced by heating a mixture of coke and boric oxide in an electric furnace, and used as an abrasive.

BORON HYDRIDES. The many hydrides of boron (often called boranes) were a strange structural puzzle to chemists until recently. The experimental facts were made plain enough by Alfred Stock in the 1920's: when magnesium boride is dropped into dilute hydrochloric or phosphoric acid, there is given off a gas consisting of hydrogen and several hydrides of boron. These hydrides may be condensed in a tube cooled by liquid nitrogen, and after careful separation by fractional distillation B_4H_{10}, B_5H_9, B_5H_{11}, B_6H_{10}, and $B_{10}H_{14}$ are obtained. Decomposition of B_4H_{10} at 100° C gives B_2H_6, which can also be prepared by the reaction of sodium borohydride with sulfuric acid:

$$2NaBH_4(aq) + H_2SO_4(aq) \rightarrow Na_2SO_4(aq) + B_2H_6(g)$$

This is really an acid-base reaction of the borohydride ions and hydrogen ions:

$$2BH_4^-(aq) + 2H^+(aq) \rightarrow B_2H_6(g)$$

The properties of the principal boron hydrides are given in Table 3–1. Notice that there is no BH_3, although ordinary rules of valence would suggest one. Moreover, all the rest of the hydrides seem to violate the ordinary rules of valence. Diborane, although the simplest hydride in the list, is not the most stable or unreactive; it decomposes readily when warmed, it ignites and burns to B_2O_3 and H_2O whenever it meets air, and it reacts rapidly with water. Pentaborane-9 is the most stable one; it may be kept indefinitely at 25° C and decomposes only slowly at 150° C. It does not react with water at room temperature, but it does ignite in air. Tetraborane does not ignite, but it does hydrolyze. Evidently there is built into every borane an inherent but selective reactivity which is much greater than that of the hydrocarbons.

How is all this to be explained? The reactivity is typical of electron-deficient compounds, which have a chronic shortage of electrons and so attract electron pairs from any substance that can donate them, such as O_2 or H_2O. Having once fastened on to O_2 or H_2O, further reaction can then take place within the complex, with the ultimate formation of B_2O_3 or H_3BO_3 and the release of much energy. The reason for the initial electron deficiency is apparent as soon as we attempt to write electron-dot formulas for the boranes—there is too much hydrogen to be bonded by ordinary shared-pair covalent bonds. Taking the simplest borane, B_2H_6, we count up three electrons from each boron atom (six for the pair), and one electron from each hydrogen atom (six for the lot), making a total of 12 electrons available for bonding. If we use two of these for the minimum bond between boron atoms, we have ten electrons left for binding six hydrogen atoms:

```
  H H
H:B:B:H
  H H
```

Inevitably, two of the hydrogen atoms are left with a single electron each, a situation we know cannot be true because there is no magnetic effect such as is always associated with unpaired electrons. Study of the geometric structure by several methods has proved that neither are the atoms deployed as just shown. Instead, there are two hydrogen atoms situated *between* the boron atoms, while the other four hydrogen atoms lie at the extremes:

```
H   H   H
  B   B
H   H   H
```

If we assign normal electron pairs to bind the four *terminal* hydrogen atoms, that uses eight and leaves us with only four electrons for all the rest of the bonding among the four central atoms. Since both the geometric structure and the chemical properties show that the two central (or "bridging") hydrogen atoms are equivalent, we must assign two of our four electrons to bonding the upper bridging hydrogen and two to bonding the lower bridging hydrogen. But *both* bridges must also connect *both* boron atoms, so we might write

```
  . H .
B       B
  . H .
```

for the central array. This cannot be correct, either, because the compound is not attracted to a magnet and cannot have unpaired electrons. We must conclude, therefore, that the two bridging hydrogen atoms are held by three-center bonds,

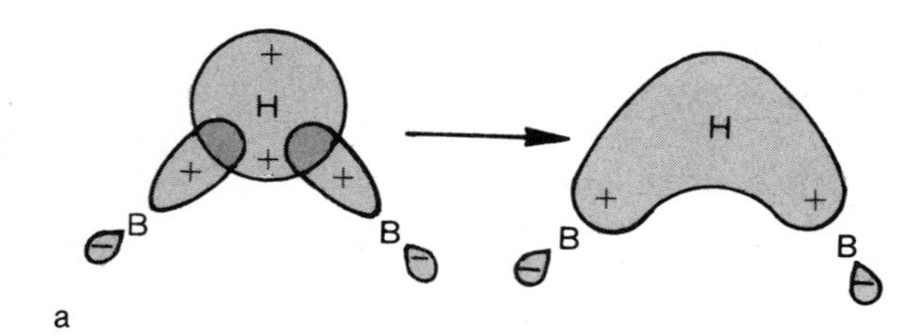

a

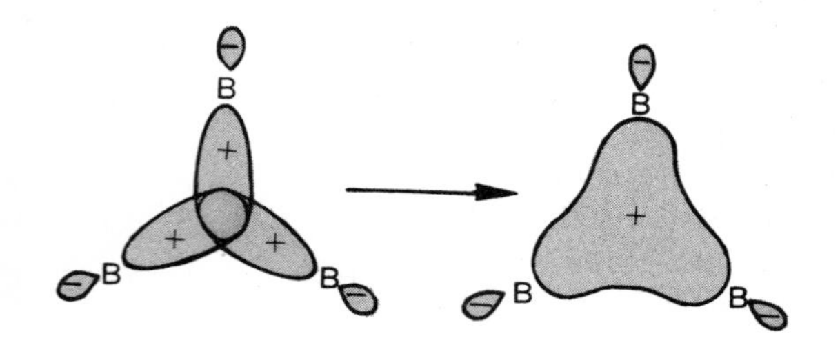

b

Figure 3–3. *a,* Formation of a three-center covalent bond between two boron atoms and a hydrogen atom by combination of the atomic orbitals to form a single bonding molecular orbital. *b,* Formation of a three-center bond by interaction of three boron atoms in similar fashion.

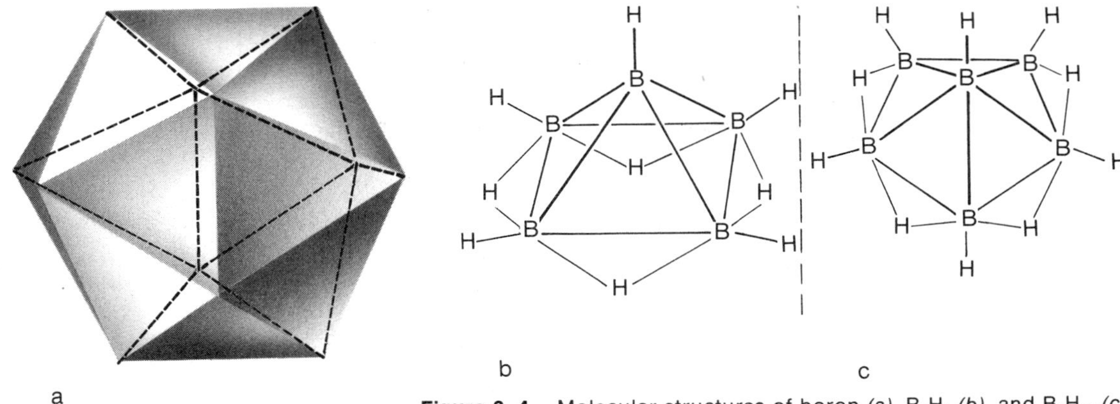

Figure 3–4. Molecular structures of boron *(a)*, B_5H_9 *(b)*, and B_6H_{10} *(c)*.

in which a pair of electrons circulates in the orbitals of *three* atoms and not the usual two atoms. The formation of such a bond from three atomic orbitals is illustrated in Figure 3–3*a*. The result is a strong but unusual bond, one limited to very small atoms with an inherent electron deficiency.*

We return briefly to the dodecahedron of boron atoms in the element itself and in the carbide $B_{12}C_3$ to point out that the boron atoms in these 12-atom cages are bound together by three-center bonds, too, as shown in Figure 3–3*b*, and that the dodecahedron is the key figure in the structure of the higher boranes (Fig. 3–4).

The boranes have no practical uses, although they were once proposed very seriously as high-energy fuels for jet engines or rockets because their heats of oxidation are considerably greater than those of the hydrocarbons. The expense, the difficulties of storage, and the somewhat toxic effect of the boric acid they would produce upon combustion all add up to a rejection of the idea.

The boranes are far more reactive toward air, water, acids, and bases than the hydrocarbons, because carbon has enough electrons to get along with conventional shared-pair bonds, and its hydrides are utterly different. The hydrides that come closest to the boranes in chemical behavior (inflammability in air, fast hydrolysis, etc.) are the corresponding compounds of *silicon*, not those of carbon or aluminum. Once again the diagonal relationship becomes apparent, and dominates other relationships.

DETECTION. Boron imparts a bright green color to the Bunsen flame, and can readily be detected in this manner provided the boron is in a sufficiently volatile form. The usual method of insuring this is to add hydrochloric acid and methanol (methyl alcohol) to the solution suspected of containing borates or other boron compounds. The acid converts borates to boric acid, which then combines with methanol to give volatile methyl borate:

$$H_3BO_3(s) + 3CH_3OH(l) \rightarrow B(OCH_3)_3(g) + 3H_2O(g)$$

Methyl borate boils at 68° C and readily carries boron into the flame, where it gives the characteristic green color.

---

*For more on boron and its hydrides, see E. G. Rochow, *The Metalloids*, Ann Arbor, Michigan: Xerox University Microfilms (1966 and 1975), Chapter 4. See also E. L. Muetterties, *The Chemistry of Boron and Its Compounds*, New York: John Wiley & Sons (1967), especially Chapter 5, pp. 223–313.

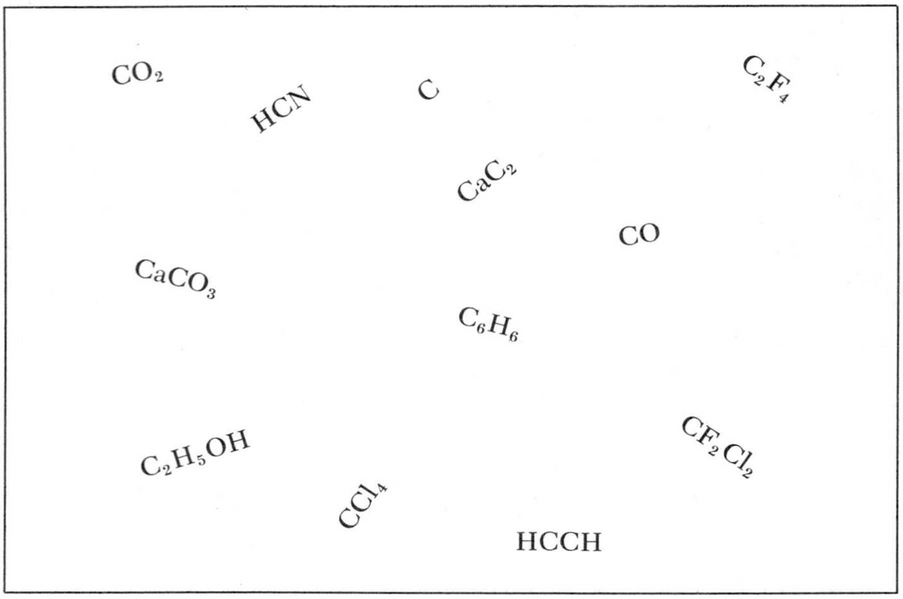

CO_2 HCN C C_2F_4

CaC_2

CO

$CaCO_3$ C_6H_6

C_2H_5OH CF_2Cl_2

CCl_4 HCCH

CARBON: ELEMENT WITH A BIG FOOT IN BOTH CAMPS

Chemistry is a large and sprawling subject that spills over into engineering, biology, medicine, home economics, geology, and metallurgy. Chemistry has many branches and the largest of them, in terms of known and catalogued compounds, is organic chemistry. This is defined by some as "the study of carbon and its compounds," but that definition would include all the mineral and geological carbonates as well as the metal cyanides and carbides, a decidedly geochemical and inorganic area. Nor is organic chemistry confined to the study of substances produced by living organisms; that definition had to be given up 150 years ago when it was discovered that many products of living matter could be synthesized from mineral substances or from the elements themselves. Today any further argument about definitions is idle, because there is no clear dividing line between organic and inorganic chemistry—we cannot even distinguish between organic and biological chemistry. For our present purposes, then, we shall consider the mineral forms of carbon and the routes by which these become the starting materials for the huge industry that manufactures fibers, polymers, plastics, paints, and pharmaceuticals. Since many students will go on to study organic chemistry as a separate subject, they can take it from there.

Carbon is not a plentiful element, as Table 1–3 has shown. It constitutes only 0.08% of our world, and about half of this is buried in the carbonate rocks—limestone, marble, chalk, dolomite, magnesite, calcite, and the rest. Consider the limestone basin of the Mississippi, the coral reefs of the Pacific, and the Dolomite Mountains of Italy, for example, to grasp the extent of this carbonate domain. Even though we quarry limestone and marble as building materials, and use limestone to make steel and cement, almost all of this carbonate rock will remain untouched, and its carbon content will remain locked up in the geologic masses. We must get along with the other half of our world supply of carbon, which goes through a slow cycle of photochemical reduction in plants (with the aid of sunlight), storage in part as coal and petroleum, oxidation to CO_2 by combustion and respiration, and then reduction of the CO_2 again by photo-

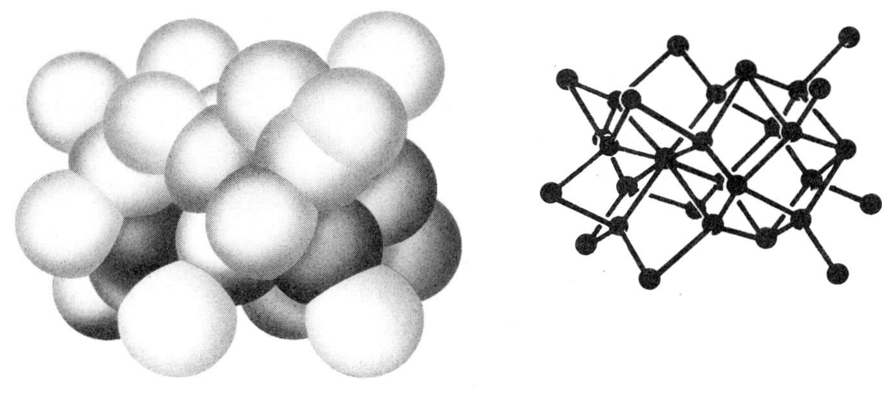

Figure 3–5. Crystal structure of diamond.

synthesis. As part of this cycle, the combustion of gasoline, oil, natural gas, coal, and wood supplies us with most of the heat and power needed by our culture; at present very little comes from water power or nuclear energy. The great question before us is how much of this carboniferous wealth we should continue to squander as an energy source, and how much we should reserve for the future manufacture of food, clothing, and household necessities.

PURE CARBON. Our discussion of carbon can begin with the two elemental forms of pure carbon, graphite and diamond. Graphite has an open layer structure, being composed of widely separated sheets of C_6 rings, but diamond has the compact cubic structure shown in Figure 3–5. The stable form under ordinary conditions is graphite, but that can be converted to diamond at extremely high pressures and temperatures, as related in most textbooks. The difference in density is especially noteworthy. Diamond, as the close-packed and thoroughly bonded form ($d = 3.51$ g/cm^3), is an electrical insulator and is chemically inert up to the temperature where it begins to oxidize (800° C). Graphite, with its open layer lattice and low density (2.25 g/cm^3), is attacked even at room temperature by acids and oxidizing agents which penetrate its layers and cause it to swell. When very finely divided, as in carbon black and charcoal, the small particles of graphite start to oxidize at 400° C. Graphite is a fair conductor of electricity because not all of its available electrons are tied down in covalent bonds; the electrons circulating in pi orbitals* are sufficiently mobile to carry an electric charge. The interaction of the same footloose pi electrons with light is responsible for the black color of graphite.

Coal has an extended layer structure like that of graphite, but with considerable hydrogen bonded to the hexagonal rings of carbon atoms. Although diamond does not react with hydrogen, and graphite reacts only under extreme conditions of temperature and pressure, coal may be hydrogenated at 475° C and a pressure of 250 atmospheres with molybdenum sulfide as catalyst. The product is a mixture of hydrocarbons resembling petroleum in its properties. It may be refined in the same way as crude oil to give gasoline, diesel oil, and lubricants. A half-million barrels of liquid fuel per day were made by this process in Germany in the 1940's, and since the United States has enormous reserves of coal, exceeding our petroleum supply many times over, we shall probably see a similar procedure for the liquefaction of coal in use in this country before long.

OXIDES. Carbon forms the oxides CO, CO_2, and some higher oxides of the composition C_nO_2. Of these, CO and CO_2 are minor constituents of the atmosphere. The solution of CO_2 in water and the subsequent dissociation and reac-

*See glossary.

tion of carbonic acid are discussed in every textbook, and need not be elaborated here. We shall consider only the monoxide, which is an active and versatile reagent as well as a very dangerous poison. The toxicity arises from the electron-donor ability of CO, a property which may be understood from its molecular structure:

$$:C \equiv O:$$

The carbon can donate a pair of electrons to any avid acceptor, such as an atom or ion of chromium, iron, nickel, or other transition metal, thereby setting up a covalent bond. Such compounds are called *metal carbonyls.* Most of them are volatile: $Ni(CO)_4$ boils at 43° C, and $Fe(CO)_5$ boils at 103° C. All are extremely poisonous. The formation of nickel carbonyl from impure nickel powder and carbon monoxide at 60° C, followed by decomposition of the carbonyl vapor at 200° C, formerly constituted a commercial method for purifying nickel.

Attachment of CO to the iron atom within a molecule of hemoglobin destroys the ability of the hemoglobin to pick up O_2 in the lungs and release it in the brain and muscle tissues of the body, simply because the CO is a more potent donor of electrons than O_2 and forms a stronger bond. Even a small proportion of carbon monoxide in the air, if breathed long enough, will incapacitate the blood and cause drowsiness, dizziness, headache, and then unconsciousness and death. Since CO is odorless, the victim may not be aware of the danger he is in. Neither will his plight be obvious to someone else, for instead of looking pale or turning blue, as in most cases of suffocation, the victim has a rosy complexion because the CO-hemoglobin compound is bright red. A physician can inject an artificial oxygen carrier, if he arrives in time; the only first-aid response available to the rest of us is to force rapid breathing of fresh air (by artificial respiration, if necessary) or to administer oxygen. A high enough concentration of oxygen in the lungs will bring about slow displacement of the CO from the hemoglobin.

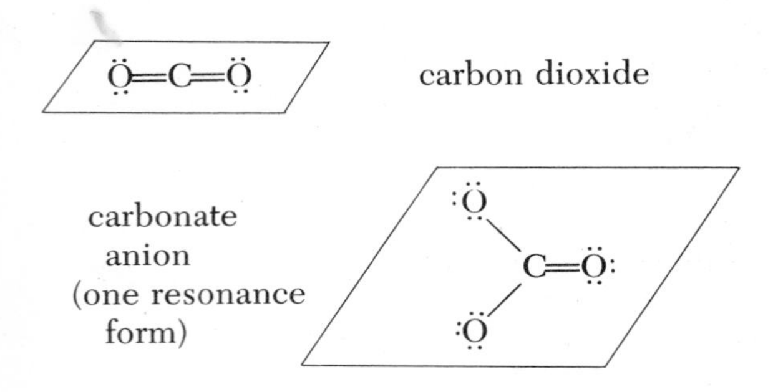

The use of carbon *dioxide* in beverages, fire extinguishers, and the manufacture of sodium carbonate is described in every textbook. Important natural carbonates are Na_2CO_3 and $NaHCO_3$; $CaCO_3$ and $MgCO_3$ in carbonate rocks; $PbCO_3$, an ore of lead; $Cu_2(OH)_2CO_3$ (malachite) and $Cu_3(OH)_2(CO_3)_2$ (azurite), both ores of copper; $ZnCO_3$, an ore of zinc; and $BaCO_3$, a common source of barium.

NITROGEN COMPOUNDS WITH CARBON. Another important class of carbon compounds is the *cyanides*, in which carbon is bonded to nitrogen. To learn something about these compounds, let us consider a sequence of reactions which provide useful products, starting with limestone and with nitrogen from the air. First the limestone is heated to give lime and carbon dioxide:

$$CaCO_3(s) \xrightarrow{850°\,C} CaO(s) + CO_2(g)$$

The CO_2 can be used in the manufacture of $NaHCO_3$ and Na_2CO_3. In an electric furnace the CaO is reduced by carbon (actually coke, made by heating soft coal) to form calcium carbide:

$$CaO(s) + 3C(s) \xrightarrow{1000°\,C} CaC_2(s) + CO(g)$$

The CO is burned as fuel gas, and the CaC_2 is used to absorb nitrogen (obtained by the distillation of liquid air):

$$CaC_2(s) + N_2(g) \xrightarrow{1100°\,C} CaCN_2(s) + C(s)$$

In former years the resulting calcium cyanamid was a source of ammonia for nitrogenous fertilizers, because it hydrolyzed readily, returning to the original calcium carbonate and liberating ammonia:

$$CaCN_2(s) + 3H_2O(l) \rightarrow CaCO_3(s) + 2NH_3(g)$$

However, calcium cyanamid is now too valuable for that. Instead, it is the source of CN groups for a long line of acrylic fibers, plastics, paints, and rubbers. It also is a good source of sodium cyanide, used in electroplating of silver, gold, and copper.

Sodium carbonate made from the CO_2 produced in the first reaction is heated with the crude residue from the nitrogen absorption reaction to obtain sodium cyanide:

$$Na_2CO_3(s) + CaCN_2(s) + C(s) \rightarrow 2NaCN(s) + CaCO_3(s) \text{ (recycled)}$$

The sodium cyanide is a source of hydrogen cyanide, which once was called prussic acid:

$$2NaCN(s) + H_2SO_4(l) \rightarrow 2HCN(g) + Na_2SO_4(s)$$

All cyanides are deadly poisons, and HCN is still more dangerous because it is so volatile (bp 26°C). It also is very useful. By reaction with acetylene, C_2H_2, it becomes relatively harmless acrylonitrile, the monomer used for the production of acrylic plastics and fibers:

$$HC{\equiv}CH(g) + HCN(g) \xrightarrow{CuCl} H_2C{=}CHCN \text{ (bp 78.5° C)}$$

Sodium cyanide is also used in the production of adiponitrile, $C_4H_8(CN)_2$, one of the two reactants used to make nylon.*

This excursion into practical organic chemistry is intended only to show how many useful and important materials can be made from ordinary cheap limestone, coal, air, and water. There are dozens of other examples that could be cited to make the same point. In order to achieve large-scale production, industrial organic chemistry must rely on starting materials from the earth, air, and water.

CARBIDES. Our last topic under carbon concerns the carbides, and especially the binary compounds of carbon with metals. These fall into three classes:

1. The ionic carbides, in which carbon is in the form of a C^{-4} or a C_2^{-2} ion balanced by a positive metal ion. Calcium carbide (p. 59) is just such a compound; it consists of an array of Ca^{+2} ions and C_2^{-2} ions in equal numbers. In general, ionic carbides are formed by metals in Group I (Li, Na, K, Rb, Cs), Group II (Be, Mg, Ca, Sr, Ba), and also by Al, Cu, Zn, Cd, Th, V, U, and some of the rare-earth elements. All react readily with water to give hydrocarbons. Those which contain C^{-4} ions (Be_2C, Al_4C_3, etc.) give methane on hydrolysis:

$$Be_2C(s) + 4H_2O(l) \rightarrow 2Be(OH)_2(s) + CH_4(g)$$

$$Al_4C_3(s) + 12H_2O(l) \rightarrow 4Al(OH)_3(s) + 3CH_4(g)$$

The other carbide ion, C_2^{-2}, has a triple bond between carbon atoms and has a complement of ten valence electrons:

$$:C\vdots C: \text{ or } :C{\equiv}C:$$

This is isoelectronic (has the same number of valence electrons) with carbon monoxide and gaseous nitrogen, N_2:

$$:C{\equiv}C:^{-2} \quad :C{\equiv}O: \quad :N{\equiv}N:$$

Carbides which contain this C_2^{-2} ion give acetylene on hydrolysis, and acetylene has the same triple-bonded structure as the ion:

$$CaC_2(s) + 2H_2O(l) \rightarrow Ca(OH)_2(s) + HC{\equiv}CH(g)$$

$$Li_2C_2(s) + 2H_2O(l) \rightarrow 2Li^+(aq) + 2OH^-(aq) + HC{\equiv}CH(g)$$

All ionic carbides are clear colorless crystals when pure, but usually are obtained as gray powders. They do not conduct electricity.

2. Covalent carbides, in which carbon is closely bound to a metal or metalloid by covalent bonds. We have already had boron carbide as an example; another good example is silicon carbide, SiC, which has the diamond structure and may be regarded as an expanded diamond in which every other carbon atom is replaced by silicon. Such carbides do not react with water or with aqueous acids or bases. They also resist oxidation at high temperatures, and may be used as refractories. They are semiconductors.

3. Interstitial carbides, which are essentially expanded metals in which the

---

*See E. G. Rochow, G. Fleck, and T. H. Blackburn, *Chemistry: Molecules that Matter*, New York: Holt, Rinehart & Winston (1974), Chapter 15, pp. 498–502.

lattice of metal atoms is expanded or distorted by the presence of carbon atoms distributed between the lattice positions. Such materials have metallic luster, are as opaque as metals, and are good conductors of electricity. All are harder and higher-melting than the pure metal itself. This hardening and stiffening effect of carbon has been known for a long time. It is iron carbide that makes steel out of iron, and swordsmiths learned many centuries ago how to harden a blade by heating it in charcoal, plunging it into cold water, and then tempering it. As the carbon diffuses into red-hot iron it forms Fe_3C, an interstitial carbide in which carbon atoms of radius 0.77 Å fit into spaces between iron atoms (radius 1.26 Å) in the lattice. If the metal is cooled quickly, the Fe_3C has no time to segregate as large crystals, and remains distributed as small grains, giving a tough, hard alloy. If too much carbon is added, the metal becomes brittle. The proper heat treatment is very important.

The suits of armor seen in museums were made in a similar way. They were hammered out as thin sheets of fairly pure wrought iron, and then were surface-hardened by heating in charcoal powder or by dipping the red-hot metal repeatedly into olive oil. Each armorer had his own secret for getting a hard, tough film on top of a flexible metal interior. Similar "case hardening" of steel gears and other machine parts is accomplished today by heating the parts (easily machined from soft steel) in a bath of melted sodium cyanide just long enough for iron carbide and nitride to develop and to form a tough, wear-resistant skin.

Interstitial carbides also find wide use in cutting and drilling tools. Lathe tools made from powdered tungsten carbide, WC, cemented by cobalt (itself a tough metal) keep their sharp edges even when red-hot. A layer of tungsten carbide on the rotating parts of aircraft jet engines greatly reduces the wear rate. Other useful interstitial carbides are TiC, ZrC, VC, TaC, and MoC, which harden the edges of cutting tools.

NITROGEN: UNWILLING BUT ESSENTIAL ELEMENT

NITROGEN AND THE FOOD SUPPLY. All living matter is organic in the chemical sense, of course, and is based on chains and networks of linked carbon atoms. But so is gasoline or asphalt. The hallmark of living protoplasm is *nitrogen*, an absolutely essential element. All proteins contain about 17% nitrogen in the form of amino acid groups linked together stereochemically. Growing plants need a suitable source of nitrogen in order to synthesize their protein, and growing animals need the plant proteins to grow on. There is no escaping this sequence.

Now that the food supply has become the foremost problem facing the world today, any discussion of the chemistry of nitrogen should properly be centered on the compounds and the processes needed to feed the ever-growing population. Improvements in the methods of "fixing" atmospheric nitrogen (converting it into agriculturally useful compounds) are sorely needed, and may well come from someone who is studying chemistry today.

The nitrogen cycle in nature is set forth in Figure 3–6. Animal metabolism of proteins results in the excretion of urea, $(NH_2)_2CO$, plus ammonium salts such as $NaNH_4HPO_4$, and various organic derivatives of nitrogen. These are all eventually hydrolyzed to ammonia and its simple salts in the soil, where the roots of growing plants can absorb and use the nitrogen in these forms. Any ammonia not used in this way is eventually oxidized to N_2 and goes into the atmosphere, which is our large reservoir of spare nitrogen.

Figure 3–6. The nitrogen cycle. 1, Lightning provides energy for the combination of some nitrogen and oxygen in the atmosphere, producing NO_2 which is washed from the air by rain which falls to the earth, entering the soil. 2, Soil contains nitrates from the NO_2, from nitrogen-fixing bacteria, and from decayed vegetation. Soil also contains $NaNH_4HPO_4$ and urea, $(NH_2)_2CO$, from animal wastes. Cultivated fields also have nitrogen, phosphorus, and potassium added as fertilizers. 3, Plants obtain nitrogen from the soil, incorporating that nitrogen into plant proteins. 4, Animals eat the plants, absorb amino acids from the plant proteins, and incorporate the amino acids into animal proteins. 5, People eat the plants and eat meat. 6, Nitrogenous wastes from plants, animals, and people return in part to the soil and in part to the atmosphere.

Unfortunately, plants cannot use molecular nitrogen for their growth because it is too unreactive. Hence nitrogen once lost from the soil to the atmosphere cannot return to the life cycle unless some outside agency intervenes and converts free nitrogen into a combined, or "fixed," form. The outside agency could be a lightning discharge (which oxidizes some nitrogen at 1000°C or more to water-soluble oxides brought down by the accompanying rain), or some special symbiotic bacteria that live on the roots of plants like clover and alfalfa, or some people who have built a factory to separate nitrogen from the air and bring it into combination as ammonia or ammonium nitrate. The lightning activity is rare, low-powered, and uncontrolled; it furnishes only about 10 lb of fixed nitrogen annually per acre in the United States. The second method, planting crops that harbor nitrogen-fixing bacteria on their roots, can provide 100 to 200 lb of fixed nitrogen per acre in a growing season, but of course this takes the land out of production for a year. The only method that can be controlled and expanded at will to insure maximum crop yield is to manufacture and distribute nitrogenous fertilizer. This has been true for the past 70 years, and so the necessary factories have been built and kept operating. They are all that keeps the world's population from starvation.

The magnitude of the problem was realized very clearly in 1898 by Sir William Crookes, who pointed out that there was only a two-week reserve supply of food in England at that time and there would be a disaster if anything cut off food imports. Something had to be done to increase productivity at home, he argued. Experiments conducted over a period of 26 years had shown that the average yield of wheat from a particular English farm was 11.9 bushels per acre when left to natural sources of nitrogen, and *36.4 bushels* when 500 lb of sodium nitrate was spread on each acre. There was no doubt that the growing plants responded to better nutrition; the question was where to get the millions of tons of nitrate required to fertilize the fields at that rate. Over the next decade three methods were devised and proved:

1. The first method, advocated by Crookes himself, was an imitation of nature's lightning discharge method. Air was blown through an electric arc and immediately into a spray of water, which absorbed the oxides of nitrogen and formed nitric acid:

$$N_2(g) + O_2(g) \xrightarrow{2000^\circ\,C} 2NO(g)$$

$$2NO(g) + O_2 \text{ from excess air} \rightarrow 2NO_2(g)$$

$$3NO_2(g) + H_2O(l) \rightarrow 2HNO_3(aq) + NO(g) \text{ (returned to process)}$$

The nitric acid was then neutralized with sodium hydroxide to give sodium nitrate equivalent to the natural nitrate used in the experiments.

2. The second method has already been described earlier in this chapter. Calcium carbide was produced from lime and coke, and the carbide allowed to absorb nitrogen to form $CaCN_2$. Hydrolysis of this cyanamid gave ammonia, half of which was oxidized to nitric acid:

$$4NH_3(g) + 5O_2(g) \xrightarrow{Pt} 4NO(g) + 6H_2O(g)$$

$$2NO(g) + O_2(g) \rightarrow 2NO_2(g)$$

$$3NO_2(g) + H_2O(l) \rightarrow 2HNO_3(l) + NO(g) \text{ (recycled)}$$

The resulting nitric acid was used to neutralize the other half of the ammonia in order to make solid ammonium nitrate, which is easier to ship than gaseous ammonia:

$$NH_3(g) + HNO_3(l) \rightarrow NH_4NO_3(s)$$

This method and the previous one fell into disuse as the cost of electric power increased, and were supplanted.

3. The third method was the last to be developed, and became the dominant procedure. Ammonia was prepared from its elements at high temperature and pressure with the help of a special iron catalyst, as described in Chapter 2:

$$N_2(g) + 3H_2(g) \xrightarrow[450^\circ\,C \\ 500\text{ atm}]{Fe} 2NH_3(g)$$

Part of the ammonia then was oxidized to nitric acid for making ammonium nitrate (see above). This method has been used almost exclusively for the past 40 years, to the current extent of the manufacture of about 230 million tons of NH_4NO_3 per year.

As we look ahead to a doubling of the world's population, serious questions arise concerning any further expansion of Method 3. It requires much hydrogen (which we now get from petroleum), it requires a great deal of expensive metal for the high-pressure equipment, and it still uses electric power for compressing the gases. The nitrogen-fixing bacteria, on the other hand, do their job quietly at ambient temperature and at 1 atmosphere pressure. We do not know how they do it, but some chemists are trying to find out. The starting point in such research is some coordination compound in which N_2 is temporarily bonded to a transition metal such as titanium. The object is to reduce the nitrogen to an NH or NH_2 group before it comes off as N_2. In the laboratory this may be accomplished with metallic sodium (scarcely a reagent employed by bacteria!), which has many electrons available for reducing action. Then the reduced nitrogen is split off in the form of ammonia by treatment with an alcohol, and the residue is processed to recover the reagents. This cycle at least works, but a more economical variation of it is required. Still better methods, more like the natural ones but capable of operation on a very large scale, are needed.

MANUFACTURED FOODS? While we are at it, the large scale manufacture of basic foodstuffs needs to be explored as an alternative to agriculture for the near future. The agricultural revolution has run its course and the time will come, rather soon, when all the arable land of the world, under cultivation with optimum fertilization but fickle weather, cannot supply the world's food requirements. Some would say we have already arrived at this point. When the tilled earth no longer produces enough food, we must turn to manufactured fats, proteins, and possibly carbohydrates. We shall be besieged like wartime Germany, where satisfactory edible fats had to be made from coal by way of synthetic glycerol and straight-chain fatty acids derived from higher alcohols. That process alone, put into adequate production, would free a great deal of farmland for grain production. Similarly, if nitrogen-bearing compounds which are intermediates in some ammonia process could be diverted into making essential amino acids (and from them nutritious proteins), more farmland would be freed for grain production and much grain now used for fattening cattle could be saved. With the bread-and-butter items coming from factories, the land could be given over to growing just the fruits and flavors and specialties necessary for good nutrition. To get ready for this we need a worldwide food policy and some farsighted implementation. The alternative will be panic programs carried out during famine and revolution.

Will the world ever run out of nitrogen for making food or fertilizer? Not likely! We live at the bottom of a sea of nitrogen and oxygen many miles thick, under a pressure of 14.7 pounds per square inch. A little calculation shows this air layer to weigh 1728 lb/sq ft, or 0.864 ton/sq ft and therefore 24.1 million tons per square mile. Four fifths of this atmosphere is nitrogen, so there are 19.3 million tons of N_2 standing and waiting above each square mile of the earth's surface. The entire resource amounts to 19.3×10^6 tons/sq mi $\times 1.97 \times 10^8$ sq mi (the area of the earth), or 38.0×10^{14} tons total—nearly four million billion tons of nitrogen waiting to be used. And all that we use gets back to the atmosphere eventually.

COMPOUNDS OF NITROGEN. Having considered the most important aspect of nitrogen from the standpoint of hungry people, we should devote some thought to a few compounds of nitrogen of current importance in other ways.

Ammonia (mp −77.7° C and bp −33.35°C) is a weak base considered in every textbook discussion of acids, bases, and aqueous equilibria. It is not the only hydride of nitrogen; it is just the most common one. The second hydride is *hydrazine*, N_2H_4, a volatile liquid which freezes at 1.8° C and boils at 113.5° C. This is made by chlorinating ammonia with sodium hypochlorite (bleaching solution) and then allowing the product to react with excess ammonia in the complete absence of transition metals (which are catalysts for the decomposition of hydrazine):

$$NH_3(g) + OCl^-(aq) \rightarrow OH^-(aq) + \underset{\text{chloramine}}{H_2NCl(aq)}$$

$$H_2NCl(aq) + 2NH_3(g) \rightarrow NH_4^+(aq) + Cl^-(aq) + N_2H_4(aq)$$

The hydrazine can be distilled off at low pressure and concentrated. Hydrazine is something like hydrogen peroxide in that it is unstable and decomposes to its elements, with the evolution of heat:

$$N_2H_4(l) \rightarrow N_2(g) + 2H_2(g) \qquad \Delta H = -12 \text{ kcal/mole}$$

It gives up its hydrogen so readily that it is an excellent reducing agent, capable of converting most metal oxides and salts to free metal. Some is used for that, but hydrazine's chief claim to fame at present is that a derivative, unsymmetrical dimethyl hydrazine, $(CH_3)_2N_2H_2$, is an excellent rocket fuel. It is oxidized instantly and energetically by dinitrogen tetroxide to liberate a great deal of gas and heat in the rocket engine:

$$(CH_3)_2N_2H_2(l) + 2N_2O_4(l) \rightarrow 3N_2(g) + 4H_2O(g) + 2CO_2(g)$$

The Lunar Excursion Module (LEM) of the Apollo space ships had pressurized tanks of $(CH_3)_2N_2H_2$ and N_2O_4 in readiness, and when the time came to blast off of the moon it was only necessary to open the valves. Dimethylhydrazine has the virtue of igniting spontaneously in an oxidizing atmosphere, and so no ignition system is necessary. There was nothing to get out of order or require repairs by local inhabitants; given an adequate supply of fuel and oxidizer, the LEM engine had to work and always did.

The *nitrogen halides* differ from each other in chemical behavior. Nitrogen trifluoride, NF_3, is a very stable gas (bp −129° C) which does not react with water and which decomposes to the elements only at 300° C. It is not made from N_2 and F_2, but by electrolyzing NH_4F in HF. Nitrogen trichloride, made by chlorinating an aqueous solution of NH_4Cl, is a very reactive liquid (bp 71° C) which decomposes explosively to the elements. It hydrolyzes rapidly to form ammonia and hypochlorous acid, rather than to the expected hydrochloric acid:

$$NCl_3 + 3H_2O(l) \rightarrow NH_3 + 3HClO(aq)$$

The action of bromine on ammonium bromide is said to give nitrogen tribromide, which decomposes so rapidly that it is hard to isolate and study. The action of iodine on ammonia gives no nitrogen triodide, but rather some unstable black crystals of $H_3N_2I_3$ which explode with a sharp report even when touched with a feather.

OXIDES. The oxides of nitrogen are gregarious compounds which change one into another so readily that it is almost impossible to get pure samples for study. They are air-polluting products of internal combustion engines, which is reason enough to study them. Due to their interchangeability they are usually lumped together as NO_x in air pollution reports.

The six principal oxides are N_2O, NO, NO_2, N_2O_3, N_2O_4, and N_2O_5. The first, N_2O, is properly called dinitrogen oxide but sometimes is still known as nitrous oxide. This is obtained by heating pure ammonium nitrate to 250° C:

$$NH_4NO_3(s) \rightarrow N_2O(g) + 2H_2O(g)$$

It has the molecular structure N=N=O, where each dash represents a pair of electrons. It is an endothermic compound, meaning that heat has to be put *into* the system to make the compound and consequently it decomposes to the elements with the liberation of heat:

$$2N_2O(g) \rightarrow 2N_2(g) + O_2(g) \qquad \Delta H = -20 \text{ kcal.}$$

For this reason N_2O is actually a supplier of oxygen, and a candle will burn in it more brightly than in air because N_2O provides one O_2 molecule for every two N_2 molecules, whereas air provides only one O_2 for every four N_2 molecules. Body temperature is *not* sufficient to initiate dissociation of N_2O, though, as Joseph Priestley found when he discovered the gas in 1771; he observed that a candle burned in the gas but a mouse could not live in it.

The fact that N_2O is an anesthetic was discovered by Humphrey Davy in 1799, and promptly gave rise to the name "laughing gas" from its effect during public demonstrations. However, 38 years elapsed before physicians got around to using it for surgical anesthesia. It was first used in surgery by C. L. Long in Jefferson, Georgia, in 1837, and again made no impression until William T. C. Morton used it repeatedly for surgical operations in Boston in 1847. Even after that it had a poor reputation for a time because some people forgot Priestley's experience and tried to administer pure N_2O without any supporting oxygen. Pure N_2O alone causes suffocation; it must always be used with oxygen.

The second oxide of nitrogen is NO, simply nitrogen oxide (or sometimes known by the obsolete name of nitric oxide). This is obtained in the laboratory from the reaction of dilute nitric acid with copper:

$$3Cu(s) + 8H^+(aq) + 2NO_3^-(aq) \rightarrow 2NO(g) + 4H_2O(l) + 3Cu^{+2}(aq)$$

It may also be made from the elements, as we have seen earlier in this chapter, but much energy is required to do so:

$$N_2 + O_2 \rightarrow 2NO \qquad \Delta H = +21.5 \text{ kcal}$$

Since it is an endothermic compound, its formation is greatly favored at high temperatures. In fact, whenever air is heated above 700° C (and especially to 1000° or 2000° C) considerable NO will form. If the gas mixture is cooled slowly, NO has a chance to decompose to N_2 and O_2, but if the gas is chilled rapidly the decomposition is too slow to be effective and the NO persists. So every automobile engine is a source of NO as air is heated in the combusion chambers and then blown out rapidly into cold air. The easiest way to reduce the emission is to run the exhaust gas from the engine into an insulated hot reaction chamber where it can remain long enough for the NO to dissociate.

Nitrogen oxide is colorless, melts at −164° C, and boils at −152° C. It has the structure $\dot{\underset{\cdot\cdot}{N}}::\ddot{\underset{\cdot\cdot}{O}}$, with one odd electron. Like all other compounds with an unpaired electron, it is paramagnetic (that is, it is attracted by a magnet because its single electron generates a magnetic field that interacts with an external magnetic field). It reacts rapidly with oxygen to form brown dinitrogen oxide, NO_2,

$$2NO(g) + O_2(g) \rightarrow 2NO_2(g)$$

as in our discussion of the oxidation of ammonia to form nitric acid. The objection to NO as an air pollutant is that it soon forms NO_2 and this reacts with organic residues from the partial oxidation of gasoline to form organic nitrates which cause a stinging sensation in the eyes and an irritation of the lungs. Two such known irritants are CH_3NO_3 (methyl nitrate) and CH_3COONO_3 (peroxyacetyl nitrate), both of which can cause plant damage as well.

Nitrogen dioxide changes spontaneously to colorless dinitrogen tetroxide, which boils at 21.2° C and freezes at −11.2° C. If the vapor is left to itself, an equilibrium mixture develops which shows these color changes with changing temperatures:

N_2O_4	$\rightleftharpoons$	N_2O_4	$\rightleftharpoons$	$2NO_2$	$\rightleftharpoons$	$2NO + O_2$
colorless solid at −11.2° C		yellow liquid at 0° C; contains NO_2		deep brown gas at 100° C		colorless gases at 620° C

The structure of NO_2 explains its dimerization (molecular doubling) to N_2O_4. Since nitrogen brings five bonding electrons into play, and oxygen brings six, there must be an odd number of electrons and one must remain unpaired. However, if two molecules of NO_2 meet, they can pair their odd electrons and form a molecule of N_2O_4:

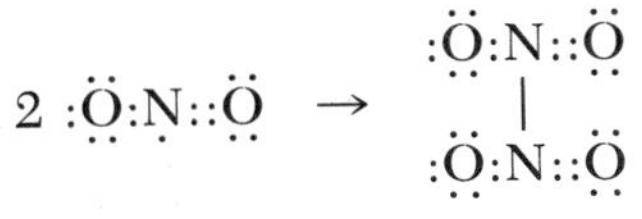

$$2\ :\ddot{O}:\dot{N}::\ddot{O} \rightarrow \begin{array}{c} :\ddot{O}:N::\ddot{O} \\ | \\ :\ddot{O}:N::\ddot{O} \end{array}$$

brown gas — colorless

There are two more oxides of nitrogen that enter the picture, N_2O_3 and N_2O_5. The first, dinitrogen trioxide, is an unstable gas (mp −103° C, bp 3.5° C) which is said to be the anhydride of nitrous acid because it reacts with water to give that weak acid:

$$N_2O_3(g) + H_2O(l) \rightarrow 2HNO_2(aq)$$

The acid itself is also unstable, and readily reverts to NO and nitric acid when it stands or is warmed. As for dinitrogen pentoxide, it is a volatile white solid that melts at 41° C and dissolves in water to form nitric acid, a very strong acid:

$$N_2O_5(s) + H_2O(l) \rightarrow 2H^+(aq) + 2NO_3^-(aq)$$

Nitric acid is not made this way, however. It is produced commercially by the oxidation of ammonia and the absorption of NO_2 in warm water, as outlined previously. Whenever N_2O_5 is required it is obtained by the dehydration of nitric acid. The molecular structures of N_2O_3 and N_2O_5 are

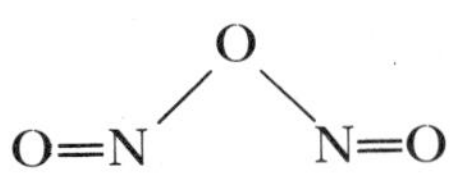

O=N–O–N=O and O=N(–O)–O–N(–O)=O

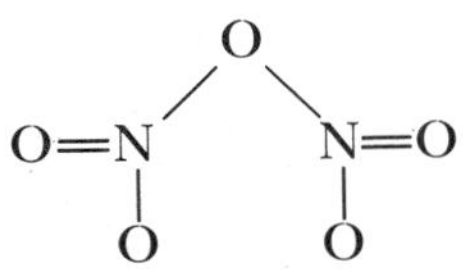

NITRIC ACID AND NITRATES. Pure nitric acid is seldom encountered, although it can be obtained by warming a mixture of sodium nitrate and 100% sulfuric acid:

$$2NaNO_3(s) + H_2SO_4(l) \rightarrow Na_2SO_4(s) + 2HNO_3(g)$$

The usual laboratory reagent called "concentrated nitric acid" is a solution of HNO_3 in water which boils at a constant temperature. It contains 68% HNO_3 and boils at 120° C. It is colorless when freshly distilled, but decomposes in sunlight and becomes yellow with dissolved NO_2. It attacks skin, wool, and all proteins to form a hard yellow substance, and its burns are slow to heal.

Nitric acid is very useful as an inorganic reagent because it is a strong oxidizing agent as well as a strong acid, and so will dissolve many metals (and their oxides) which are not soluble in hydrochloric acid. All metal nitrates are water-soluble, too, which helps matters a great deal. So nickel, copper, and silver are dissolved by 50% HNO_3:

$$Ag(s) + 2H^+(aq) + NO_3^-(aq) \rightarrow Ag^+(aq) + NO_2(g) + H_2O(l)$$

Similarly, the spent fuel rods from nuclear power reactors are dissolved in nitric acid in order to separate the fission products and recover the uranium and plutonium. Even though the fission of uranium produces about 35 different metallic elements (in addition to the gases krypton and xenon), all of them dissolve in nitric acid and stay dissolved.

Because of their solubility in water, nitrates are favorite salts to use in laboratory experiments and in many manufacturing processes. Some important nitrates are $NaNO_3$ (used in preserving meats), KNO_3 (used in gunpowder), and $AgNO_3$ (used to prepare photographic film and paper).

Nitric acid is useful not only as an inorganic acid and as the source of ammonium nitrate for fertilizer, but also as a nitrating reagent for introducing nitrogen into a large variety of organic compounds. If the nitration is carried far enough, the organic derivative may be useful as an explosive, as in these examples:

1. Glycerol trinitrate ("nitroglycerine") is made by nitrating glycerol, the trihydric alcohol obtained from edible fats when soap is made from them:

$$C_3H_5(OH)_3 + 3HNO_3 \xrightarrow{H_2SO_4} C_3H_5(NO_3)_3 + 3H_2O$$

Here sulfuric acid acts as a dehydrating agent, accelerating the nitration. Pure glycerol trinitrate detonates so easily that it is sensitive to shock, and so it is not used in that form but is soaked up in some porous absorbent material to render it less sensitive. The dry solid combination is called dynamite. When a detonator cap sets it off, the glycerol trinitrate decomposes to nitrogen, water, carbon dioxide, and surplus oxygen:

$$4C_3H_5(NO_3)_3 \rightarrow 10H_2O + 12CO_2 + 6N_2 + O_2$$

This extra oxygen can oxidize more carbon and hydrogen, so the absorbent material is usually lightly nitrated cotton or some similar substance calculated to make the oxygen balance come out right.

2. Trinitrotoluene, TNT, is made by nitrating toluene, a cyclic hydrocarbon:

$$C_6H_5CH_3 + 3HNO_3 \xrightarrow{H_2SO_4} C_6H_2(NO_2)_3CH_3 + 3H_2O$$

TNT is a yellow solid which stands a great deal of shock and requires a primer and a booster charge to detonate it:

$$2C_6H_2(NO_2)_3CH_3 \rightarrow 3N_2 + 7CO + 5H_2O + 7C$$

The products show an oxygen deficiency, and so the explosive force is increased by adding an oxygen supplier such as ammonium nitrate.

Commercial explosives have been a mainstay in mining and excavating for 75 years, but now they are being displaced by a much cheaper explosive, a simple mixture of 95% ammonium nitrate and 5% fuel oil. This mixture is more difficult to detonate than dynamite, but that is its virtue; any unexploded portions remaining in the drill holes are not dangerous when the mining machinery encounters them. The new explosive has the form of little spheres ("prills" of fertilizer-grade NH_4NO_3 that have been tumbled with the required amount of fuel oil. The mixture is simply poured into a drill hole and set off with a special blasting cap.

OXYGEN, THE LIFE-GIVER, ACID-FORMER, AND WATER-MAKER

Oxygen, our most abundant element on earth, is one of the few elements discussed in some detail in elementary textbooks. It is the component of the atmosphere that sustains life and supports combustion, it is the chief constituent of water and the earth's rocks and minerals, it is an essential part of all fats, carbohydrates, and proteins, and it is the major constituent of the common oxyacids H_2SO_4, HNO_3, H_3PO_4, and so on. (For a long time oxygen was considered the essential part of *all* acids; the German name for it is *sauerstoff*, which means "acid material.") The chief danger of such detailed discussion is not that it can be overdone, but that an oversimplified impression of oxygen may result. It is a very versatile element as well as an all-pervasive one, and so a few special remarks about it may be in order here.

It is best to think of oxygen as making up most of the material world, including us. An enormous cage of oxygen and silicon atoms forms the solid foundation of the earth, the moon, and (so far as we know) all the inner planets and satellites of the solar system. Embedded in this oxygen-silicon cage structure are ions of metals such as aluminum, iron, calcium, magnesium, and the alkali elements. Assorted other elements are imprisoned there, too, in smaller amounts. Above this foundation of crust and mantle ride the hydrosphere (the oceans, lakes, and rivers, with their combined oxygen) and the atmosphere (carrying more oxygen). Gaseous molecular oxygen dissolves in the water, providing the necessary O_2 for fish, and it pervades the rocks and soil, bringing about the slow changes we associate with oxidation and with the decay of plant and animal material.

Oxygen is also the great life-giver, and we cannot imagine life on another planet or another solar system without it. To realize in how many ways we depend upon our oxygen-bearing atmosphere, consider the situation encountered by the men who landed on the moon: they had to seal themselves in space suits inflated with oxygen to 5 lb/sq in, they had to carry elaborate and unavoidably awkward life-support systems that provided more oxygen and removed carbon dioxide and water from exhaled air, they had to control the temperature inside the suit and, since there was no air to transmit sound, they had to communicate entirely by radio, even if only two feet apart.

OZONE. The atmosphere contains not only O_2, of course, but also some ozone at the higher levels. There is renewed interest in the oxygen-ozone transformation since the ultraviolet-absorbing ozone layer in our upper atmosphere is now said to be threatened by the chlorofluorocarbons used in spray cans. Ozone itself has long been known to be an endothermic substance, and consequently O_2 can be converted to O_3 only by injecting energy in some suitable form. Sunlight accomplishes this in the upper atmosphere, and an electric discharge does the same thing in an earthbound laboratory. When pure oxygen* is passed through a silent high-voltage discharge between the concentric tubes of an ozonizer, 3 or 4% of the O_2 is converted to O_3. If we think of the injected energy as a reagent, then

$$3O_2 + 68 \text{ kcal} \rightarrow 2O_3 \text{ (mp } -193^\circ \text{C, bp } -112^\circ \text{C)}$$

It follows that when ozone reacts, say in bleaching paper or purifying water or killing bacteria, the O_3 molecules carry their burden of energy into the fray. The ozone dissociates exothermally to oxygen atoms, and these do the work of oxidizing the offending material. Hence ozone is a more active and energetic form of oxygen, and that is why it gets the oxidizing jobs to do.

OXIDES, PEROXIDES, AND SUPEROXIDES. It is widely believed that in ordinary combustion the O_2 of the air is converted solely to covalent oxides such as H_2O and CO_2, and that the oxidation of metals always gives oxide ion, O^{-2}. This is not so. If an oxyhydrogen flame is directed on a block of ice, the condensed water is found to contain some hydrogen peroxide, H_2O_2. If sodium is allowed to burn in an excess of air, it gives a mixture of the normal oxide, Na_2O, and the peroxide, Na_2O_2, with about 20% of the latter. If the temperature of the mixture exceeds 400° C, it is all converted to the peroxide. Indeed, this is the commercial process used for making about 12 million pounds of Na_2O_2 per year in the U.S. It is used in water solution to bleach textiles and wood pulp. The solution does not contain Na_2O_2, of course; all alkali-metal peroxides react with water to give hydrogen peroxide and the hydroxide of the metal, so Na_2O_2 yields a solution of H_2O_2 and NaOH:

$$Na_2O_2(s) + 2H_2O(l) \rightarrow 2Na^+(aq) + 2OH^-(aq) + H_2O_2$$

When potassium burns in air it yields mostly the peroxide, and so do rubidium and cesium. Only lithium gives the simple oxide Li_2O. With pure oxygen instead of air, even higher oxides prevail. Thus potassium burns in pure oxygen to the superoxide, KO_2, an orange crystalline solid. If oxygen is bubbled through melted potassium hydroxide at 300° C it also produces some KO_2, a fact very important to the operation of high-capacity fuel cells which use KOH as an electrolyte. The superoxides are even more potent oxidizers than the peroxides, but they have no practical application in the pure form.

The superoxide ion may be thought of as an O_2 molecule which has gained one electron, while the peroxide ion may be thought of as an O_2 molecule which has gained two electrons:

$:\dot{\underset{\cdot\cdot}{O}}:\dot{\underset{\cdot\cdot}{O}}:$ molecular oxygen, O_2

$:\ddot{\underset{\cdot\cdot}{O}}:\dot{\underset{\cdot\cdot}{O}}:$ superoxide ion, O_2^-

$:\ddot{\underset{\cdot\cdot}{O}}:\ddot{\underset{\cdot\cdot}{O}}:$ peroxide ion, O_2^{-2}

*Oxygen is used instead of air in order to avoid the formation of oxides of nitrogen.

Hydrogen peroxide itself is a very interesting compound. Ordinarily we see only a 3% solution of it in water, sold in drugstores as a germicide and bleach. Such a solution is made by electrolyzing cold concentrated sulfuric acid to convert it to peroxydisulfuric acid, $H_2S_2O_8$, and then hydrolyzing that acid and distilling off the peroxide:

$$H_2S_2O_8(l) + 2H_2O(l) \rightarrow H_2O_2 + 2H_2SO_4(l)$$

Alternatively, an organic peroxide is produced by the action of oxygen on an ether or a hydroxy compound, and this is hydrolyzed to H_2O_2. In either case the H_2O_2 is recovered by distillation and concentrated to 30% for chemical use, or diluted to 3% for pharmaceutical use. Pure H_2O_2 is a pale blue, somewhat viscous liquid which freezes at $-0.89°$ C and boils at 152° C. It is an endothermic compound, and decomposes rapidly and energetically:

$$2H_2O_2 \rightarrow 2H_2O + O_2 \qquad \Delta H = -24 \text{ kcal}$$

This dissociation is catalyzed by iron and manganese salts, and becomes so violent that highly concentrated solutions of H_2O_2 are dangerous to store and use unless kept exceedingly pure.

Group II metals (with the exception of Be) and some transition metals also form peroxides, some of which are slightly water-soluble and all of which yield H_2O_2 when treated with an acid. Some examples are CaO_2, BaO_2, ZnO_2, and CdO_2,* all of which contain the O_2^{-2} ion. The peroxides should not be confused with *dioxides,* which may be oxidizing agents but which contain oxide ions rather than peroxide ions, and which do not yield H_2O_2 on treatment with acid. Examples of true dioxides are PbO_2 (the oxidizing agent in the lead storage battery) and MnO_2 (the oxidizing agent in the ordinary dry cell battery).

FLUORINE, THE CHAMPION ELECTRON-SNATCHER

FREE FLUORINE, F_2. Suppose we are given some sodium fluoride or potassium fluoride, and are asked to extract free fluorine from it. How shall we proceed?

Those who think and act by analogy will consult the methods for obtaining chlorine, whereby hydrochloric acid is oxidized by manganese dioxide, or a saturated water solution of NaCl is electrolyzed to produce chlorine, hydrogen, and NaOH. However, neither of these methods works for fluorine because fluorine is a stronger oxidizing agent than oxygen. Free F_2, if it were formed, would oxidize water to O_2 and HF. Free F_2 would also convert most oxides and oxyacid salts to fluorides, the reverse of what we want. Electrolysis of an aqueous solution of KF produces *oxygen,* not fluorine, because O_2 is discharged at a lower voltage than F_2. So all these methods are unsuitable, and so are those for preparing free bromine and iodine.

If we forsake the watery world, the problem is much simpler. Hydrogen fluoride, HF, is a liquid that boils at 19° C and is a good solvent for most substances, organic and inorganic. It also dissolves glass, so we need to keep it in a container of iron or nickel or copper. If we immerse a pair of nickel electrodes

*For more on peroxides, see I. I. Volnov, *Peroxides, Superoxides, and Ozonides,* New York: Plenum Press (1966).

in a saturated solution of KF in anhydrous HF, and apply 10 or 15 volts to the electrodes (keeping them separated by a diaphragm so that the products cannot mix) we get a little oxygen given off at first, until all traces of water are driven out of the solution. Then fluorine begins to evolve copiously from the positive electrode while hydrogen is evolved at the negative electrode. If the two gases meet, there is a sharp explosion as they combine to form HF again. If kept separate, the fluorine can be conducted through copper tubing to a metal trap cooled in liquid nitrogen, where it condenses to a light green liquid which boils at −188° C. The gas can be stored at moderate pressure in nickel cylinders. Copper and nickel are not exempt from attack by fluorine, but the coating of metal fluoride which forms on the surface protects the underlying metal, in the same way that aluminum oxide protects an aluminum pot in air.

The spontaneously igniting combination of F_2 with H_2 is awesome to behold, and the heat of reaction is enormous: 129 kcal per mole of F_2. This heat of combination is so high that fluorine will remove hydrogen from all its compounds other than HF. Wood, paper, and hydrocarbons burst into flame when exposed to a stream of fluorine, and so will sulfur, charcoal, and powdered metals such as zinc dust. Free fluorine is used to prepare uranium hexafluoride, a volatile compound (bp 56.2° C) used for the separation of the isotopes of uranium. It is also used for the preparation of SF_6, an inert gas employed as the insulating fluid in high-voltage transformers and x-ray equipment. To some extent fluorine is also used as an oxidizer in experimental rocket engines.

HYDROGEN FLUORIDE. Calcium fluoride, CaF_2, occurs as the mineral fluorite. The action of sulfuric acid on this produces hydrogen fluoride, a colorless corrosive gas which condenses just below room temperature:

$$CaF_2(s) + H_2SO_4(l) \rightarrow CaSO_4(s) + 2HF(g)$$

In keeping with its name, fluorine is said to "lend wings to the other elements," and it might therefore be expected that hydrogen fluoride would be very low-boiling, probably under −100° C. Instead, it boils at +19.5° C. This abnormal behavior is caused by strong association of the HF molecules through hydrogen bonding, a phenomenon in which a covalently bonded hydrogen atom acts as a bridge between its own fluorine atom and the fluorine atom of a neighboring HF molecule. The H—F bond distance is 1.0 Å, and the HF—HF bond distance is 1.5 Å. At 0° C the average molecular weight of the HF aggregates corresponds to $(HF)_6$.

Liquid HF comes close to being a universal solvent. Being a highly polar liquid, it dissolves salts even better than water does. Being a very reactive liquid as well, it dissolves many oxides which are insoluble in water, converting them to fluorides in the process:

$$UO_2(s) + 4HF(l) \rightarrow UF_4(s) + 2H_2O(l)$$

Fluorides of the metalloids are all volatile, so their oxides disappear during attack by HF:

$$SiO_2(s) + 4HF(l) \rightarrow SiF_4(g) + 2H_2O(l)$$

$$B_2O_3(s) + 6HF(l) \rightarrow 2BF_3(g) + 3H_2O(l)$$

Furthermore, most organic compounds which are insoluble in water will dissolve in liquid HF. This is especially true of organic substances which contain oxygen, but even hydrocarbons will dissolve to some extent in HF. The solutions conduct electricity, so they may be electrolyzed with unreactive electrodes (such as those of carbon or platinum) to produce fluorine-substituted derivatives. Experimenters should be very cautious about using HF, however, because the anhydrous liquid is exceedingly destructive to skin and tissues, and even a 40% aqueous solution of HF can produce painful burns which are slow to heal.

FLUOROCARBONS AND CHLOROFLUOROCARBONS. Freons are trademarked chlorofluorocarbons which are made by a method more economical than direct fluorination. The simplest common one, Freon 12, has the composition of CF_2Cl_2 and is made by replacing half the chlorine atoms in CCl_4 by stepwise reaction with antimony chlorofluoride:

$$SbCl_5(s) + HF(l) \rightarrow SbCl_4F(s) + HCl(g)$$

then

$$CCl_4(g) + SbCl_4F(s) \rightarrow CCl_3F(g) + SbCl_5(s)$$

and

$$CCl_3F(g) + SbCl_4F(s) \rightarrow CCl_2F_2(g) + SbCl_5(s)$$

Some chlorofluorocarbons are nontoxic and unreactive to metals, so they are used extensively in refrigerators as the working fluid for heat transfer, and also as the propellant in spray cans which dispense everything from paint to perfume. The possible danger that these unreactive gases might persist and collect eventually in the upper atmosphere is being investigated, for their presence there might interfere with the formation of ozone and thus allow harmful ultraviolet radiation to reach the earth's surface.

HEXAFLUORIDES AND MAXIMUM COVALENCY. Two useful hexafluorides, SF_6 and UF_6, have been mentioned, and it would be hard to find two substances more different in chemical behavior. Sulfur hexafluoride does not hydrolyze or decompose, and may even be breathed without any toxic or irritating effects. It withstands acids and alkalies, and may be bubbled through melted sodium without losing its fluorine. Uranium hexafluoride, on the other hand, is very sensitive to water, with which it reacts to form uranium oxide and HF. It fumes in moist air, liberating choking clouds of HF. It reacts violently with aqueous alkalies, and even fluorinates organic compounds. Why are these two compounds of similar composition so different?

The principle of maximum covalency provides the answer. For sulfur, in the third period of the Periodic Table, the maximum covalency is six. There is no room for more than six atoms or groups around the comparatively small sulfur atom, nor are there any more electrons for further chemical bonding beyond six partners. Hence SF_6 cannot bond water or oxygen as a prelude to further reaction. Uranium, on the other hand, is a very heavy element situated in the seventh period, where the maximum covalency is 8 and sometimes 12. It follows that UF_6 can pick up and hold a molecule or two of H_2O on its surface, where further reaction brings hydrogen and fluorine together and leaves oxygen attached to the uranium.

FLUORIDE SALTS VS. COVALENT FLUORIDES. While fluorine does indeed "lend wings to the elements" in providing volatile covalent fluorides, fluoride salts are very high-melting. For example, alumimum fluoride, AlF_3, has a melting point of 1040° C, while $AlCl_3$ melts at 190° C. This shows that fluorine is so much more electronegative than chlorine that its compounds with some metals are decidedly ionic, while the chlorides of the same metals may be covalent.

Aluminum and sodium fluorides combine to make Na_3AlF_6 (mp 1000° C), the electrolyte used in aluminum production. Calcium fluoride, CaF_2, is also a high-melting white solid (mp 1360° C), with a very low solubility in water (only 0.0016 g dissolves in 100 ml of cold water). Fluoride ion in concentrations over 100 ppm in human blood plasma is toxic because the ionic balance of Na^+, K^+, and Ca^{+2} ions is very delicate, and severe malfunction of body control results if this electrolyte balance is disturbed by the precipitation of CaF_2. On the other hand, a concentration of 1 or 2 ppm of fluoride ion in drinking water is beneficial because it converts some of the acid-sensitive basic calcium phosphate ($Ca_5(PO_4)_3OH$, which goes by the strange name of apatite) in growing teeth to nonbasic fluoroapatite, $Ca_5(PO_4)_3F$. This has the same crystal form as apatite and performs the same function in teeth, but without the same tendency toward corrosion and cavities caused by acids from food decay.

Table 3–2 lists the physical properties of all the elements in the sequence from lithium to fluorine.

TABLE 3–2 PHYSICAL PROPERTIES OF THE ELEMENTS FROM LITHIUM TO FLUORINE

Atomic No.	Symbol	Atomic Weight	Melting Point, °C	Boiling Point, °C	Density g/cm^3	State at Room Temperature
3	Li	6.939	186	1326	0.534	Silvery metal
4	Be	9.0122	1283	2970	1.848	Hard gray metal
5	B	10.811	2300	2550	2.34	Black lustrous crystals
6	C	12.011	3570	sublimes	2.25	Black graphite
					3.51	Diamond
7	N	14.007	−210	−196	1.25 g/liter	Colorless gas
8	O	15.999	−218	−183	1.43 g/liter	Colorless gas
9	F	18.998	−220	−188	1.81 g/liter	Pale yellow gas

SUMMARY

In this chapter we have documented the entire progression of a typical period from a strongly electropositive metal to a strongly electronegative nonmetal. The period begins with lithium, an alkali metal that reacts with water to form a strong base, LiOH. The next element, beryllium, is decidedly a metal and finds use as such, but its compounds are mostly covalent because the small size and double charge of the Be^{+2} ion draws pairs of electrons to itself and converts its compounds to covalency. Boron, the third element, shows none of the properties of a true metal; it forms very volatile halides and hydrides, and its oxide forms a very weak acid. Carbon, the middle element, is given over to covalency and especially to chains of carbon atoms; its ionic chemistry is represented only by carbonates and a few ionic carbides. Nitrogen, the fifth element, is electronegative. Its hydrides are bases because they are electron donors, but its oxides are acid-formers and nitric acid is one of the strongest acids. The sixth element, oxygen, is the archetype acid-former and oxidizing agent. Fluorine, the seventh element, is the most potent electron-seeker in the Periodic Table; it will take electrons even from oxygen as it forms a long series of covalent and ionic fluorides.

A few special aspects of each element were considered as the chapter unfolded. Due to its small size and the high heat of hydration of its ion, lithium generates the highest voltage of all the elements when used as the negative electrode of a battery. Its hydride, LiH, contains hydrogen in the very uncommon state of H^- ions. Beryllium is a rare but strong metal with low neutron absorption. Both lithium and beryllium provide good reasons for examining the diagonal relationships of elements in the Periodic Table. Boron is exceedingly hard and strong, and so are its carbide and nitride, which are special abrasives. The hydrides of boron do not follow the usual rules of valency, but are held together by three-center covalent bonds in which one pair of electrons circulates in the orbitals of *three* atoms, not two. The oxide B_2O_3 is a glass-former, and is used with SiO_2 and a decreased proportion of sodium and calcium oxides to make Pyrex and other borosilicate glasses. The halides (and especially BF_3) led to a discussion of some good examples of Lewis acids and their reactions.

The major part of carbon chemistry is left to courses in organic chemistry, but the special behavior of carbon monoxide in forming metal carbonyls and the lethal carboxyhemoglobin was pointed out. Carbon forms inert covalent carbides such as SiC and B_4C, reactive ionic carbides (Al_4C, CaC_2, Li_2C_2, etc., which give hydrocarbons upon hydrolysis), and unreactive metallic interstitial carbides (Fe_3C, WC, etc., useful in alloys). Nitrogen is important principally because of ammonia and its relation to the world food supply. The oxides of nitrogen are air pollutants which lead to nitric acid and nitrates. The organic chemistry of nitrogen leads to proteins, nylon, cyanides, and explosives.

Oxygen is the basis of rocks, mountains, soil, water and hundreds of ionic and covalent oxides. It also is important as ozone, peroxides such as H_2O_2 and Na_2O_2, the superoxide KO_2, and many oxyacids.

The preparation of free fluorine provides an illustration of the unique characteristics of that element. Combination of hydrogen and fluorine is especially vigorous, and HF is a good solvent for almost everything except transition metals. Among the other covalent fluorides, the reactive UF_6 and the very unreactive SF_6 were singled out to illustrate the workings of the principle of maximum covalency.

GLOSSARY

alcohol: any organic compound in which an OH group is linked to a carbon atom which bears no other oxygen. Examples: methyl alcohol, CH_3OH; ethyl alcohol, CH_3CH_2OH.

borane: a boron hydride; any binary compound of boron and hydrogen, so named in keeping with the alkane series of hydrocarbons (methane, ethane, etc.).

carbonyl: a binary compound of any transition metal (elements 21–20, 39–48, and 73–80) with carbon monoxide, characterized by volatility and reversible bonding of the CO.

chlorofluorocarbon: a ternary compound of carbon, fluorine, and chlorine, such as CCl_2F_2.

coordinate covalent bond: a donor-acceptor bond in which one ion or molecule (a Lewis base) donates a pair of electrons to an acceptor (a Lewis acid) to establish a covalent bond.

electrode potential: the potential (voltage) registered by a given half-cell versus a standard hydrogen electrode, usually at standard conditions of molar concentration, at 25° C and 1 atm.

electrolyte: the ionic medium (usually liquid) which surrounds the electrodes of an electrochemical cell.

electronegativity: the attraction of an atom for electrons in a compound.

endothermic compound: a compound which requires *absorption* of heat to prepare it from its elements, contrary to the usual *evolution* of heat as compounds form.

ether: an organic oxide of the type ROR′, in which R and R′ are the same or different hydrocarbon groups. Example: $C_2H_5OC_2H_5$, diethyl ether.

fuel cell: an electrochemical primary cell in which the substance oxidized is not a metal, but a substance normally considered to be a combustible fuel, such as coal or natural gas. The oxidant is usually air, but sometimes pure oxygen or H_2O_2.

halogen: an element of the group fluorine, chlorine, bromine, iodine, and astatine.

hydration: the attraction and bonding of water molecules *as such* by any element or compound, and especially by positive ions.

hydrocarbon: a binary compound of carbon and hydrogen, such as CH_4 or C_2H_2.

hydrolysis: the reaction of a substance with water, in which water molecules are destroyed and new reaction products result.

hydrosphere: the watery covering of the earth, comprising the oceans, seas, lakes, and rivers.

lithosphere: the rocky covering of the earth, comprising sand, soil, rocks, and minerals.

metalloid: a solid element which resembles a metal in physical appearance but is not completely metallic in chemical or physical properties; an element of electronegativity between 1.8 and 2.2.

ozonizer: an apparatus for passing a high-voltage electrical discharge through a stream of oxygen to produce ozone, O_3.

pi orbital: in molecular orbital theory, a region not symmetrical with a line connecting two bonded atoms but available for electron occupancy.

refractory: as a noun, any material which will keep its form and identity at high temperatures (say above 1000° C), such as fire brick, alumina, magnesia, silicon carbide, etc.

three-center bond: a special variety of covalent bond in which a pair of electrons with opposite spin is shared by *three* very small atoms capable of combining one atomic orbital each for this purpose.

trihydric alcohol: an organic compound containing three hydroxyl (OH) groups in its molecular structure and capable of alcohol-type reactions on the part of all three.

unsaturated: as used in organic chemistry, any compound containing one or more multiple bonds between carbon atoms.

EXAMINATION QUESTIONS

1. Name all the elements in the second period, and write their symbols. Which of these elements are discussed in this chapter?

2. Give three examples of diagonal relationship (close similarity of elements located

on a diagonal, rather than within one group, of the Periodic Table) encountered so far, and explain briefly the reasons for the similarity in terms of atomic structure and dimensions.

3. Why does lithium deliver a higher electrode potential in an aqueous electrochemical cell than the more electropositive elements sodium and potassium?

4. How can lithium be separated from sodium and potassium?

5. What is the source of beryllium, and how is it obtained? What uses are there for the metal?

6. How plentiful is boron, and where is it found? How is the pure element obtained, and what are its properties?

7. Describe the physical and chemical properties of the boron halides BX_3. How is BF_3 used, and in what form? How does it accomplish its action in this use?

8. What three solid compounds described in this chapter are exceedingly hard and unreactive, and are competitors of diamond dust as high-quality abrasives? What features of composition and structure account for these properties?

9. List as many boron hydrides as you can, and describe their chemical behavior. What is different about the chemical bonding in these compounds vs. the hydrocarbons?

10. What three general types of carbides are there, and what chemical reactions are typical of each class? Give balanced equations as illustrations.

11. What important products can be made from calcium carbide? Give balanced equations to show how they are obtained, and how the carbide itself is produced.

12. What is the nitrogen cycle, and how does it operate? Why is nitrogen essential to living organisms?

13. By means of balanced equations, outline the three principal methods by which ammonia and ammonium nitrate, or sodium nitrate, may be produced on a large scale.

14. What drawbacks are involved in present-day production of nitrogenous fertilizers? What can (and eventually must) be done to overcome these handicaps?

15. Differentiate between oxides, dioxides, peroxides, and superoxides, giving examples of each. Under what conditions is each formed?

16. Describe the preparation of free fluorine, giving all necessary conditions. In what way does this preparation differ from the methods used for chlorine and bromine?

17. What are fluorocarbons, and how are they produced commercially? Where are they used? (If it is any easier, use chlorofluorocarbons as examples.)

18. Why do some tetrafluorides and hexafluorides hydrolyze rapidly, whereas others are unreactive to water? Give examples to substantiate your answer.

19. Name some fluorides which are *not* volatile or low-melting, and explain why they are so.

20. Illustrate the progression of chemical properties of the elements in the second period, from lithium through fluorine, by giving a brief summary of the chemical behavior of each element.

"THINK" QUESTIONS

A. It is explained in the text that dynamite is seldom used anymore in mining ore, coal, or salt below ground. Instead, a mixture of ammonium nitrate and fuel oil is detonated with a special blasting cap. The mixture has two advantages: it is safer (since unexploded charge is harmless when struck), and it is much cheaper. The text says that only 5% of fuel oil is used with the NH_4NO_3. That is not even enough to wet the prills of nitrate; the mixture looks dry, although it smells of fuel oil. Is 5% of oil enough to use up all the oxygen available in the nitrate? Is it possibly too much, leading to the danger of incomplete oxidation of the carbon in the oil and hence to carbon monoxide poisoning of the miners? Take $C_{12}H_{26}$ as a good approximation of the composition of fuel oil, set up a balanced equation for its reaction with NH_4NO_3 to give N_2, H_2O, and CO_2, and then find out exactly what weight of fuel oil is required to react with 95 g of NH_4NO_3. This will give an unequivocal answer to both questions.

B. The Principality of Lower Phosphoria requires 51,000 tons of ammonia per year for the production of ammonium nitrate as agricultural fertilizer. Finding no country with a surplus to export, they must make their own.

a. By means of balanced equations, outline a method for the preparation of ammonium nitrate from atmospheric nitrogen.

b. What weight of atmospheric nitrogen must be combined annually to produce this amount of ammonia?

c. How many tons of ammonium nitrate can be made from this much ammonia?

C. If lithium would give the highest possible voltage, as measured by standard electrode potential, when used as reducing agent in the negative electrode of a battery, what element would contribute the highest voltage when used as the oxidizing agent in the positive electrode? What practical difficulties do you foresee in using this oxidizing agent?

D. Oxygen forms a difluoride, OF_2, which can be prepared by the electrolysis of a very concentrated solution of HF and KF in water. It is a yellow gas (bp $-145°$ C) which reacts rather slowly with liquid water but explodes when mixed with steam. What products would you expect from the reaction with steam? Write an equation for the hydrolysis.

E. The other well-established fluoride of oxygen is O_2F_2, which is an orange solid at very low temperatures (it melts at $-163°$ C) and decomposes to the elements when its vapor warms to $-50°$ C. It is extremely reactive, and serves as both an oxidizing agent and a fluorinating agent. Write an electron-dot formula for O_2F_2 with what you think is the most likely bonding between the four atoms, and from this predict its molecular configuration. What very common compound of oxygen does your model resemble? How do the molecular structure you have written and the resemblance you have deduced explain the instability and the reactivity of dioxygen difluoride?

F. Over and over, an old chestnut arises in rumor and in the press. It goes something like this: A man drives into a service station, disdains gasoline, and asks that his gas tank be filled with water. Into the water he pours a little bit of white powder, and then drives off, leaving the impression that he is using water (or a dilute aqueous solution of something) as fuel. Oil companies are denounced then by the teller for suppressing a valuable invention.

In terms of the teachings of chemical principles in your textbook and the material in the present chapter of this book, criticize or comment upon the legend from the standpoint of

a. the heat energy available from the oxidation of gasoline vs. any possible heat available from the oxidation of water or of the aqueous solution of a mysterious white powder.

b. the possible mechanical methods or tricks by which a man could actually do exactly what the story-teller says, without violating any chemical principles.

c. the proven use of fuels other than gasoline for propelling automobiles in other countries, stressing what these fuels have in common with gasoline.

G. Ammonium nitrate used for the preparation of N_2O must be very pure. Why?

PROBLEMS

1. How much lithium hydride can be made from 13.88 g of lithium, and what would the product look like? How will it behave toward water?

2. What volume of oxygen is needed to burn 10 liters of C_2H_2 to CO_2 and H_2O? (All volumes are considered to be at standard conditions.)

3. An organic compound containing only carbon and hydrogen is found to be 92.3% carbon. At 80° C and 192.5 mm pressure, 200 ml of the gas is found to weigh 0.1353 g.

   a. What is the empirical formula of the compound?

   b. What is its molecular weight?

   c. What is the true molecular formula?

4. Tabulate all the weight and volume relationships that can be deduced from the equation

$$SiO_2 + 4HF \rightarrow SiF_4 + 2H_2O$$

given the atomic weights Si = 28.08, O = 16.00, H = 1.008, and F = 19.00.

5. Given the heat of formation of BF_3 (−265.4 kcal/mole), the heat of formation of H_2O (−57.80 kcal/mole), and the information in this chapter, calculate the heat of reaction of 1 mole of BF_3 as it hydrolyzes:

$$BF_3 + 3H_2O \rightarrow 3HF + H_3BO_3$$

6. Balance the following equations:

   a. $H^+ + NO_3^- + Zn \rightarrow Zn^+(aq) + N_2O + H_2O$

   b. $MnO_4^- + H_2O_2 + H^+ \rightarrow Mn^{+2} + H_2O + O_2$

   c. $H^+ + NO_3^- + PbS(s) \rightarrow Pb^{+2}(aq) + SO_4^{-2}(aq) + NO + H_2O$

7. Eighteen grams of sorbitol (a sugar with the empirical composition $C_6H_{12}O_6$) is completely oxidized in the human body to CO_2 and H_2O.

   a. How many grams of CO_2 are formed?

   b. What volume of air is required for the oxidation, assuming that the air is 20% oxygen and is at 20° C and 1 atm?

   c. How much heat is required to heat all the air from 20° C to body heat (37.5° C), if the specific heats are: oxygen, 0.219, and nitrogen, 0.249, cal/(g °C)?

4 SODIUM TO CHLORINE: MORE WELL-KNOWN ELEMENTS

We turn now to the third period, and especially to the first seven of its eight elements, the sequence from sodium (atomic number 11) through chlorine (atomic number 17). As before, we shall leave the last element in the period to a special discussion of the zero group gases in Chapter 10, and consider only the chemically active elements in this second period of eight. This will complete a survey of all the active light elements which make up Periods 1, 2 and 3.

SIMILARITIES AND DIFFERENCES

First, what do the elements sodium (Na, for the Latin *natrium*), magnesium (Mg), aluminum (Al), silicon (Si), phosphorus (P), sulfur (S), and chlorine (Cl) have in common? They all have 3*d* orbitals, along with the 3*s* and 3*p* orbitals being filled in this period, and so pairs of electrons from some good donor can be parked there. This allows a maximum covalency of *six*, instead of the maximum of four permitted to the period Li-F above. The effect of this increase on reaction mechanism and on reactivity itself is seen in the difference between $SiCl_4$ and CCl_4 when water is added: $SiCl_4$ fumes, heats up, and soon boils and spatters violently as it reacts with ready-bonded water molecules, while CCl_4 will sit quietly for months in contact with water for lack of a mechanism to react with it. Further evidence of hexacovalency is found in the ions AlF_6^{-3} and SiF_6^{-2}, and in the neutral compound SF_6, a gas. Lastly, and perhaps most convincingly, there are the hexacovalent chelate compounds. The term "chelate" comes from the Greek word for a crab's claw *(chela)*, and refers to the molecular structure of some substances that can attach themselves at two places to a metal ion or atom. For example, the compound acetylacetone has two oxygen atoms spaced and oriented in just the right way to form two covalent bonds to a metal:

$$\begin{array}{ccccc} & \ddot{O}H & & \ddot{O} & \\ & | & & \| & \\ CH_3{-} & C & & C & {-}CH_3 \\ & & \diagdown\!\!\diagdown \quad \diagup & & \\ & & \underset{H}{C} & & \end{array}$$

acetylacetone

This is a weak acid, and dissociates to H^+ ion and the uninegative ion:

$$\begin{array}{ccccc} & \ddot{O}^- & & \ddot{O} & \\ & | & & \| & \\ CH_3{-} & C & & C & {-}CH_3 \\ & & \diagdown\!\!\diagdown \quad \diagup & & \\ & & \underset{H}{C} & & \end{array}$$

When this is mixed with a solution of beryllium nitrate, the compound BeA_2 is formed (where A represents the acetylacetonate ion above). The compound has a tetrahedral structure in which the beryllium is 4-covalent, as shown in Figure 4–1*a* (where the solid dots represent oxygen atoms). However, when magnesium, aluminum, and silicon combine with acetylacetone, they combine with *three* chelate molecules each to form MA_3 compounds: first the ion MgA_3^-, then the neutral compound AlA_3, and then the positive ion SiA_3^+. In all three instances the structure is octahedral, with *six* equivalent covalent bonds about the central atom, as shown in Figure 4–1*b*.

The point is that the Mg, Al, and Si atoms are large enough to accommodate three bulky adducts, each held by two covalent bonds, whereas the elements of the second period can hold only two such adduct molecules, fastened by two covalent bonds each in tetrahedral configuration.

There are many other electron-donating molecules capable of forming chelate compounds. One of them is ethylene diamine

$$\begin{array}{ccc} \ddot{N}H_2 & & \ddot{N}H_2 \\ | & & | \\ CH_2 & {-} & CH_2 \end{array}$$

which can contribute the two pairs of electrons shown by dots. It and acetylacetone are *bidentate* reagents, meaning literally that they have two sets of teeth for forming a chelate compound. There are also some tridentate and quadri-

Figure 4–1. Structures of beryllium and magnesium chelates, showing tetrahedral (*a*) and octahedral (*b*) arrangements.

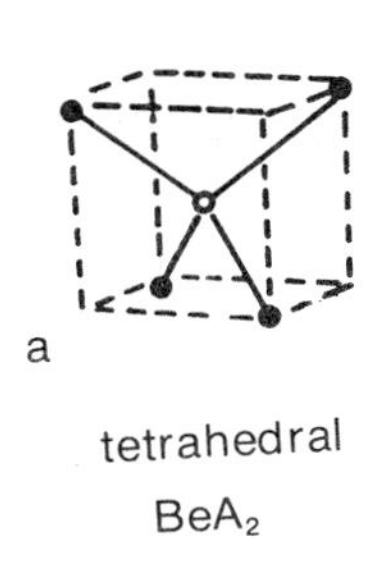

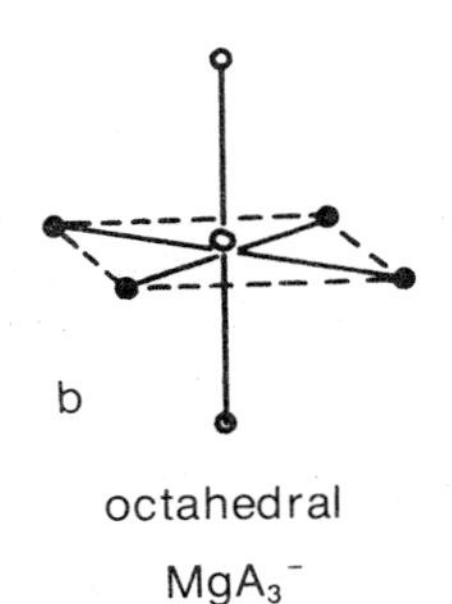

TABLE 4–1 TRENDS IN THE THIRD PERIOD, SODIUM THROUGH CHLORINE

Element	*Atomic Number*	*Atomic Radius, Å*	*Ionic Radius, Å*	*Electro-negativity*	*Standard Reduction Potential, Volts*
Na	11	1.86	(+1) 0.95	0.9	−2.71
Mg	12	1.60	(+2) 0.65	1.2	−2.37
Al	13	1.43	(+3) 0.50	1.5	−1.66
Si	14	1.17	(+4) 0.41	1.8	−0.102
P	15	1.10	(−3) 2.12	2.1	−0.06
S	16	1.04	(−2) 1.84	2.5	+0.14
Cl	17	0.99	(−1) 1.81	3.0	+1.36

dentate reagents which are important in analytical chemistry and coordination chemistry. In all their compounds with metals the Maximum Covalency Rule is observed.

In what ways do the elements in the sequence Na-Cl differ? In all their chemistry, of course. They also show gradual changes in atomic radii and electronegativity, as given in Table 4–1. Their ionic radii decrease rapidly from 0.95 Å for Na^+ to 0.41 Å for Si^{+4}, and then rise abruptly as the change from positive-ion to negative-ion formation occurs (Table 4–1). The standard electrode potential shows an abrupt decrease for chlorine. The maximum potential difference for a Na-Cl cell is 4.07 volts, compared with 5.92 volts for a Li-F cell.

SOME ASPECTS OF SODIUM

SALT. To most people, sodium means salt and salt means sodium, and that's it. They have an instinctive craving for salt because its chloride content is es-

Important Salts

Chemical Preservatives
Sodium sulfite
Sodium propionate
Sodium sorbate
Potassium bisulfite

Miscellaneous Food Additives
Magnesium hydroxide
Sodium acetate
Sodium chloride
Sodium bicarbonate

Nutrients and Dietary Supplements
Magnesium oxide
Manganese sulfide
Potassium chloride
Sodium phosphate
Potassium iodide

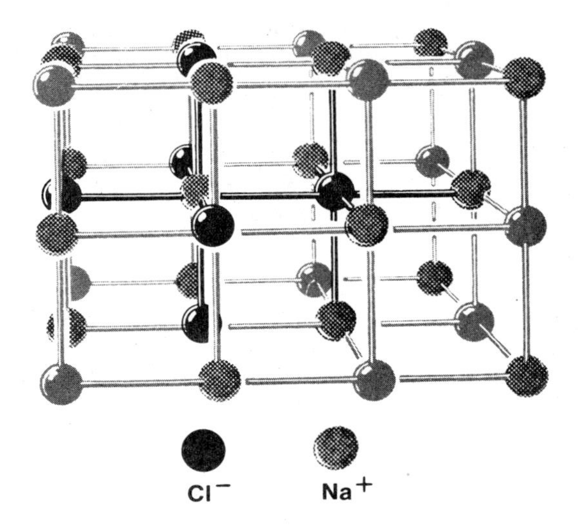

The structure of sodium chloride, a three-dimensional network of bonded ions. Each Na^+ ion is bonded to 6 Cl^- ions and each Cl^- ion is bonded to 6 Na^+ ions. Actually, the ions are very closely packed.

sential for the continuous supply of gastric juice (which is 0.1 N HCl), and hence is essential for digesting our food. Herbivorous wild animals will travel great distances to find a salt lick, and domestic animals are fed salted hay. The most precious cargo of camel caravans for centuries has been dried sea salt for the camels themselves, and the most valuable item in the Roman military baggage was salt for the soldiers (*salarium,* from which our word "salary" is derived). Like most goods in our culture, though, this one has been abused by too much purification. In our intense desire to have clean, white, free-flowing table salt, 100% pure, we remove the potassium, calcium, and magnesium chlorides and iodides that were in sea salt, and then run the risk of ionic imbalance in the blood plasma from the consumption of too much pure sodium chloride. Crude sea salt is better nutritionally, and at the very least some sodium iodide should be put back into purified salt to provide the essential iodine.

As a consequence of having cheap, reliable sources of sodium chloride, we use too much of it. Too much salt in the diet promotes high blood pressure, dizziness, and dehydration. Too much salt on city streets and highways destroys concrete, rusts away automobiles before their time, kills roadside shrubs and lawns, and pollutes streams with the runoff.

The Age of the Oceans. The saltiness of ocean water has fascinated people throughout the ages. Was the sea always salty? Is it getting saltier? Does seawater harm skin or hair? Why can't we drink it? Why does it corrode iron and steel (and most other metals) so much faster than fresh water? Is there any material impervious to it? Answers to all these questions would fill a book larger than this one, but a few items can be settled. The rivers of the world return 6500 cubic miles of water to the ocean each year, and analyses show that this water carries about 160 million tons of sodium chloride down to the sea (along with many other dissolved and suspended materials, of course). Assuming that the ocean was once fresh water, early in the history of the earth, and assuming that the rate of accumulation of salt has remained constant (neither assumption being justified), we can divide the present content of salt in the sea by the annual input and arrive at an age of the ocean. This turns out to be 256 million years, a period much smaller than the age of the earth as measured by the decay of radioactive isotopes. Some explanation for the discrepancy can be found in the periodic removal of salt in the form of accumulated deposits that remain from evaporated ocean basins of long ago. And of course the earth may have been hot for a very long time before it cooled enough for water to condense and accumulate.

Corrosion by Seawater. A 3.5% solution of sodium chloride is an excellent electrolyte for galvanic action of the same sort that takes place in batteries, so any dissimilar metals joined together and immersed in seawater will set up a voltaic cell and generate electricity. A current of only 1 ampere, continued for one year, will dissolve 22 lb of steel under this electrolytic corrosion. Since most metal chlorides are soluble (those of silver and lead being exceptions), no insoluble protective coating is formed. The metals most susceptible to seawater corrosion are magnesium, zinc, and aluminum, with cast iron and steel coming next; brass and bronze resist fairly well, and the least corroded metals are silver, gold, and platinum. Silver bars and gold coins from the wrecks of Spanish galleons are still found intact off the coast of Florida, after 300 years of immersion in seawater.

Manufactures from Salt. Sodium chloride is the raw material for making metallic sodium, sodium hydroxide, sodium carbonate and bicarbonate, and sodium sulfate. The processes for these manufactures are described in every textbook and need not be repeated here, but a few additional points should be noted:

1. Sodium is obtained by electrolysis of a molten mixture of 40% NaCl and 60% $CaCl_2$ by weight, not from pure NaCl. The mixture melts at 580° C, vs. 801° C for pure NaCl, and so the cell can be operated at a lower temperature and lasts longer. Sodium ions are discharged preferentially in the melt

$$Na^+ + e^- \rightarrow Na(l)$$

and the liquid sodium rises to a collector pipe. Any particles of calcium that may form at the negative electrode are insoluble in the liquid sodium and soon settle out of it, only to be dissolved by the molten salt. Chlorine collects at the positive electrode

$$2Cl^- \rightarrow Cl_2(g) + 2e^-$$

and is cooled, compressed, and liquefied. Most of the chlorine is used for water purification, but some goes into bleaching operations, industrial organic chemistry, and various inorganic chlorides such as $AlCl_3$, $TiCl_4$, and PCl_3.

Metallic sodium is still used for making tetraethyl lead (see next chapter) for ethyl gasoline, but this product is slowly being phased out to reduce pollution of the air by the lead in exhaust gas. Considerable sodium is used to reduce animal fats in order to make detergents, and some is used to reduce titanium and zirconium from their chlorides:

$$TiCl_4(g) + 4Na(l) \rightarrow 4NaCl(s) + Ti(s)$$

Sodium hydride and amide are also made from metallic sodium:

$$2Na(l) + H_2(g) \rightarrow 2NaH(s)$$

$$2Na(l) + 2NH_3(g) \rightarrow 2NaNH_2(s) + H_2(g)$$

Sodium cyanide can be made from sodium, but is more economically made from calcium cyanamide:

$$CaCN_2(s) + C(s) + Na_2CO_3(s) \rightarrow 2NaCN(s) + CaCO_3(s)$$

The proposed large-scale use of metallic sodium as a heat-transfer liquid in fast-breeder reactors for a new generation of nuclear power plants arouses opposition from many chemists who are familiar with the violent reaction of sodium with water

$$2Na(l) + 2H_2O(l) \rightarrow 2NaOH(aq) + H_2(g)$$

and the spontaneous ignition or explosion of the resulting hydrogen, and who look upon the large quantity of hot sodium circulating through a water boiler as a huge chemical bomb, poised for disaster.

2. For over half a century sodium bicarbonate (and from it the carbonate) have been made by the Solvay process, in which CO_2 is forced into a concentrated solution saturated with salt and ammonia:

$$CO_2(g) + NH_3(aq) + H_2O(l) + Na^+ + Cl^- \rightarrow NaHCO_3(s) + NH_4^+ + Cl^-$$

$$2NaHCO_3(s) \rightarrow Na_2CO_3(s) + H_2O(g) + CO_2(g)$$

Recently, however, methods have been developed for the extraction of $NaHCO_3$ from natural deposits of alkali in dried-up lakes, while the increasing cost of ammonia and the difficulty of disposing of chloride wastes have made the Solvay process less attractive. The market continues to increase, as more and more sodium carbonate is needed to make glass, paper, and detergents.

3. Sodium hydroxide is the country's most important alkali, and some five million tons of it are made annually by the electrolysis of a water solution of salt:

$$2NaCl(aq) + 2H_2O(l) \xrightarrow{\text{Electrolyzed}} 2NaOH(aq) + H_2(g) + Cl_2(g)$$

Most of the sodium hydroxide is shipped as a concentrated solution ("caustic soda") which is used in making soap, rayon, cellophane, paper, and dyes.

4. Sodium sulfate ("salt cake") was formerly a by-product from the manufacture of hydrochloric acid:

$$2NaCl(s) + H_2SO_4(l) \rightarrow Na_2SO_4(s) + 2HCl(g)$$

Now it is the end-product of hundreds of industrial operations in which residual sulfuric acid used for processing is neutralized by sodium hydroxide. For a long time it had very little value, but now it is a mainstay in the paper industry, which makes 640 lb of paper for each man, woman, and child in the U.S. each year. The specific process which requires Na_2SO_4 is the kraft process for making brown wrapping paper and corrugated boxes. Digestion of wood chips or sawmill waste in a very hot alkaline solution of Na_2SO_4 dissolves the brown resinous component of wood, called lignin, which cements the fibers together. This liberates the cellulose fibers as a pulp which then goes to the papermaking screens, while the solution is evaporated until it can be burned, thereby providing steam for the plant and heat for the evaporation. The fused sodium sulfate and hydroxide that survive the flame are used over again in the process. Indeed, most paper can be recycled, too, by throwing it in with the wood chips of the next batch. A modern paper mill burns its own odors, uses its own waste water over and over, and even uses the water in incoming green logs. It is not necessary that a paper mill pollute the surrounding air and water; all the chemical technology required to keep it self-contained and to consume its own waste is there.

5. Sodium hydride is made from sodium and hydrogen, in the same manner as lithium hydride. It is another salt-like hydride, a white crystalline solid made up of Na^+ and H^- ions, and is used to make sodium borohydride ($NaBH_4$) for bleaching paper and textiles and for recovering silver and mercury from waste water.

SOME ASPECTS OF MAGNESIUM

EXTRACTION. In the discussion of seawater in Chapter 2 we saw that magnesium could be recovered from it. This is done by precipitating magnesium hydroxide, converting this to the chloride, and electrolyzing the molten anhydrous chloride. One seaside plant uses oyster shells as the source of hydroxide ions to precipitate $Mg(OH)_2$. First the oyster shells are calcined at red heat:

$$CaO_3 \text{ in oyster shells} \xrightarrow{\text{Heat}} CaO(s) + CO_2(g)$$

Then the quicklime is slaked with water:

$$CaO(s) + H_2O(l) \rightarrow Ca(OH)_2(s) \text{ (solubility 1.85 g/liter)}$$

The resulting calcium hydroxide is then added to the seawater to increase its pH to a level where magnesium hydroxide precipitates, leaving the very soluble $CaCl_2$ in solution:

$$2OH^-(aq) + Mg^{+2}(aq) \rightarrow Mg(OH)_2(s)$$

Magnesium hydroxide has a solubility of only 0.009 g/liter at room temperature, so it separates as a milky suspension which slowly settles and is collected. In order to produce metallic magnesium from it, the wet hydroxide is dissolved in hydrochloric acid obtained from the chlorine recovered in the final step:

$$Mg(OH)_2(s) + 2H^+(aq) + 2Cl^-(aq) \rightarrow Mg^{+2}(aq) + 2Cl^-(aq) + 2H_2O(l)$$

The water is driven off by heat, the dry magnesium chloride is melted at 708° C, and the melt is electrolyzed:

$$Mg^{+2} + 2e^- \rightarrow Mg(l)$$

$$2\,Cl^- \rightarrow Cl_2(g) \text{ (used over again)} + 2e^-$$

Magnesium oxide is the source of metallic magnesium by the silicothermic process, used in Canada and the Pacific Northwest. Magnesite is calcined (heated to red heat) to convert it to the oxide, and then this is mixed with an alloy of silicon and iron, pressed into pellets, and heated to 1200° C in a vacuum furnace. At that temperature silicon and iron remove oxygen from magnesium, driving off the metal:

$$2MgO(s) + Si(s) \rightarrow SiO_2(s) + 2Mg$$

and

$$3MgO(s) + 2Fe(s) \rightarrow Fe_2O_3(s) + 3Mg$$

Since magnesium melts at 650° C and boils at 1110° C, it is vaporized at the furnace temperature and condenses in a cooled arm of the reaction vessel as shiny crystals of pure metal. This method requires no electric power, but does require considerable fuel.

THE METAL. Pure magnesium metal is the starting point for a long list of structural light alloys containing zinc, aluminum, and manganese. These are used for aircraft wheels and other die-cast parts, ladders, luggage, shovels, portable tools, truck bodies, and even cameras. Undoubtedly there will be greater use in the automotive field (and elsewhere) as good iron ore becomes more rare and steel becomes more expensive. Good workable deposits of iron ore do run out, but it seems impossible that we shall ever run out of magnesium for structural use. It is our eighth most abundant element, and is available at any seacoast. It has been estimated that if we took 100 million tons of magnesium out of the sea each year, and kept that up for a million years, the concentration of Mg in the sea would drop from the present 0.13% to 0.12% (assuming that no Mg was washed back into the ocean in the meantime). Given a cheap, compact source of electrical energy (a packaged nuclear power plant?), it is even possible to conceive of magnesium breeder ships, all programmed to extract more magnesium from the sea, to melt it, alloy it, roll it, and build more ships from it, outfitting these to come home upon command to be scrapped for use as structural metal on land.

The principal advantage of magnesium as a structural metal is easily seen from its density, 1.74 g/cm^3 vs. 2.70 g/cm^3 for aluminum and 7.87 g/cm^3 for iron. For equal strength, the best aluminum alloys weigh 35% as much as steel, while the best magnesium alloys weigh 24% as much as steel.

Magnesium is an active metal, as its electrode potential (−2.37 volts, just below that of sodium) shows. It might be expected to decompose water the way sodium does, but normally magnesium is protected by a film of oxide and it does not react with water. A boiling solution of $MgCl_2$ attacks the oxide, however, and then the metal does decompose the water:

$$Mg(s) + 2H_2O(l) \xrightarrow{MgCl_2} Mg(OH)_2(s) + H_2(g)$$

In its other chemical properties magnesium acts much like lithium and calcium. It forms a soluble chloride like LiCl and $CaCl_2$, and its sulfate is also very soluble, whereas $CaSO_4$ is not. The Mg^{+2} ion is strongly hydrated in solution, due to its small size and double charge; hydrated salts such as $MgCl_2(H_2O)_6$ and $MgSO_4(H_2O)_7$ in which the magnesium ion is hexahydrated and hence hexacovalent) are common. Magnesium also forms many organometallic compounds which are covalent and are valuable reagents for organic synthesis. The reaction of magnesium chips with ethyl bromide in ether solution provides an example:

$$Mg(s) + C_2H_5Br(l) \xrightarrow{\text{Ether}} C_2H_5MgBr \text{ (soluble in ether)}$$

Such organomagnesium halides are called Grignard reagents.

COMPOUNDS. Some important magnesium compounds are the oxide, hydroxide, chloride, and sulfate. When magnesium hydroxide is precipitated, filtered, and washed, it can be stirred with a limited amount of water to form a thick white suspension called milk of magnesia. This is a very weak base, but weight for weight it can neutralize 1.37 times as much acid as sodium hydroxide and 2.85 times as much acid as sodium bicarbonate, as a little calculation of acid-base equivalent weights will show. Hence milk of magnesia is a favorite digestive antacid, not as alkaline and hence not as irritating as sodium bicarbonate but much more effective.

Magnesium oxide has two sources: it is obtained by heating the hydroxide to dull red heat, driving off water,

$$Mg(OH)_2(s) \rightarrow MgO(s) + H_2O(g)$$

or else it is obtained by heating natural magnesium carbonate (the mineral *magnesite*) to 700° C:

$$MgCO_3(s) \rightarrow MgO(s) + CO_2(g)$$

The first method can give a purer product because the hydroxide can be purified beforehand; oxide from the second method still contains the natural impurities (such as Ca, Al, and Si) present in the magnesite. Impure fluffy magnesium oxide is used for thermal insulation because it is unreactive and heat-stable, and because it traps a large volume of air that does the actual insulating. A much purer grade is fused in an electric arc furnace to obtain hard, colorless, crystalline magnesium oxide with the mineral name of periclase. This has the same crystal structure as sodium chloride, but with the corners of the cube occupied by alternate Mg^{+2} and O^{-2} ions instead of Na^+ and Cl^- ions. Because the ions all have double charges, without much change in spacing, the lattice energy is much higher than in NaCl, so the melting point is much higher, 2800° C instead of 801° C.

Clear crystalline magnesium oxide is a very unusual material, for it is a good conductor of heat but a very poor conductor of electricity, even at high temperatures. Because of this anomaly, crystalline MgO is the substance that makes the modern electric kitchen range possible; it conducts the heat rapidly from a very hot coil of resistance wire to a grounded metal sheath only a tenth of an inch away, yet insulates that 220-volt coil from the grounded sheath, even at 1000° C.

SOME ASPECTS OF ALUMINUM

SOURCES. Aluminum is our third most abundant element, and it is widespread in the form of thousands of different aluminosilicates and oxides. It is almost impossible to classify these by chemical composition, but they can be thought of in terms of relatively few structural types:

1. Oxides which contain no silicon. The oxide Al_2O_3 itself (corundum and the gemstones, ruby and sapphire) is rather rare, but the *spinels*, in which some

dipositive ion teams up with the Al^{+3} ion in a mixed oxide, are very widespread. Some examples are $MgAl_2O_4$, $FeAl_2O_4$, and $ZnAl_2O_4$. Clear crystalline magnesium spinels, especially when colored red (ruby spinels) are prized as gemstones. They have a hardness of 8 on the mineralogist's Mohs scale, vs. 9 for corundum (including ruby and sapphire) and 10 for diamond.

2. Silicates which contain aluminum and other metals, as in beryl, $Be_3Al_2Si_6O_{18}$, and in the layer minerals (mica, talc, and the clays). Here the hexagonal rings of silicon and oxide ions are joined together to form chains, sheets, or layered structures, with positive ions and OH^- ions between the layers. In this structure aluminum can substitute for silicon in the rings because it has about the same size, but since Al^{+3} has one less charge than Si^{+4}, it must be compensated for by an additional Na^+ or K^+ or Li^+ ion between the layers. The clay minerals, such as halloysite and kaolinite, are all aluminosilicates. All contain structural water and all undergo further slow weathering and metamorphosis until they reach the ultimate end-product bauxite, a hydrated oxide of aluminum containing considerable iron and other impurities. Bauxite is the raw material for the commercial production of aluminum.

3. In addition to the natural minerals which contain aluminum as a structural constituent, there are many artificial mineral building materials which depend upon it. Portland cement, the basis of concrete, is an aluminosilicate made by calcining clay or shale with limestone, sand, and the glassy slag from iron-making blast furnaces. The clinker that results is mixed with some calcium sulfate and ground to a very fine powder, which contains 3 to 8% Al_2O_3, 3 to 6% Fe_2O_3, 15 to 20% SiO_2, and 58 to 64% CaO. When mixed with water, the aluminosilicate powder slowly reacts (with evolution of heat) and crystallizes as an interleaved mass of hydrated minerals. Similarly, brick and tile are composed of aluminosilicates because they are made of fired clay, and every piece of chinaware or earthenware has a considerable aluminum oxide content. High-temperature porcelain is mostly a mass of sillimanite crystals, Al_2SiO_5. Window glass usually contains only a small amount of aluminum oxide because that raises the softening point and makes fabrication of the glass more difficult. However, laboratory glassware contains some Al_2O_3 in order to raise the softening point and to improve chemical resistance to reagents.

THE METAL. The production of metallic aluminum by the electrolysis of Al_2O_3 from bauxite, dissolved in melted Na_3AlF_6, is well-known. Every textbook points out that the process was discovered by an Oberlin College student, Charles Hall, and that it has made aluminum cheap and plentiful, a tremendous boon to our aircraft industry and to the manufacture of household goods and automobiles. As we look to the future, however, we see aluminum as a highly energy-intensive material (it takes 1 faraday of electrical energy, or 96,500 coulombs, to produce only 9 grams of aluminum!) dependent on a starting material which is becoming increasingly scarce (bauxite, a natural hydrated Al_2O_3). It is time for some other student to devise a new method for extracting aluminum from its myriad sources, in order to produce the metal cheaply without electrolysis. In the meantime, all those four billion aluminum beer cans made each year need to be returned to the foundry as scrap, both to save energy and to prevent our drowning in a sea of beer cans. They will not disappear by themselves.

Pure aluminum itself is a rather soft and weak metal. What is even worse, it loses strength rapidly above 300° C, even though it does not melt until 660° C. It would be impossible to build a modern airplane out of pure aluminum, because its wing spars, its girders, and its skin would not stand the strain. The great advances in aluminum metallurgy have been in the direction of light alloys that are strong and stiff, and are easy to fabricate yet keep their strength under all reasonable conditions of use. The favorite method of accomplishing this is to dissolve a limited amount of copper, silicon, magnesium, and manganese in hot aluminum, chill the metal, fabricate the part, and then heat-treat the metal at such a temperature and time that tiny particles of $CuAl_2$, Mg_2Si, and other insoluble compounds precipitate out. These dispersed particles of insoluble material harden and stiffen the metallic matrix in much the same way as dispersed particles of iron carbide harden and stiffen iron and transform it into steel. A typical heat-treatable alloy will contain about 4.4% copper, 0.8% silicon, 0.4% magnesium, and 0.8% manganese. Softer and more corrosion-resistant alloys for window frames, sheathing, and cooking utensils omit the Cu, Si, and Mg and contain only about 1.2% Mn.

Pure aluminum does have one important property that leads to direct use, and that is its high electrical conductivity. By volume, aluminum can conduct 64% as much electric current as copper; that is, a wire of the same diameter has 64% of the conductivity of a pure copper wire. But, weight for weight, aluminum will carry twice as much current as copper because of its low density (2.70 for Al, 8.96 for Cu). Wherever bulk does not matter, as in long-distance power transmission lines, the advantage is with aluminum. As copper becomes more and more scarce and expensive, aluminum very likely will take over the electrical jobs.

The electrochemical equivalent weight of a metal or of any element is the weight of it that is deposited or liberated by exactly 1 faraday of electricity. It is the atomic weight divided by the number of electrons removed or acquired by the action of the electric current. The equivalent weight of lithium is 6.94 g, of magnesium 12.16 g, and of aluminum only 9.0 g (the atomic weight, 26.98, divided by 3 for the change $Al^{+3} + 3e^- \rightarrow Al$). This is why aluminum metallurgy was stated earlier to be so excessively energy-intensive—aluminum is a real energy hog. But, by the same token, once we have metallic aluminum on hand it becomes a concentrated source of chemical energy which can deliver 1 faraday of electricity for every 9 g consumed in a suitable battery. Just what constitutes "a suitable battery" is the problem. We do not yet know how to make an aluminum electrode function in a battery, because in aqueous electrolytes it likes to coat itself with an insulating layer of Al_2O_3 and stop reacting. If a suitable nonaqueous electrolyte could be found, aluminum could be incorporated in a cheap and light battery for automobiles. The voltage developed by an aluminum electrode, 1.66 volts on the standard scale, cannot match that developed by lithium (3.045 V) or even by magnesium (2.37 V), but the reactivity of aluminum might be more controllable, and of course it has an enormous weight advantage over zinc (equivalent weight 32.7 g) and lead (103.6 g), which are the materials now in use. Perhaps the ultimate would be an electric automobile with a battery that consumed beer cans.

Aluminum is an active metal (see Table 4–1) which is ordinarily protected by a tenacious film of oxide. It is an amphoteric element, and the clean metal will dissolve in strong acids and strong bases:

$$2Al(s) + 6H^+(aq) \rightarrow 2\ Al^{+3}(aq) + 3H_2(g)$$

$$2Al(s) + 60H^-(aq) \rightarrow 2AlO_3^{-3}(aq) + 3H_2(g)$$

Among the compounds of aluminum that are important are the oxide, the chloride, and the sulfates (plural because there are several useful ones). The oxide Al_2O_3 has already been encountered in the impure powdery form as bauxite. In the crystalline form of corundum it is an important abrasive for grinding wheels and "sand" paper; as ruby or sapphire it is a gemstone. Sapphire is also used to make pivot bearings for watches and instruments—it is so hard it does not wear away. The supply of natural sapphire crystals is now inadequate for all its uses, so pure Al_2O_3 is fused in an oxyhydrogen flame to grow crystals of suitable size and quality.

Aluminum chloride, Al_2Cl_6, is a cheap and effective Lewis acid catalyst for many reactions in industrial organic chemistry. It has a bridged structure similar to that of diborane:

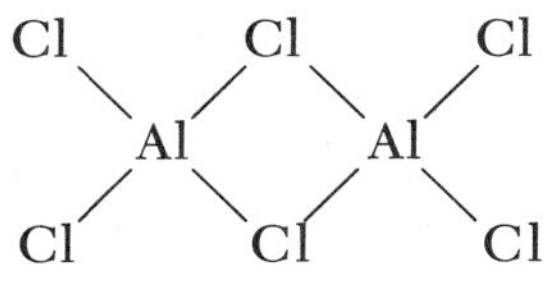

```
Cl        Cl        Cl
  \      /  \      /
    Al        Al
  /      \  /      \
Cl        Cl        Cl
```

It is made by chlorinating scrap aluminum or a red-hot mixture of bauxite and coke. Pure Al_2Cl_6 sublimes at 178° C, so it leaves the scene of reaction and condenses in a cooled receiver as colorless volatile crystals. These are conveniently soluble in such diverse organic solvents as alcohol, chloroform, ether, and even benzene, but they react vigorously with water to revert to the oxide:

$$Al_2Cl_6(s) + 6H_2O(l) \rightarrow 2Al(OH)_3(s) + 6HCl(g)$$

Aluminum sulfate, $Al_2(SO_4)_3$, is prepared by the action of sulfuric acid on hydrated aluminum oxide obtained in the bauxite purification process. It crystallizes from the reaction mixture as $Al_2(SO_4)_3(H_2O)_9$, sometimes called "papermaker's alum" because huge quantities are used in the paper industry. If paper were made only from the cellulose fibers derived from wood, all paper would be blotting paper. In order to make a strong, nonporous sheet with a smooth writing surface, paper needs to be "filled" with an opacifier and "sized" with a surface coating, then rolled to make it smooth and dense. To accomplish all this, fillers such as clay and chalk are first added to the wet pulp, and then a milky suspension of rosin and sodium soaps of rosin acids is added as sizing agent. To fix the rosin to the paper fibers, aluminum sulfate is added. Since this is acidic, it neutralizes the highly alkaline sodium soap and deposits rosin on the fibers. The other product of reaction, hydrated aluminum oxide, acts as further filler and opacifier. The mass is then poured on the screens of a papermaking machine, and the resulting web is washed, dried, and passed between heated polished rolls to obtain a smooth sheet of paper like this one.

In the correct sense, a true alum is a double sulfate of aluminum and lithium, sodium, potassium, or ammonium ions. Potassium alum, the best known one, is $KAl(SO_4)_2(H_2O)_{12}$; sodium alum is $NaAl(SO_4)_2(H_2O)_{12}$, and so on. Not only can the unipositive ions Li^+, Na^+, K^+, and NH_4^+ substitute for each other, but Fe^{3+} and Cr^{3+} ions can substitute for the Al^{3+}, giving rise to a large number of possible combinations. All these alums are isomorphic (that is, they have the same crystal pattern and very similar lattice dimensions), so they can be deposited in layers on one seed crystal. It is fun to grow large crystals from saturated solution in a constant-temperature room, and the alums give the beginner his best chance for beautiful results. For example, a 1-inch layer of colorless potassium alum can be deposited on a 4-inch purple crystal of chromium alum, and so on. The best way to learn about crystals is to grow them.

Like magnesium, aluminum forms a long series of very reactive organometallic compounds with Al—C bonds. Some of these, notably triethyl aluminum, $(C_2H_5)_3Al$, have become important catalysts for polymerization of hydrocarbons to make high-strength plastics. They are called Ziegler catalysts, and one of their products is polypropylene, used for carpets and ropes.

SILICON, THE PLANET-BUILDER

GENERAL. Silicon is a shiny, silvery, blue-gray, brittle element that looks every bit as metallic as chromium, but doesn't always act like a metal. For example, all metals conduct electricity by flow of their free electrons according to the principles of metallic bonding, but in silicon the atoms are all linked together by shared-pair covalent bonds, and so electrical conduction is on a much lower level because it can only result when an electron is torn away from such a bond. In a true metal the flow of free electrons is hampered only by the positive metal ions that get in the way. As the temperature is raised, the vibrations of positive ions about their mean positions become more extensive, resulting in greater hindrance to the flow of free electrons and a higher resistance to electric current in the metal. In silicon, on the other hand, higher temperatures and more vibration of silicon atoms cause a *loosening* of the covalent bonds and a greater availability of electrons as carriers of current, so the electrical resistance of silicon *decreases* markedly with rise of temperature. This is but one of several ways in which a metalloid differs from a metal, and silicon is a typical metalloid.

SEMICONDUCTOR THEORY. The fundamental differences between metals and semiconductor metalloids can be made clear by resorting to energy-level diagrams such as have already been encountered in discussions of atomic spectra. For a single isolated atom the allowable energy levels corresponding to the principal quantum numbers n are drawn in at appropriate heights on the diagram. In true metals there is massive overlap of the atomic orbitals in the multiatom lattice, resulting in such a profusion of low-lying allowable energy levels that the few available electrons can circulate freely in these levels. The lines of the atomic energy-level diagram therefore become *bands* when a solid is depicted, and in a metallic solid the bands are numerous but are very sparsely populated because of the chronic shortage of electrons.

When we turn to the metalloids we have more electrons available per orbital because the metalloids are situated further to the right in the Periodic Table. In a semiconductor such as silicon (or in diamond or gray tin) the electrons are localized in covalent bonds, and therefore when we assign electrons two at a time to the available energy levels (in accordance with the Pauli exclusion principle) we find that the lower levels are completely filled (Fig. 4–2). At the same time there are some higher energy levels which are completely empty. Allowing for the usual broadening out of lines into bands, then, we find that solid silicon may be represented by a sequence of filled and empty energy levels (Fig. 4–3). Moreover, since diamond, silicon, germanium, and gray tin all have the same crystal structure, they all have a similar pattern of energy-level bands (Figs. 4–2, 4–3, 4–4, and 4–5). The only difference lies in the separation of filled and empty bands and, of course, in the number of bands. In diamond the spacing is wide, and so a great deal of excitation energy is necessary to promote some electrons from the uppermost filled band to the empty band, so that they can migrate freely throughout the lattice and conduct an electric charge. The promotional energy is so large, in fact, that diamond is an insulator at ordinary temperatures and becomes a semiconductor only at red heat. Silicon, on the other hand, does not hold on to its electrons so strongly, and a few can bridge the

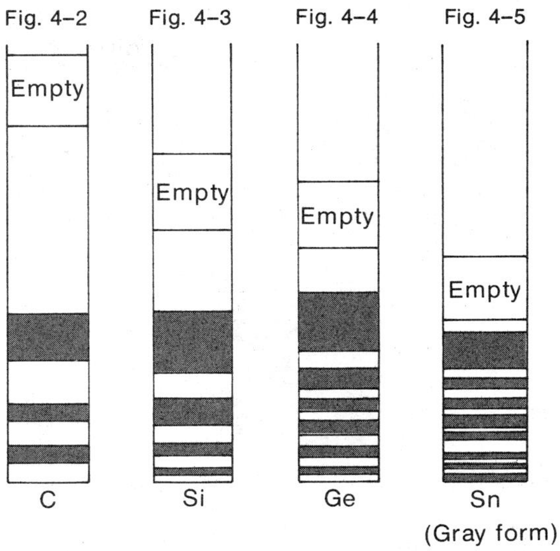

Fig. 4–2 Fig. 4–3 Fig. 4–4 Fig. 4–5

Empty Empty Empty Empty

C Si Ge Sn (Gray form)

Figures 4–2 to 4–5. Band theory diagrams for diamond (Fig. 4–2), silicon (Fig. 4–3), germanium (Fig. 4–4), and gray tin (Fig. 4–5).

gap and be present in the empty bands even at room temperature. It follows that at elevated temperatures more electrons will jump the gap to the empty band, and so there will be more carriers and more conductance in the silicon (Fig. 4–3). Moreover, when substantial numbers of electrons are promoted from a filled band to a hitherto empty band, electron vacancies will be left in the "filled" bands. Electrons from other covalent bonds can move to these vacancies under the stress of an electric field, and other electrons can move into the new "holes." Hence considerable current can be carried by presumed motion of the "holes," and the conductance rises more.

In germanium the spacing between filled and empty bands is still smaller (Fig. 4–4), and so at ordinary temperatures even more conductance electrons have already been promoted to the "empty" bands. Hence the intrinsic conductivity of germanium is greater than that of silicon at 25° C and becomes still higher at elevated temperatures. The trend continues in the allotropic form of tin which has the diamond structure (usually called gray tin to distinguish it from the bright metallic form, white tin). In gray tin the energy gap between filled and empty bands is so small that electrons are promoted to the conductance bands in large numbers at room temperature (Fig. 4–5), making the substance a much better conductor than silicon or germanium at that temperature. Indeed, it requires only a moderate rise in temperature to bridge the gap entirely, so that there is no longer any distinction between filled and empty bands and gray tin begins to conduct like a metal.

Consequently, semiconductance must always be considered in terms of temperature, and comparisons can be made only at a specified reference temperature. When cooled sufficiently, all semiconductors are insulators, and when heated sufficiently, their electrical behavior merges with that of the metals.

SILICA, SILICATES, AND CERAMIC CHEMISTRY. When an atom of silicon meets a molecule of oxygen and combines with it, much heat is evolved (amounting to 198 kcal/mole) and the very stable dioxide, SiO_2, is formed. Since the crust of the earth is made up of 49.5% oxygen and 25.7% silicon (see Chapter 1), there is more than enough oxygen to combine with all the silicon as well as all the iron and aluminum in the world, and so we see oxygen as the natural and pervasive partner of silicon everywhere. Before concentrating on pure SiO_2 and the many

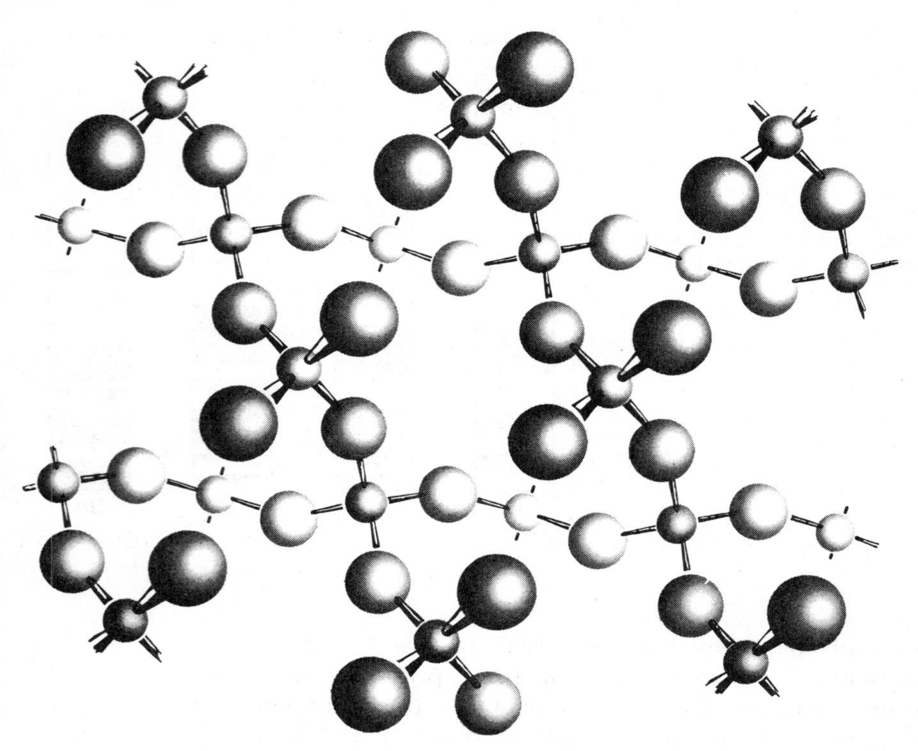

Crystal structure of quartz, SiO_2.

In SiO_2 every silicon atom (small) is bonded tetrahedrally to four oxygen atoms (large)

products made from it, however, it would be well to consider the role of silica and silicates in anthropology, archeology, and the plastic arts.

A little early history will help us to get oriented. It has been shown that some birds and mammals are capable of using tools to get what they want, but these efforts are far exceeded by the ability of *Homo sapiens* to find and devise tools to accomplish his aims. It is man's ability to make tools and machines to extend his limited strength that accounts for the rapid rise in human civilization. Early in human evolution, rounded flint pebbles became the first tools and weapons. It is very likely that such fist-sized stones shaped the evolution of the human hand,* which differs so much from the forepaws of the primates. These simple tools were so plentiful that many were left with the bones and debris of Early Stone Age settlements, and the dates of their use can be fixed at about 600,000 years ago. Flint is an uncrystallized variety of silicon dioxide (often still called silica), so the technology of silicon and silicates can be said to have begun at that time. Later the same kind of stones were fastened in handles to make hammers and axes for hunting and building. The art of *shaping* flint by chipping or flaking it came much later, of course; flint tools with sharp edges go back only 20,000 or 30,000 years. Nevertheless, the history and migrations of the American Indian have been traced by the different shapes and sizes of arrowheads he left behind. He was Stone Age Man, and flint was his favorite stone. The Latin word for flint is *silex*, gen. *silicis*, and from it come our words silica, silicon, silicate, silicone, and so on.

Another continuous and reliable thread that runs through human history is traced by pottery. The art of shaping vessels from a siliceous mass of wet clay, drying them, and then hardening them in fire goes back at least 10,000 years, both in China and in Egypt and the Near East. Wherever settlements flourished,

*See John Napier, "Evolution of the Human Hand," *Scientific American*, **207**:56 (1962).

or traders went, or peoples migrated, shards of broken pottery remained behind as a trail that can now be uncovered and read. Artifacts of wood or cloth rot away, and most metal objects corrode and disintegrate, but pottery shards resist all passage of time and bear with them a silicate history of the regional composition of the clay, the telltale marks of their method of fabrication, and of course whatever artistic development prevailed at the time. Such fragments are a keystone of archeology.

Out of the art of decorating and glazing pottery came the art of brightly colored vitreous enamels and the technique of applying them to metal surfaces and melting them in place so that they stayed bonded to the metal. Beautiful examples of these silicate enamels were found in the tomb of the Pharaoh Tutankhamen, who died in 1350 B.C. Glass bottles also were known to the ancient Egyptians, so they were well versed in at least three branches of ceramic art. The artists and workmen who made these objects had to carry out a great deal of research into silicate chemistry in order to reproduce colors and achieve the desired effects from the starting materials on hand, so it became a science. Ceramic chemistry remains a thriving branch of science, a branch so extensive that we cannot do it justice in a book like this. (The interested student will find some helpful articles in the Encyclopedia Britannica and the McGraw-Hill Encyclopedia of Science and Technology.)

The silicate chemistry of silicon, embracing the vast area of natural silicate rocks and minerals as well as the artifacts of ceramic chemistry, still represents only one aspect of silicon. In addition to this we have: the element itself, with its elaborate refinement to "hyperpure" silicon for use in solid-state electronic devices such as transistors, rectifiers, and solar batteries; the binary silicides of the metals and their use in metallurgy; the covalent compounds of silicon (gases, liquids and solids); the organic derivatives of silicon; and the industrially important organosilicon polymers, called silicones. Silicate minerals were considered in the previous section, and these other aspects will now be discussed in the numbered sections to follow.

OTHER ASPECTS OF SILICON

1. The Element. Silicon of 98% purity is obtained on a large scale by heating a mixture of purified silica sand and petroleum coke (which contains very little ash) to 3000° C in an electric arc furnace:

$$SiO_2(s) + 2C(s) \rightarrow Si(l) + 2CO(g)$$

The reactants are charged continuously at the top of the furnace, and the carbon monoxide escapes and burns. Molten silicon (mp 1414° C) is tapped from the bottom of the furnace, and cools to a shiny blue-gray polycrystalline mass. This material is pure enough for metallurgical use in aluminum and magnesium alloys, and for making silicone polymers. Ferrosilicon, an alloy of silicon with 25% or 50% iron, is made in the same way from a mixture of SiO_2,Fe_2O_3, and coke, and is used for Duriron corrosion-resistant laboratory drain pipes and for the magnetic cores of AC motors and transformers. Very pure silicon for transistors and other electronic devices is produced by the reduction of silicon tetrachloride or trichlorosilane, which are volatile liquids obtained by chlorinating scrap silicon or ferrosilicon:

$$Si(s) + 2Cl_2(g) \rightarrow SiCl_4 \text{ (bp 57.6° C)}$$

or

$$Si(s) + 3HCl(g) \rightarrow SiHCl_3 \text{ (bp 31.8° C)} + H_2$$

These are distilled exhaustively to free them from the deleterious chlorides of boron, aluminum, and arsenic, and then are reduced by very pure distilled magnesium from the carbothermic process:

$$SiCl_4(g) + 2Mg(s) \rightarrow 2MgCl_2(s) + Si(s)$$

or

$$2SiHCl_3(g) + 3Mg(s) \rightarrow 3MgCl_2(s) + H_2(g) + 2Si(s)$$

Magnesium chloride is very soluble, so it is washed out with hot water. The powdered silicon so obtained is melted, cast into bars, and then repeatedly zone-refined until it contains less than one part per *billion* of boron, aluminum, or arsenic. This is accomplished by supporting the bar at both ends, melting just a narrow zone of it by inducing a high-frequency current in it (the liquid stays in place by surface tension), and moving the melted zone along slowly to that pure silicon crystallizes out of the melt and the impurities are carried along in the liquid. When carried out repeatedly, this process concentrates the impurities at one end of the bar. Only the purified region is used for semi-conductor devices.*

2. Covalent Compounds. For 10,000 years people thought of silicon only in terms of the solid ionic silicates, and then 200 years ago a Swedish chemist named Scheele discovered the first covalent compound of silicon—the tetrafluoride, SiF_4, a gas at room temperature. He obtained it by the action of hydrofluoric acid on silica or silicates (even glassware!):

$$SiO_2(s) + 4HF(aq) \rightarrow SiF_4(g) + 2H_2O(l)$$

It was another 50 years before silicon tetrachloride was discovered, and this turned out to be a volatile liquid, as noted above. Twenty-six years after that, the first hydride was made (SiH_4, monosilane, a gas with mp −185° C and bp −112° C), and covalent silicon could no longer be ignored. An entire series of silane hydrides followed—Si_2H_6, Si_3H_8, Si_4H_{10}, etc.—with physical properties that parallel those of the methane hydrocarbons but with very different chemical behavior. The silanes are very reactive; as ordinarily prepared they ignite spontaneously in air, leaving a white smoke:

$$SiH_4(g) + 2O_2(g) \rightarrow SiO_2(s) + 2H_2O(g)$$

Organosilicon compounds, in which organic groups are bonded directly to silicon with C—Si bonds, were first made in the 1860's, and now there are over 6,000 of them known. They are the basis of silicone polymers, which are described in a later section. At about the same time it was discovered that $SiCl_4$ reacts with alcohol to give a colorless, pleasant-smelling liquid which boils at 165° C. Mendeleev was the first to figure out that this was ethyl silicate, $Si(OC_2H_5)_4$, in which four ethyl groups derived from the alcohol were linked through oxygen to silicon. Today there is a sizeable industry based on ethyl silicate and its analogs, which are used as sources of pure silica for the light-emitting powders inside fluorescent lamps, and as a vehicle for rustproofing paints.

3. The Oxide SiO_2. Silicon dioxide deserves special consideration because it is present everywhere throughout the mineral, animal, and vegetable worlds. The pure substance is known in 22 different forms or phases: alpha

*For an explanation of how transistors, diodes, and solar cells operate, see E. G. Rochow, *The Metalloids*, Ann Arbor, Michigan: Xerox University Microfilms (1966 and 1975), pp. 55–94.

quartz, beta quartz, the two forms of cristobalite, the two forms of tridymite, fused silica or "quartz glass," and the high-pressure vitreous form known as coesite, just to name a few. The impure forms may mean more to the reader: rock crystal, rose quartz, amethyst, citrine, smoky quartz, and "Glens Falls diamonds" as crystalline varieties, and flint, onyx, jasper, carnelian, chalcedony, obsidian, and opal as noncrystalline varieties. Silica dissolves in water as silicic acid, $Si(OH)_4$, and so furnishes the silica for petrified wood, for the quartz crystals that grow inside geodes (and for all well-crystallized quartz, in fact), and for the skeletons of tiny one-celled animals called diatoms from which we recover diatomaceous earth.

Dissolved silica influences the growth of plants and concentrates itself in the shiny coating of wheat straw. Within animal bodies it forms complex organic derivatives which influence the growth of hair and feathers, and again becomes concentrated in them. The amount of silica inside the living organisms of the world at any one time is estimated at millions of tons, and yet we understand very little about its exact form or function. Pure SiO_2, on the other hand, has been investigated more than any other pure substance except H_2O.

Ordinary quartz (alpha or low-temperature quartz) crystallizes as hexagonal prisms with either left- or right-handed configuration of the end faces. These two varieties rotate the plane of polarized light correspondingly to the left or the right. Both varieties are piezoelectric; that is, the quartz generates an electric potential when pressed in a particular direction and, conversely, the crystal contracts if an external voltage is applied in the same way. Since the contraction is a function of thickness in the required direction, a polished plate cut from a quartz crystal will vibrate in exact resonance with an applied alternating current of the appropriate frequency. This is the basis of the crystal-controlled radio transmitter in every police car and every citizens band installation, and also the basis for the so-called quartz watch, which relies on a precise crystal-controlled oscillator frequency as the standard of time.

The resulting demand for high-grade quartz crystals has been so great that the natural supply is inadequate, and the necessary crystals are now grown artificially. Vitreous silica ("quartz glass"), made by melting pure silica sand, is more soluble in very hot water than is quartz,* so pieces of vitreous silica are put in the bottom of a heavy steel bomb, alkaline water is added, and a tiny seed crystal of quartz is suspended from the top. The bomb is heated above the critical temperature of water for days, and the seed crystal grows at the expense of the silica glass, which dissolves. The process is a good lesson in the properties of water and the functioning of equilibria.

4. The Nitride Si_3N_4. During the reduction of SiO_2 to Si in the electric arc furnace, the temperature is high enough to dissociate N_2 molecules of the air to nitrogen atoms, and these combine freely with hot silicon vapor to produce a white fluffy mass of Si_3N_4. This substance resembles SiO_2 in being unreactive and refractory, but it has no use. A great deal of it is available for anyone who finds out what it is good for.

5. The Halides. The compounds SiF_4, $SiCl_4$, $SiBr_4$ and SiI_4 are compared in Table 4–2 because they are typical covalent compounds of silicon that serve as starting materials for many other covalent compounds. The figures are just for comparison, not memorization. The formation of SiF_4 from silica has been described in section 2 above, and it is the only halide of silicon that can be made in this way because the bond energy of forming Si—F bonds is higher than

*The free energy of formation of silica glass is −190.9 kcal/mole, while that of alpha quartz is −192.4 kcal/mole, so the latter is more stable and will grow from solution at the expense of the silica glass fragments.

TABLE 4–2 THE TETRAHALIDES OF SILICON

	SiF_4	$SiCl_4$	$SiBr_4$	SiI_4
Melting point, °C	−90.3	−70.4	+5.4	+120.5
Boiling point, °C	Sublimes −96	57.0	155	287.5
Si—X bond energy, kcal/mole	135	91	74	56
Heat of formation, kcal/mole	−370	−153	−95.1	−31.6
Bond length, Å	1.54	2.01	2.15	2.43

that of forming Si—O bonds. The bond energies of Si—Cl, Si—Br, and Si—I are all lower than that of Si—O, so HCl, HBr, and HI do not act on SiO_2. Hence $SiCl_4$, $SiBr_4$, and SiI_4 must be made by the action of Cl_2, Br_2, or I_2 on elementary silicon or on the carbide SiC. Conversely, silicon tetrachloride, bromide, and iodide hydrolyze energetically to the oxide:

$$SiCl_4(l) + 2H_2O(l) \rightarrow SiO_2(s) + 4HCl(aq)$$

$$SiBr_4(l) + 2H_2O(l) \rightarrow SiO_2(s) + 4HBr(aq)$$

$$SiI_4(s) + 2H_2O(l) \rightarrow SiO_2(s) + 4HI(aq)$$

But SiF_4 reaches an equilibrium, because as soon as any HF accumulates to a concentration of a few per cent, it dissolves the SiO_2 product, which reverts to SiF_4:

$$SiF_4 + 2H_2O \rightleftharpoons SiO_2 + 2HF$$

The equilibrium mixture is also found to contain a new species, called fluorosilicic acid:

$$SiF_4 + 2HF \rightarrow H_2SiF_6(aq)$$

This acid forms hexafluorosilicate salts that are potent insecticides and rodenticides. The ion SiF_6^{-2} is seen to be an excellent example of hexacovalent silicon, a stable arrangement of a central silicon surrounded by six equidistant fluorines in an octahedral pattern. There are no equivalent compounds with Cl, Br, or I because the ions of these three heavier halogens are too large to accommodate six of them around the small silicon atom.

All the silicon tetrahalides can be reduced to the hydride SiH_4, monosilane, by an alkali-metal hydride or (better still) by lithium aluminum hydride:

$$SiCl_4 + LiAlH_4 \xrightarrow[\text{solvent}]{\text{Ether}} SiH_4(g) + LiCl(s) + AlCl_3(s)$$

In fact, any Si—Cl bond can be reduced to an Si—H bond by this reagent, so there is a ready connection between the halides and hydrides of silicon. At the same time any hydride of silicon, as well as any chloride, bromide, or iodide, will revert to the corresponding oxide by reaction with air or water, if given a chance. The heat of formation of SiO_2 (and hence the bond energy of forming an Si—O bond) is so high that oxygen ends up with silicon at every opportunity. Knowing this, any student can predict the result of the reaction of $SiCl_4$ with ethyl alcohol

$$SiCl_4(l) + 4C_2H_5OH(l) \rightarrow \underset{\text{ethyl silicate}}{Si(OC_2H_5)_4(l)} + 4HCl(g)$$

or the reaction of $SiCl_4$ with the previously mentioned organometallic reagent CH_3MgCl in ether

$$SiCl_4 + 2CH_3MgCl \xrightarrow[\text{solvent}]{\text{Ether}} 2MgCl_2 + (CH_3)_2SiCl_2$$

followed by hydrolysis of the product

$$(CH_3)_2SiCl_2(l) + H_2O(l) \rightarrow \underset{\text{as polymer}}{(CH_3)_2Si{-}O{-}}\ (l) + 2HCl(g)$$

The last two equations show an old-fashioned way of making silicone polymers from $SiCl_4$. The polymeric product consists of a flexible chain of alternate silicon and oxygen atoms, with each silicon atom bearing two methyl (CH_3) groups linked to it. Here again, the silicon atom finds its preferred partner in oxygen, and the resulting dimethyl silicon oxide is an unreactive and heat-stable but flexible material.

6. The Hydrides (Silanes). The hydrides of silicon are worth a separate book,* but a few interesting points can be made here. The first such hydride, SiH_4, was prepared in 1851 by Friedrich Wöhler, the same man who first made a typical organic compound (urea) from purely inorganic reagents. Wöhler wanted to find out whether the newly emerged covalent chemistry of silicon would lead to a complete parallel with organic chemistry, with silicon taking the place of carbon in alcohols, ethers, organic acids, fats, carbohydrates, proteins, and even completely siliceous living creatures. Just the spontaneous inflammability of SiH_4† and the violence of its reaction with alkaline water convinced Wöhler that there was no chemical parallel with CH_4 and the rest of organic chemistry. But suppose we take a look at just the boiling points for a moment, considering them as typical physical properties. For the simplest members of the two series, SiH_4 boils at −111.9° C or 161° K, while CH_4 boils at −161.3° C or 112° K. The ratio of their boiling points in °K is 161/112, or 1.44. For the next pair, Si_2H_6 boils at −14.5° C or 259° K, C_2H_6 boils at −88.7° C or 185° K, and the ratio of absolute boiling points is 1.40. For the following pair, Si_3H_8 boils at 52.9° C or 326° K, C_3H_8 boils at −44.5° C or 229° K, and the ratio is again 1.42. Lastly, Si_4H_{10} boils at 109° C or 382° K, C_4H_{10} boils at 0.5° C or 274° K, and the ratio is once more nearly the same, 1.39. Anyone who relied solely on physical properties to establish similarity would be impressed by the constant parallel. But Si_4H_{10} barely stays together, and the higher silanes are very difficult to make, while stable pure hydrocarbons of the methane series are known as far as $C_{100}H_{202}$. Obviously the strength of the C—C bond and the lack of a *d*-orbital

*See A. Stock, *The Hydrides of Boron and Silicon*, Ithaca, New York: Cornell University Press (1933).

†Recent research has indicated that very pure SiH_4 does not ignite in air by itself, but requires traces of higher hydrides to get it started. It is unwise, however, to rely on such high purity as a safeguard; the real world is full of circumstances and substances that will cause a mixture of SiH_4 and air to ignite or explode.

mechanism for hydrolysis or oxidation favor the hydrocarbons over the silicon hydrides, and an imaginary "silicon man" (even if he were to evolve on a dry planet) would be no threat in our watery world.

Just because of their reactivity, silicon-hydrogen bonds are useful functional groups in the synthesis of organosilicon compounds and in the vulcanizing of silicone rubber. They also have a strong reducing action, and will precipitate metals such as silver from waste water containing their ions, thereby aiding in the recovery and recycling of scarce materials.

7. The Carbide. Silicon carbide, SiC, is a very hard, tough, refractory solid which has the diamond structure. It was mentioned previously as a typical covalent carbide, in effect a diamond in which alternate carbon atoms are replaced by silicon. The pure carbide forms colorless crystals, although the commercial variety is glossy black because of the presence of iron and titanium as impurities. It is made on a large scale by the reaction of an excess of coke with silica sand in an electric resistance furnace (not an arc furnace, but one in which the charge is heated only by its own resistance to passage of electric current):

$$SiO_2(s) + 3C(s) \xrightarrow{1950°\,C} SiC(s) + 2CO(g)$$

The resulting crystals of SiC are crushed, and the sharp-edged fragments are used as an abrasive. Their hardness on the Mohs scale is 9.5, relative to diamond as 10. The fragments can be mixed with some wet clay and pressed into discs which become grinding wheels after they are fired. Other clay-bonded and fired shapes are used as refractory supports in the firing of china dinnerware, because SiC does not oxidize or sag at the kiln firing temperature. It does not melt at all, but decomposes at 2700° C. It is a semiconductor and is used in resistance elements for electric furnaces. A more dense beta form is made by heating ultrapure silicon in a graphite crucible at 1500° C in vacuum, and is used as a semiconductor in solid-state electronic devices that must withstand high temperatures, as in space vehicles.

8. Silicones. The silicone polymers are actually organic-substituted silicon oxides, and are best considered as relatives of SiO_2. They are polymeric for the same reason that SiO_2 is solid and polymeric while CO_2 is always monomeric and gaseous: the carbon atom is small enough to form double bonds to oxygen by overlap of its *sp* hybrid orbitals, giving rise to the structure O=C=O, but the larger silicon atom forms no double bonds to anything, and must satisfy its bonding capabilities by four covalent bonds to separate oxygen atoms:

```
 —O     O—
    \  /
     Si
    /  \
 —O     O—
```

At the same time, each oxygen atom must bond itself to two separate silicon atoms, leading to the arrangement Si—O—Si. Hence SiO_2 is a solid network of interlocking chains of alternate silicon and oxygen atoms (Fig. 4–6*a*), while CO_2 retains individual molecules and exists as a gas.

Silicone polymers are not all silicon and oxygen, however; each silicon atom bears one or more organic groups, and usually two such groups. Each covalent bond that links such a group to silicon necessarily uses one of the silicon atom's four bonding electrons, and correspondingly limits that atom's ability to form bonds to oxygen. Therefore a silicon atom bearing two organic groups can bond to only two oxygen atoms. The result is a single chain of alternate

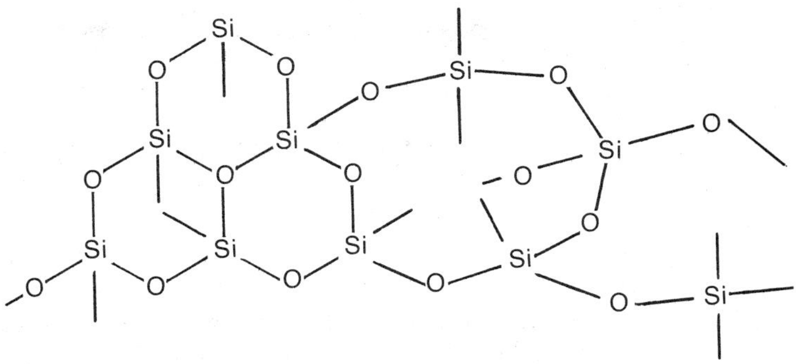

Figure 4–6. *a*, Random structure of silica glass. *b*, Structure of a silicone polymer. (R represents any hydrocarbon group attached directly to silicon, such as a —CH_3 group or a —C_6H_5 group.)

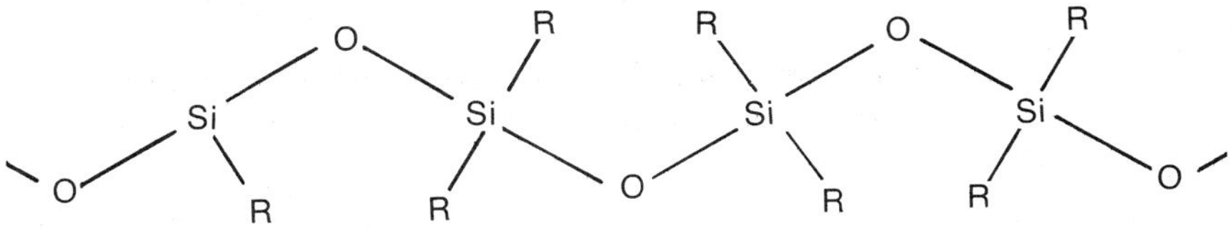

silicon and oxygen atoms (a siloxane chain), with two organic groups on each silicon. It is as though we had removed one long siloxane chain from a quartz crystal, and festooned it with all the organic groups it will hold (Fig. 4–6*b*). Such a chain is flexible and has properties in between those of quartz and glass, on the one hand, and organic plastics and rubber on the other. That is, the silicone polymer retains much of the thermal stability and chemical inertness of its inorganic siloxane backbone, but acquires the flexibility that comes from easy rotation about single carbon bonds and from low intermolecular attraction between neighboring hydrocarbon groups.

The siloxane backbone of a silicone polymer need not remain a single siloxane chain. If occasional silicon atoms bear only one organic group, then such atoms can form *three* bonds to oxygen, and since only two are required to propagate a chain, the third oxygen atom can form a bridge or cross-link to a similar silicon atom in a neighboring siloxane chain. The more such cross-links are introduced, the more the resulting network will resemble a quartz crystal, and the less flexible it will be. Silicone resins used for coatings and adhesives have a limited number of such siloxane cross-links; silicone rubber has very few, and silicone oil has none.

The favorite kind of organic group used in silicones is a methyl group, —CH_3. This carries no C—C bonds, and minimizes the carbon and hydrogen content of the polymer so that it can retain maximum inorganic character. A polymer made up only of —$(CH_3)_2Si$—O— units contains only 32.4% carbon and 8.16% hydrogen, while the siloxane backbone makes up 59.45% of its weight. A polymer with a substantial proportion of siloxane cross-links will contain still less carbon and hydrogen, of course.

Silicone polymers may be made by the older and rather difficult method indicated in section (5) above, or they can be made directly from elementary silicon by reaction with methyl chloride:

$$Si(s) + 2CH_3Cl(g) \xrightarrow[Cu]{300°C} (CH_3)_2SiCl_2(g) + \text{other products}$$

$$(CH_3)_2SiCl_2(l) + H_2O(l) \rightarrow \underset{\text{as cyclic polymers}}{—(CH_3)_2SiO—} + 2HCl(g)$$

The resulting liquid cyclic polymers are separated and purified, and then are used to make silicone oil and rubber. The HCl by-product is recycled to produce more methyl chloride from methanol, so the chlorine is not wasted:

$$HCl(g) + CH_3OH(l) \rightarrow CH_3Cl(g) + H_2O(l)$$

The methanol comes from carbon monoxide and hydrogen, which in turn come from the reaction of red-hot coal with steam:

$$C(s) + H_2O(g) \xrightarrow{800°C} CO(g) + H_2(g)$$

$$CO + 2H_2 \xrightarrow[500\ atm]{450°C} CH_3OH(g)$$

So the starting materials are actually coal, water, and silica sand, all readily available and not dependent on petroleum.

INORGANIC PHOSPHORUS

Phosphorus is both a mineral element and an element essential to all living organisms. On the mineral side, calcium phosphate sediments from age-old oceans are compacted and overlaid with sand and soil to form phosphate rock, which is mostly $Ca_3(PO_4)_2$ and $Ca_5(PO_4)_3F$. This rock is mined in Florida and in North Africa, treated with sulfuric acid to make it slightly soluble in water, and returned to the biological domain by spreading it on the soil used to grow grain. Phosphate ions absorbed by the growing plants become attached to the simple sugars formed from CO_2 and water by action of the sunlight during photosynthesis. Phosphate groups also find their way into the plant's proteins as these are synthesized, and thence to any animal that eats the plant. In the human body an organic phosphate (adenosine triphosphate, or ATP) transforms chemical energy into mechanical energy in the muscles as we breathe and move about, and basic calcium phosphate or fluorophosphate makes up about a quarter of our bones and teeth. Phosphorus from urine and from dead tissue is leached from the soil and carried by rivers to the sea, where it joins the skeletons of marine animals and leads to sediment as the cycle is repeated. Hence phosphorus is an essential part of us and of all living matter, and so it is necessary to help it along through its cycle.

THE ELEMENT. Free phosphorus is produced on a very large scale by heating phosphate rock with sand and coke in an electric furnace. The phosphate is reduced and P_4 vapor escapes, leaving the calcium behind as a silicate slag:

$$2Ca_3(PO_4)_2(s) + 10C(s) + 6SiO_2(s) \rightarrow 6CaSiO_3(s) + 10CO(g) + P_4(g)$$

The phosphorus vapor is very inflammable, but can be condensed under water where it hardens to a waxy white solid which melts at 44° C. The P_4 molecules that make up this solid are tetrahedral in shape, with each phosphorus atom covalently bonded to three phosphorus atoms at the other corners of the tetrahedron. This tetrahedral arrangement is retained in expanded form in many compounds of phosphorus, such as the sulfide P_4S_3 and the oxides P_4O_6 and P_4O_{10}.

White phosphorus must be kept under water or it will oxidize so vigorously that it ignites and burns. When kept under water for a long time it turns yellow and then develops a reddish coating as the unstable white phase changes slowly to the more stable allotropic form, a brick-red nonvolatile polymer which does not ignite in air and can be kept indefinitely. If red phosphorus is heated to 600° C it depolymerizes to P_4 molecules which can be condensed again in the waxy white form. The white form is soluble in carbon disulfide; the red form is not. The white form is poisonous, the red form much less so.

Most of the phosphorus produced from phosphate rock is converted to P_4O_{10} by controlled burning under steam boilers

$$P_4(s) + 5O_2(g) \rightarrow P_4O_{10}(s) \qquad \Delta H = 720 \text{ kcal/mole}$$

and the oxide is used to make phosphoric acid for fertilizers.

OXIDES AND OXYACIDS. We shall consider only P_4O_6, P_4O_{10}, and the acids H_3PO_2, H_3PO_3, and H_3PO_4. The lower oxide, P_4O_6, has an expanded tetrahedral structure with oxygen atoms between the phosphorus atoms on each of the six edges (Fig. 4–7*a*). It hydrolyzes to phosphorous acid:

$$P_4O_6(s) + 6H_2O(l) \rightarrow 4H_3PO_3(aq)$$

The salts of this acid are called phosphites, and they are easily oxidized to phosphates. The higher oxide, P_4O_{10}, has the same expanded tetrahedral structure as P_4O_6 but with four more oxygen atoms sticking out from the corners of the tetrahedron (Fig. 4–7*b*). It absorbs moisture avidly from the air or from anything else, and hisses if it is dropped into water. With limited water it forms a metaphosphoric acid, HPO_3, but with sufficient water it forms a colorless crystalline solid of the composition H_3PO_4. This dissolves in more water to form a syrupy colorless liquid which is the ordinary phosphoric acid. This has neither reducing nor oxidizing properties; its salts (the phosphates) are similarly stable.

The lowest oxyacid of the three considered here, hypophosphorous acid (H_3PO_2), is more difficult to make. White phosphorus reacts slowly with a boil-

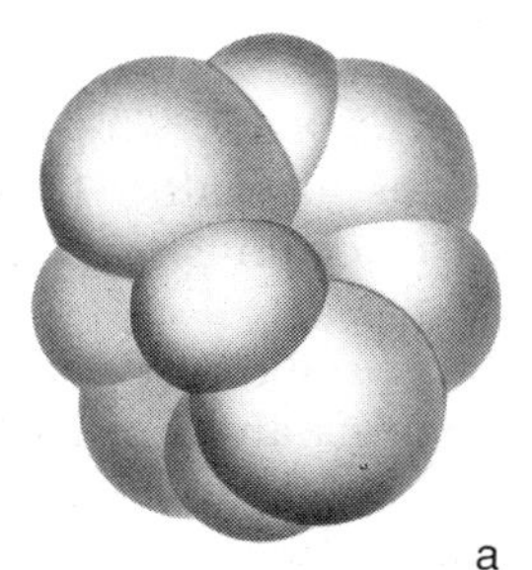

P—O—P
O O O O
P P
O

a

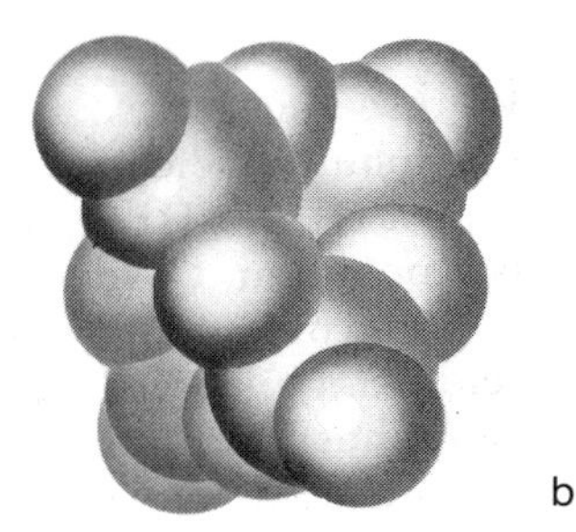

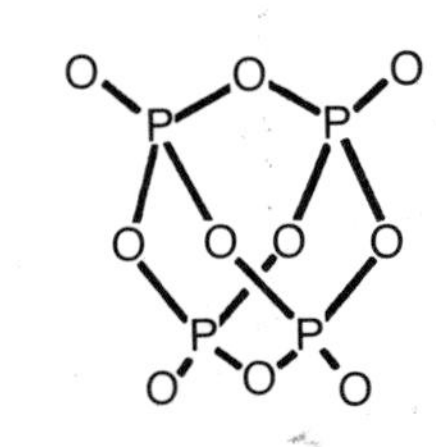

b

Figure 4–7. Structures of *a*, P_4O_6 and *b*, P_4O_{10}.

ing solution of barium hydroxide to form barium hypophosphite, $Ba(H_2PO_2)_2$, and phosphine, PH_3:

$$2P_4(s) + 3Ba(OH)_2(aq) + 6H_2O(l) \rightarrow 3Ba(H_2PO_2)_2(aq) + 2PH_3(g)$$

The phosphine is a gas which escapes and ignites as it meets the air. If just the right amount of sulfuric acid is added to the remaining solution of barium hypophosphite, slowly and with much stirring, insoluble barium sulfate precipitates and a solution of the desired hypophosphorous acid is left.

As indicated by the formula for its barium salt given above, hypophosphorous acid is a monoprotic acid (that is, only one of its three hydrogen atoms can be neutralized by a base or displaced by a metal). The other two hydrogen atoms are bonded directly to phosphorus; they function strongly as reducing agents but do not react when the acid is titrated with a standard base. The structure

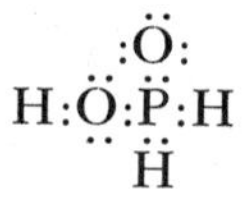

:Ö:
H:Ö:P:H
Ḧ

indicates these features. The reducing action of the two P—H bonds is shown by the reduction of most transition-metal ions to free metal, as with cobalt:

$$H_3PO_2 + CoCl_2 + H_2O \rightarrow H_3PO_3 + 2HCl + Co$$

The oxygen-bonded hydrogen is the acidic one, with a dissociation constant of only 10^{-9}.

The insertion of one more oxygen atom into the structure gives phosphorous acid

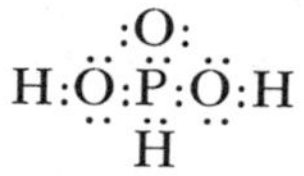

:O:
H:Ö:P:Ö:H
Ḧ

which is a dibasic acid with two neutralizable hydrogen atoms and two dissociation constants, 7×10^{-3} and 2×10^{-5}. Phosphor*ic* acid has all three hydrogen atoms linked through oxygen, and hence is tribasic and without any reducing action:

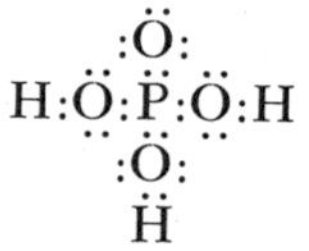

:Ö:
H:Ö:P:Ö:H
:Ö:
Ḧ

Its three acid dissociation constants are 1.1×10^{-2}, 7.5×10^{-8}, and 4.8×10^{-13}. Phosphoric acid forms three corresponding series of salts, illustrated by NaH_2PO_4, Na_2HPO_4, and Na_3PO_4. The last one, trisodium phosphate, or TSP, is a potent and widely sold cleansing agent.

The oxyacids of phosphorus condense in much the same way as do the silicic acids and silanols, yielding polymers containing chains and networks of alternate phosphorus and oxygen atoms. The sodium salts of these complicated condensed acids are valuable laundry detergents, particularly sodium hexametaphosphate, $Na_6P_6O_{18}$. Formerly it was general practice to load household detergents with large proportions of phosphate, but since none of this was consumed in the laundering operation, it was all drained into sewers whence it got into streams and lakes and fertilized the growth of unwanted weeds there. Now polyphosphates are substituted in part by sodium carbonate and sodium silicate.

HYDRIDES. The hydrides of phosphorus are fewer in number than those of silicon, but are just as reactive. Phosphine, PH_3, is best produced by dropping calcium phosphide (made from the elements) into water in a closed flask:

$$Ca_3P_2(s) + 6H_2O(l) \rightarrow 2PH_3(g) + 3Ca(OH)_2(s)$$

As the phosphine emerges from a delivery tube it catches fire spontaneously and burns to a white smoke:

$$4PH_3(g) + 8O_2(g) \rightarrow P_4O_{10}(s) + 6H_2O(g)$$

It's fortunate that this happens, for phosphine itself is very poisonous. It has a very bad odor which serves to warn people when it is around. What smells bad in a chemical laboratory usually *is* bad for you; PH_3 and H_2S are two prime examples.

A second hydride, P_2H_4, is an unstable and easily oxidized liquid, like N_2H_4 (hydrazine) which we met in the previous chapter.

HALIDES. There are two series of phosphorus halides, of the types PX_3 and PX_5. The trihalides like PCl_3 (mp −112° C, bp 74° C) hydrolyze energetically to phosphorous acid and the corresponding hydrohalogen acid:

$$PCl_3(l) + 3H_2O(l) \rightarrow H_3PO_3(aq) + 3HCl(g)$$

We saw that NCl_3 behaves very differently

$$NCl_3(g) + 3H_2O(l) \rightarrow NH_3(aq) + HOCl$$

which illustrates the big difference between nitrogen and phosphorus. The pentahalides such as PCl_5 (mp 148° C, bp 160° C) hydrolyze to phosphor*ic* acid:

$$PCl_5(s) + 4H_2O(l) \rightarrow H_3PO_4(aq) + 5HCl(g)$$

The pentafluoride PF_5 is a fluorinating agent, and both PCl_5 and PCl_3 are used in organic chemistry to convert alcohols and acids to the corresponding chlorides.

SULFUR: FAMILIAR AND COMMON, BUT WORTH A SECOND LOOK

Sulfur was highly valued by the alchemists, who strove valiantly and unsuccessfully to transfer its yellowness to lead in order to obtain gold. Sulfur is a constituent of several essential amino acids, so it is part of the protein structure of our bodies. Similarly, it is in the protein of eggs, giving rise to the odor of hydrogen sulfide in decaying eggs. Sulfur in animal tissue is oxidized to sulfuric acid and excreted, and sulfates collect in ground water and seawater. Much more sulfate ion gets into the sea from the oxidation of common sulfide minerals such as FeS_2 (pyrite), PbS (galena), ZnS (zinc blende), and Cu_3FeS_3 (bornite). Such sulfide minerals are the principal ores for obtaining copper, zinc and lead. The accumulated sulfates of sodium, calcium, and magnesium become concentrated as a landlocked body of seawater evaporates, so Na_2SO_4 and $MgSO_4$ are prominent in Great Salt Lake and crystallize along its edges during cold dry weather. Calcium sulfate is less soluble (only 2.4 g/liter) so it crystallizes out as deposits of gypsum, $CaSO_4(H_2O)_2$, from which we make plaster of paris and wallboard.

Sulfur is studied extensively in general science and in high-school chemistry, and its compounds are considered throughout college textbooks. Indeed, H_2SO_4 is probably the first formula learned after H_2O. Very little more about the element and its compounds need be added here, but the ubiquitous nature of sulfur should be stressed while we face a shortage of elementary sulfur on the one hand and struggle to cope with too much SO_2 and SO_3 as pollutants in air on the other hand.

THE ELEMENT. Ordinary sulfur does not consist of isolated atoms or of diatomic molecules like O_2, but instead forms puckered rings of six or eight atoms. The two common lemon-yellow crystalline forms of sulfur (orthorhombic, mp 112° C, and monoclinic, mp 119° C) consist of S_8 molecules, as does the melt. When liquid sulfur is heated above 150° C, the S_8 rings begin to break open and the resulting chains become tangled, causing an increase in viscosity. At the same time the color darkens, because the chain ends have unpaired electrons. At 180° C the viscosity is at a maximum and the color is dark brown. The mass will scarcely pour. Heated still more, the chains are ruptured and the viscosity decreases. At the boiling point, 444° C, the liquid is thin again.

If liquid sulfur at 300° C is poured into cold water, a rubbery brown mass of uncrystallized polymer (called lambda or plastic sulfur) is obtained. This is a supercooled liquid; if left at 25° C for several days it reverts to brittle crystalline sulfur.

A much less common allotrope of sulfur is the orange-brown rhombohedral crystalline form, which is made up of S_6 rings. This is more soluble in organic solvents than are the S_8 varieties, but is unstable to light and reverts readily to S_8 even in solution unless kept in the dark.

SOURCES, DESIRABLE AND UNDESIRABLE. There are at least four present-day sources of sulfur available in this country: the deposits of elementary sulfur along the Gulf shore, laid down by bacterial reduction of gypsum (and mined by melting the sulfur with high-pressure steam and hot water and blowing the liquid to the surface with hot water); the SO_2 emitted by sulfide ores of copper, lead, and zinc during smelting operations; the sulfur in petroleum, present as a result of its biological origin and carried along as complicated organic compounds; and the sulfur in coal, retained from the plants that formed the coal and present as FeS_2 and cyclic organic compounds. Elementary sulfur is mined at the rate of about 7.5 million tons per year, most of which goes into the manufacture of sulfuric acid. If *all* of it were converted to sulfuric acid at 100% yield, the acid produced would come to

$$7.5 \times 10^6 \text{ tons S} \times \frac{\text{mol. wt. } H_2SO_4}{\text{at. wt. S}} = 7.5 \times 10^6 \times \frac{98.1}{32.1} = 22.9 \times 10^6 \text{ tons } H_2SO_4$$

But over 30 million tons of H_2SO_4 are produced in this country each year, so it is obvious that some is being made from other materials. For example, some of the smelting operations that formerly released sulfur dioxide to the air now divert it to a sulfuric acid plant, and some petroleum refineries are equipped to burn H_2S and other waste sulfur compounds to SO_2, which is then recovered as H_2SO_4. The key environmental question is, how far can chemical engineering go toward extracting or recovering all the unwanted sulfur in our fuels and ores so that it could be marketed as H_2SO_4 instead of being released to pollute the atmosphere?

We begin with a consideration of quantities. About 600 million tons of coal are mined in the U.S.A. each year, and 400 million tons are used in the production of electric power. With about 5% sulfur in the coal, this means that we have 34 million tons of sulfur in the coal coming out of the mines each year, and 20 million tons of sulfur present in the coal burned in power plants alone. While some attempts have been made to separate sulfur minerals from powdered coal before burning, most power plants burn the coal as it is, and the sulfur ends up as SO_2 and SO_3 in the flue gas. If we want to remove the oxides of sulfur, the most promising method that springs to mind is to wash the flue gas thoroughly by a water spray system, using some alkaline solution to trap the acid oxides. The cheapest base for the absorbing would be calcium hydroxide, derived from limestone:

$$CaCO_3 \xrightarrow{750°\,C} CaO \xrightarrow{H_2O} Ca(OH)_2$$

A suspension of $Ca(OH)_2$ (it is soluble only to the extent of 1 g/liter in hot water) would absorb the acid oxides, if enough is used and if the contact is long enough

$$SO_2(g) + Ca(OH)_2(s) \rightarrow CaSO_3(s) + H_2O(l)$$

and then

$$2CaSO_3(s) + O_2(g) \rightarrow 2CaSO_4(s)$$

because sulfites are good reducing agents and are easily oxidized. Any SO_3 present in the flue gas would also be absorbed

$$SO_3(g) + Ca(OH)_2(s) \rightarrow CaSO_4(s) + H_2O(l)$$

and undoubtedly a great deal of CO_2 would be absorbed at the same time:

$$CO_2(g) + Ca(OH)_2(s) \rightarrow CaCO_3(s) + H_2O(l)$$

There is much more CO_2 than SO_2 in the flue gas, and so the last reaction might predominate. Let us concentrate only on the calcium sulfate, however. The formula weight of $CaSO_4$ is 136.14, of which 32.06 represents sulfur. The complete absorption of 20 million tons of sulfur from power-station flue gas therefore would produce $20 \times 136.14/32.06 = 84.9$ *million tons of* $CaSO_4$ to dispose of each year, in addition to whatever $CaCO_3$ forms. It could be filtered off and shipped back (wet?) to the coal mines to be dumped where the coal came from, but this is not going to beautify the landscape. What with the mining of the limestone, its shipping cost, the fuel required to calcine it, and then the return freight on a quantity of $CaSO_4$ amounting to at least one fourth of the coal tonnage used, the cost of trapping the sulfur this way would about double the cost of burning coal in the power station. In addition, there would be other troubles related to disposal, as we have seen.

A little calculation like this enables even a beginning student of chemistry to consider various proposals and newspaper accounts of "environmental breakthroughs" in a rational light, so that he can vote intelligently and can advise his less knowledgeable friends. In the matter of using CaO to absorb SO_2, obviously something better has to be devised, purely from the chemical standpoint. In addition, from the engineering standpoint, no power plant stack in the country will operate if we introduce cold flue gas (cooled by washing with a water spray) at the bottom, so there would have to be other means of forcing draft through the furnaces and washers. How could we keep the flue gas hot and still absorb the SO_2 and SO_3? Well, we could add ground limestone to the coal and rely upon the combustion to convert $CaCO_3$ to CaO, and then let the dry hot particles of CaO absorb the oxides of sulfur. The difficulty here is that the contact between solid and gas is poor and the time is short. Moreover, while SO_3 might be removed reasonably well because it is so acid, the more prevalent SO_2 is much less acid and will be absorbed much less readily. We should have to use excess CaO, and then we should have the extreme difficulty of removing millions of tons of $CaSO_4$ and extra CaO *as fine dust* from the hot flue gas. We should seem to be trading a gas pollutant for a dust problem, and still have all the difficulties of disposal of a huge quantity of $CaSO_4$. Perhaps it or the wet $CaSO_4$ from the spray process could be used to make wallboard or some other marketable product right at the power station site.

An alternative to the wet or dry neutralization scheme would be to adsorb the oxides of sulfur from the hot flue gas on some porous solid material such as silica gel* or porous Al_2O_3 or charcoal, and then drive off the collected SO_2 and SO_3 by heating the adsorbent in a separate vessel. This could give a *concentrated* stream of SO_2 and SO_3 which could be oxidized with air over the usual V_2O_5 catalyst and then converted to H_2SO_4 for sale. The adsorbent could then be returned to the furnace flue to collect more sulfur oxides, and endlessly recycled this way. Such a procedure would not result in 100% recovery of the sulfur, of course, but assuming that even 70% of it converted to H_2SO_4 we should then have 43 million tons of sulfuric acid available for industrial use. This would expand the supply by 43% and would put off the exhaustion of our deposits of elementary sulfur.

H_2SO_4. Sulfuric acid is the world's most important industrial chemical, so it is worth an historical perspective. The acid was well known to the alchemists of the sixteenth century, who made it by heating crystals of "green vitriol," a hydrated sulfate of iron:

$$FeSO_4(H_2O)(s) \rightarrow H_2SO_4(l) + FeO(s)$$

It remained an alchemical nostrum until Joshua Ward developed a new method for making it: he burned "brimstone" (a corruption of the German word for sulfur, *brennstein,* the stone that burns"), mixed with some "nitre" ($NaNO_3$), in a large glass globe. The nitrate gave oxides of nitrogen which served to oxidize the SO_2 to SO_3, which was absorbed by a little water in the globe:

$$NO_2(g) + SO_2(g) \rightarrow SO_3(g) + NO(g)$$

$$SO_3(g) + H_2O(l) \rightarrow H_2SO_4(l)$$

Ward sold the acid in London for two shillings per pound in 1740. John Roebuck built large boxes of sheet lead to replace Ward's glass globes, and then in 1827 Gay-Lussac in France found he could recover the NO from the product acid in a stream of warm air, which served also to oxidize the NO back to NO_2 so it could be used over again. Later the heat of burning sulfur was used in the recycling of oxides of nitrogen, and by 1860 the price was down to one penny per

*Silica gel is an extremely porous form of SiO_2 made by adding hydrochloric acid to a solution of sodium silicate, and then washing and drying the gelatinous mass.

pound. There it stayed for a century; as the operating costs for the lead-chamber process rose, the new "contact method" for oxidizing SO_2 to SO_3 over a contact catalyst came into use around 1910 and eliminated the lead chambers. Further engineering improvements in each new plant kept the price down during successive inflations, until all possible economies in the process were exhausted and the price began to rise in the early 1970's. What is needed now is a fourth major change, breaking away from two centuries of dependence on elementary sulfur as the principal raw material and turning to recovery of inevitable sulfur in fossil fuels as H_2SO_4.

The word "inevitable" is used advisedly. Crude oil contains anywhere from 0.2 to 7% sulfur, with 3% as a reasonable figure to work with now that special low-sulfur crudes have become scarce. World production of petroleum is about 18 *billion* barrels per year, of which the U.S. uses about one fourth, or 4.5 billion barrels (about 614 million tons). At 3% S, this amounts to 18.4 million tons of sulfur to get rid of, each year, during refining or after combustion of the petroleum products. Refining operations remove most of the sulfur, but residual oil burned in power plants and ships still contains 1 to 2% S, which presumably could be recovered from the flue gas as described above. The oil burned in electric generating stations constitutes only 13% of the petroleum products; gasoline and home heating oil are the high-volume items, and these are desulfured by "hydrofining," where hydrogenation reduces the organic sulfur compounds to H_2S which is subsequently oxidized to marketable sulfur or converted to H_2SO_4.

Gasoline still contains some sulfur, however, and automobiles put about a million tons of sulfur oxides into the air each year. Formerly this was in the form of SO_2, but with the advent of the catalytic converter this is converted to SO_3, a much more dangerous substance which damages lungs and attacks clothing. Clearly there is as much potential for recovering sulfur from petroleum and its products as there is from coal, and there is as much necessity to do so from the environmental standpoint. It is just that the task of removing the sulfur from petroleum is easier than its removal from coal, and most of the job is already done in the refineries, so we have much less to cope with later.

The reader may well wonder where all the 30 million tons per year of H_2SO_4 goes to now, and whether there would be a place for 50 or 100% more of it if most of the sulfur in our mineral fuels were converted to H_2SO_4. In the first place, it would take quite a few years to develop processes and equipment for the removal of SO_2 and SO_3 and for the recovery of H_2SO_4, and our economy is expanding all the time, so the demand for H_2SO_4 will rise. The demand has increased at a rate of 3.2% per year for several decades. In the second place, we might well leave the elementary sulfur in the ground while we strive to use the unwanted sulfur in our fuels. As for utilization, at present 27% of the H_2SO_4 produced goes into the manufacture of phosphate fertilizers, 8.4% into making pigments such as TiO_2, 8% into making chemicals, 6.9% into removing oxide scale from steel plate, 5% for making $(NH_4)_2SO_4$ as fertilizer, 4.7% for absorbing ammonia from coke ovens, 3.7% for making synthetic fibers, 3% for making petroleum products, 2% for making alum, and the remaining 31.3% for thousands of other uses.

With that, we shall leave the environmental aspects of SO_2, SO_3, and H_2SO_4, and turn to some points about a few other compounds of sulfur.

HYDROGEN SULFIDE, H_2S. This gas is familiar to chemists for its bad odor. It is formed when sulfur oxidizes a hydrocarbon such as hot paraffin wax, or reacts with any other hydrogenous reducing agent:

$$C_{15}H_{32}(l) + 2S_8(s) \rightarrow 16H_2S(g) + 15C(s)$$

It also results from the treatment of almost all sulfides of metallic elements with a strong acid:

$$FeS(s) + H_2SO_4(aq) \rightarrow H_2S(g) + Fe^{+2}(aq) + SO_4^{-2}(aq)$$

It is a gas which condenses at −62° C and freezes at −83° C, far below the corresponding melting and boiling points of H_2O, which is much more associated due to hydrogen bonding. Hydrogen sulfide burns in air to SO_2 and H_2O, and it is oxidized by SO_2 in the presence of water to give free sulfur in colloidal form, used as a fungicide:

$$16H_2S(g) + 8SO_2(g) \rightarrow 16H_2O(l) + 3S_8(s)$$

Hydrogen sulfide is a very weak acid, with a first dissociation constant of 10^{-7} and a second of 10^{-15}. The dissociation into H^+ and S^{-2} ions and the manipulation of that equilibrium by adding acids and bases formerly were very important to analytical chemistry, but analysis for metal ions is now carried out much more conveniently by spectrographic instrumental methods.

Hydrogen sulfide is very poisonous, and should be avoided even in very low concentrations. One part per thousand in air causes unconsciousness in a few minutes, and two parts per thousand is fatal.

HALIDES OF SULFUR. The fluorides and chlorides include S_2F_2, SF_2, SF_4, S_2F_{10}, SF_6, S_2Cl_2, SCl_2, SCl_4, and the oxychlorides $SOCl_2$ and $SOCl_4$. There is also a bromide, S_2Br_2, but no iodides. The lower fluorides can be made by fluorinating sulfur with AgF, but SF_6 is made from the elements. All fluorides but SF_6 react with water to give HF and an oxide of sulfur:

$$SF_4(g) + 2H_2O(l) \rightarrow 4HF(aq) + SO_2(g)$$

$$S_2F_{10}(l) + 5H_2O(l) \rightarrow 10HF(aq) + SO_2(g) + SO_3(g)$$

It might be expected, then, that SF_6 would hydrolyze to $6HF + SO_3$, certainly a horrendous mixture to breathe because both compounds destroy lung tissue rapidly. Yet, as we saw in Chapter 3, SF_6 can be inhaled with impunity (provided oxygen is present to sustain life), and is actually administered sometimes during x-ray examination of the lungs in order to see greater detail. This surprising behavior was explained earlier in terms of the principle of maximum covalency. The octahedral structure involves six sp^3d^2 hybrid orbitals, and with six electrons from sulfur plus one electron from each of six fluorine atoms, these orbitals will be completely filled. Sulfur hexafluoride therefore is chemically unreactive for the same reason neon is—all its available orbitals are already filled.

The chloride S_2Cl_2 is manufactured by the action of chlorine on hot sulfur, and the liquid product (bp 138° C) dissolves considerably more sulfur. The solution is used to vulcanize rubber.

CHLORINE, A VERSATILE ACID-FORMER

The word *chlorine* is familiar to the man in the street and to every schoolchild. There is only one chlorine, though, and that is a green gas which has a choking odor. What is used in swimming pools and called chlorine is not chlorine at all, but a solid chlorite or hypochlorite salt which kills bacteria as efficiently as gaseous chlorine but is more convenient to store and to use. Elementary chlorine, the real thing, *is* used in treatment of municipal water supplies to control the bacterial level, but a water treatment plant has equipment to handle the necessary steel cylinders and control the pressure of the gas. Chlorine condenses to a liquid at −34° C and the liquid exerts a pressure of only 6.6 atm at 20° C, so it can be shipped in steel tank trucks and pumped into steel storage cylinders. About ten million tons of it are made each year, because it is needed not only for water treatment but for bleaching wood pulp and textiles, and for making plastics, insecticides, dry-cleaning solvents, and a host of chemical products.

THE ELEMENT. Free chlorine is made in the laboratory by oxidizing hydrochloric acid or a chloride with any strong oxidizing agent, such as a permanganate or a dichromate or manganese dioxide:

$$4H^+ + 2Cl^- + MnO_2(s) \rightarrow Mn^{+2} + 2H_2O(l) + Cl_2(g)$$

or

$$14H^+ + 6Cl^- + Cr_2O_7^{-2}(aq) \rightarrow 2Cr^{+3} + 7H_2O + 3Cl_2$$

The ten million tons of commercial chlorine are not made this way, however; they come from the electrolysis of melted NaCl or of a saturated aqueous solution of NaCl, as described in the section on sodium.

Chlorine combines vigorously with hydrogen, as can be demonstrated by inflating soap bubbles with a stream of equal volume of the two gases. The bubbles explode upon ignition:

$$H_2 + Cl_2 \rightarrow 2HCl(g) \qquad \Delta H = -22 \text{ kcal/mole}$$

Similarly, chlorine extracts hydrogen from many compounds to form hydrogen chloride, and therefore replaces hydrogen in hydrocarbons:

$$C_6H_6(l) + Cl_2(g) \rightarrow HCl(g) + C_6H_5Cl(l)$$

benzene — chlorobenzene, a solvent and dye intermediate

Another hydrocarbon, turpentine, bursts into flame if it is warmed and introduced into chlorine:

$$C_{10}H_{16}(l) + 8Cl_2(g) \rightarrow 16HCl(g) + 10C(s)$$

a black smoke

Chlorine is also a strong oxidizing agent in other ways, as when it reacts with water to form strong hydrochloric and weak hypochlorous acids:

$$Cl_2(g) + H_2O(l) \rightarrow H^+ + Cl^- + HOCl(aq)$$

When chlorine dissolves in water, it is the hypochlorous acid which kills bacteria or bleaches colored substances, accomplishing both by oxidation as the HOCl decomposes to stable HCl and liberates atoms of oxygen:

$$HOCl(aq) \rightarrow H^+ + Cl^- + O$$

OXIDES. In contrast to phosphorus and sulfur, chlorine forms oxides which are unstable, explosive, and of no practical importance—but they still form acids. The chief oxides are Cl_2O, ClO_2, Cl_2O_6, and Cl_2O_7. All are highly endothermic, and cannot be obtained from the elements. The first, Cl_2O, is an orange gas produced by the action of chlorine on mercury oxide:

$$2Cl_2(g) + 2HgO(s) \rightarrow Cl_2O(g) + Hg_2OCl_2(s)$$

Its only claim to fame is that it reacts with water to form hypochlorous acid:

$$Cl_2O(g) + H_2O(l) \rightarrow 2HOCl(aq)$$

This is not a practical way to obtain hypochlorous acid or hypochlorites, of course. To do so on a large scale, a solution of sodium chloride is electrolyzed while being stirred rapidly, so that the chlorine produced at the positive electrode has a chance to react with the sodium hydroxide produced at the negative electrode:

$$Cl_2(g) + 2OH^-(aq) \rightarrow Cl^- + OCl^- + H_2O$$

The resulting solution of sodium hypochlorite is the familiar household bleach, a potent germicide as well as a popular laundry bleaching agent.

The second oxide, ClO_2, is an evil-smelling dark yellow gas that can be made (with extreme danger) by the action of sulfuric acid on chlorates (if they don't explode in the process). It reacts with water to give two oxyacids:

$$2ClO_2(g) + H_2O(l) \rightarrow HClO_2(aq) + HClO_3(aq)$$

The next, Cl_2O_6, is obtained by ozone oxidation of chlorine dioxide:

$$2ClO_2 + 2O_3 \rightarrow Cl_2O_6 + 2O_2$$

It too hydrolyzes to give two oxyacids:

$$Cl_2O_6 + H_2O \rightarrow HClO_3 + HClO_4$$

The highest oxide, Cl_2O_7, is obtained by brave people who dehydrate $HClO_4$ with P_4O_{10}. It is a violently explosive oily liquid which will, of course, recombine with water to form perchloric acid:

$$Cl_2O_7 + H_2O \rightarrow 2HClO_4$$

Hence all the oxides of chlorine are acid formers, and the resulting acids are very important.

ACID STRUCTURE VS. ACID STRENGTH. We have seen that chlorine forms a variety of oxyacids, and it is worthwhile to look into their structures and their characteristics in order to get some idea of what makes an acid tick, so to speak. What governs the strength of an acid? Which acids are also oxidizing agents, and which are reducing agents? In answer to the first question, the strength of an acid increases with increasing electronegativity of the central atom and is enhanced if other electronegative atoms, such as oxygen, are also attached to the

TABLE 4–3 PRINCIPAL ACIDS OF CHLORINE

Name	*Structure*	*Dissociation Constant*
Hydrochloric acid	H:C̤̈l:	1.0 (very strong)
Hypochlorous acid	H:Ö̤:C̤̈l:	2.95×10^{-9} (very weak)
Chlorous acid	H:Ö̤:C̤̈l:Ö̤:	1.1×10^{-2} (weak)
Chloric acid	:Ö: H:Ö̤:C̤̈l:Ö̤:	0.20 (moderately strong)
Perchloric acid	:Ö: H:Ö̤:C̤̈l:Ö̤: :Ö̤:	1.0 (very strong)

central atom. This is borne out by HCl and the four oxyacids of chlorine, the structures and dissociation constants of which are given in Table 4–3.

The actual bond energy or bond strength of the hydrogen bond can often temper the conclusion about the effect of electronegativity. Given the generalization just quoted, and the stated values of electronegativities for the halogen elements (F 4.0, Cl 3.0, Br 2.8, and I 2.5), a student would conclude that HF should be the strongest binary acid, while HI should be the weakest. This is not so; the actual measured strengths are in the reverse order, with HI strongest! Why is this so?

In an acid HX we are dealing with the breaking of the covalent H:X bond under the solvolytic action of water, with the release of H^+ and X^- ions. Naturally the strength of the original bond, as measured on the pure HX compound, has a lot to do with it. Secondly, the ions become hydrated; in terms of energy, that is why water attacks the HX bond. Pure hydrogen chloride, either as liquid or gas, contains no ions. It conducts no electric current. It is soluble in liquid hydrocarbons, but the solution conducts no current. Dissolve the same amount of HCl in the same volume of *water*, however, and the solution is strongly conducting and strongly acidic. The reason is that H^+ ions give off energy as they become hydrated, and the solution becomes warm. This heat of hydration is what provides the incentive for dissociating the HX bond.

Turning to the series of hydrogen halides, we see that all provide H^+ ions and that the heat of hydration of H^+ will be the same for any molecule of HX that comes apart. But how easily do the molecules come apart? We need only look at their bond energies to find out: the bond energy of HF is 135 kcal per mole, while that of HCl is 103 kcal; for HBr it is 88 kcal, and for HI it is only 71 kcal. Hence HI comes apart easiest, and is the strongest acid; HBr is next, HCl next, and HF is by far the weakest acid, with the hydration effect struggling quite ineffectually to tear apart the very strong H—F bonds. In 0.1 M solution only 10% of HF exists as ions, compared with 92.6% of HCl, 93% of HBr, and 95% of HI.

OXIDIZING ABILITY. As for oxidizing properties, the HX acids have none because they have no oxygen to give up. Of the oxyacids of chlorine, HClO has the strongest oxidizing action as it gives up its lone oxygen atom and reverts to stable HCl. Chlorous acid, $HClO_2$, has less oxidizing tendency but has twice as much oxygen to give up to an eager acceptor. Chloric acid, $HClO_3$, and the chlorates are still weaker oxidizing agents, but at the same time have more oxygen to give up when they are induced to do so. Perchloric acid, $HClO_4$, is considered a very stable acid, and perchlorates are thought to be weak oxidizing agents, but once activated they provide a very concentrated source of free oxygen atoms. For that reason perchloric acid and all perchlorates should be considered potentially very dangerous. In dilute solution they may be all right, but in concentrated form they are hazardous. For example, there was a small Western electroplating plant that blew up and destroyed the entire building because clips of organic plastic (easily oxidizable!) held the metal parts to be plated in a bath of warm perchloric acid. Similarly, in an Eastern medical school laboratory, a sample of food was being prepared for analysis for toxic metals by heating it in concentrated perchloric acid, but unfortunately the exothermic reaction got out of hand, resulting in a devastating explosion and tragic loss of life.

OXYACID SALTS. You may well wonder where chlorates and perchlorates come from, if the oxides of chlorine are made from the oxyacids or their salts, and not vice versa. Hypochlorites come from the electrolysis of water solutions of the corresponding chlorides, as explained. Further electrolytic oxidation of chlorides under appropriate conditions of temperature, concentration, and design of the electrolytic cell will produce chlorites, chlorates, and even perchlorates. In the laboratory, potassium chlorate can easily be made by the action of chlorine on a hot concentrated solution of potassium hydroxide:

$$3Cl_2 + 6KOH \rightarrow 5KCl + KClO_3 + 3H_2O$$

or

$$3Cl_2(g) + 6OH^-(aq) \rightarrow 5Cl^-(aq) + ClO_3^-(aq) + 3H_2O(l)$$

Potassium chlorate is only one fifth as soluble as potassium chloride in cold water so it can be crystallized from the cooled mixture and then purified by recrystallization. Notice that in the reaction some of the chlorine is oxidized to chlorate ion and some is reduced to chloride ion. In terms of electron transfer, one chlorine atom gives up five electrons to provide five other chlorine atoms with one electron each, so that they may become Cl^- ions. This is a good example of *autooxidation*, a reaction in which one substance serves simultaneously as an oxidizing agent and as a reducing agent. Autooxidation is one form of *disproportionation*, a general term applied to any reaction in which one substance rearranges itself to form two unequal products.

TABLE 4–4 PHYSICAL PROPERTIES OF THE ELEMENTS FROM SODIUM TO CHLORINE

Atomic No.	*Symbol*	*Atomic Weight*	*Melting Point, °C*	*Boiling Point, °C*	*Density g/cm³*	*State at Room Temperature*
11	Na	22.99	97.5	889	0.97	silvery metal
12	Mg	24.31	650	1120	1.74	gray metal
13	Al	26.98	660	2327	2.70	light gray metal
14	Si	28.09	1414	2355	2.33	bluish metalloid
15	P	30.97	44	280	1.83*	waxy white solid
16	S	32.06	112	444	2.07†	yellow solid
17	Cl	35.45	−101	−34	3.21 g/liter	green gas

*Properties are those of white phosphorus. Red phosphorus melts at 590°C under 45 atm pressure, and has a density of 2.20 g/cm³. Black phosphorus has a density of 2.70 g/cm³ and does not melt.

†Properties are those of orthorhombic S_8. The monoclinic form melts at 119°C and has a density of 1.96 g/cm³; it is stable only in the range 96° to 119°C.

Potassium perchlorate can be made by heating $KClO_3$ very gently and carefully:

$$4KClO_3(s) \rightarrow 3KClO_4(s) + KCl(s)$$

(This reaction is in itself an example of disproportionation.) Perchloric acid can then be derived from its salt by heating with successive small amounts of sulfuric acid, so that nonvolatile $KHSO_4$ remains behind while $HClO_4$ is evaporated off:

$$KClO_4(s) + H_2SO_4(l) \rightarrow KHSO_4(s) + HClO_4$$

Perchloric acid is a commercial strong acid, used where sulfuric acid might be reduced to sulfur dioxide.

SUMMARY

The active elements of the third period, sodium through chlorine, range from an exemplary alkali metal which forms a strong base to an exemplary nonmetal which forms strong acids. All are abundant and well-known, and all are capable of hexacovalency. The range of electronegativity is less than that in the second period, but exceeds that in all subsequent periods. The physical properties of these elements are summarized in Table 4–4.

All the elements in the sequence Na-Cl form reactive hydrides. Sodium hydride reacts violently with water to release hydrogen and to form the strong base, NaOH. Magnesium hydride hydrolyzes vigorously to form the weak base, $Mg(OH)_2$. The substituted aluminum hydrides hydrolyze vigorously to form the corresponding substituted aluminum hydroxides, which are amphoteric. The silicon hydrides hydrolyze readily, especially in alkaline solution, to give the exceedingly weak acid H_4SiO_4. Phosphine, PH_3, is a very weak acid itself, but phosphorus forms a series of oxyacids of increasing acid strength, culminating in H_3PO_4, a moderately strong acid. Hydrogen sulfide is a weak acid, but the fully oxidized oxyacid H_2SO_4 is a strong acid. Hydrochloric acid is a very strong acid itself, and chlorine forms a series of oxyacids of increasing acid strength culminating in $HClO_4$, the strongest acid known to chemistry.

A comparison of electron-dot diagrams reveals clearly two reasons for the sharp increase in acidity among the oxyacids H_4SiO_4, H_3PO_4, H_2SO_4, and $HClO_4$:

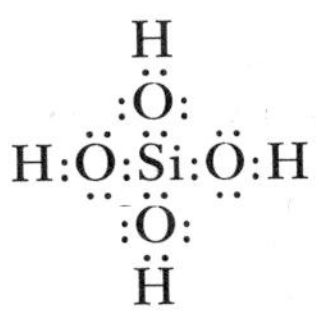

```
      H
     :Ö:
  H:Ö:Si:Ö:H
     :Ö:
      H
```

silicic acid,
a very weak acid

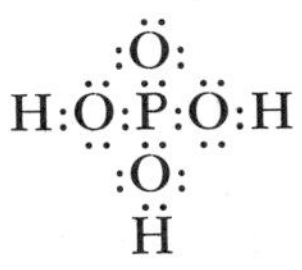

```
     :Ö:
  H:Ö:P:Ö:H
     :Ö:
      H
```

phosphoric acid,
a moderately strong acid

```
     :Ö:
  H:Ö:S:Ö:H
     :Ö:
```

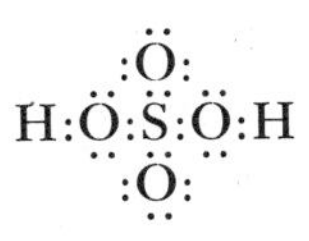

sulfuric acid,
a strong acid

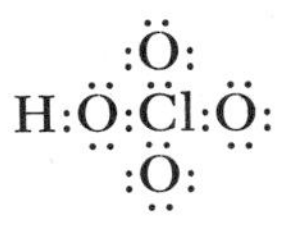

```
     :Ö:
  H:Ö:Cl:Ö:
     :Ö:
```

perchloric acid,
the strongest acid known

Each element in the series Si, P, S, Cl has one more electron than the last, so one less hydrogen atom is bound. As the electronegativity of the central atom rises, it withdraws electrons more firmly from the O:H bond or bonds, enabling an H^+ ion to be released more easily. This makes release of an H^+ ion from perchloric acid easiest. Furthermore, the combined electron-withdrawing effect of four oxygen atoms *and* the central atom is concentrated on one O:H bond in $HClO_4$, whereas it is spread out among four O:H bonds in H_4SiO_4 and is correspondingly diluted. The two effects act together to produce the marked change in acidity.

Sodium and its major industrial compounds NaOH, $NaHCO_3$, Na_2CO_3, NaCN, Na_2SO_4, and NaH are all produced from NaCl, which is therefore a very important commercial raw material as well as an essential nutrient. Sodium itself is a reducing agent used in the manufacture of detergents, dyes, and ethyl gasoline; the metal is also the intended heat-transfer fluid in the new fast-breeder power reactors (a use which gives most chemists the shudders). The alkaline compounds of sodium are used to make glass and soap, and for industrial processing.

Magnesium is obtained from inland brines and from seawater, and is of great value as a structural metal. The important compounds of magnesium are its oxide (thermal insulation and for electric range units), its hydroxide (milk of magnesia), its carbonate (magnesite, dolomite), and its organometallic derivatives (Grignard reagents).

Aluminum is also a structural metal, obtained from Al_2O_3 by electrolysis. It is reactive but amphoteric, and is an important constituent of the clay minerals and igneous rocks. Its most important compounds are Al_2O_3 (ruby, sapphire, corundum), Al_2Cl_6 (a Lewis acid catalyst for organic synthesis), and $Al_2(SO_4)_3$ and the mixed sulfates (alums).

Silicon is our most abundant electropositive element, and teams with oxygen to form the silicate rocks that cover our globe. Silicates (often with aluminum incorporated) are also important as the ceramic products glass, porcelain, chinaware, bricks, tile, cement or concrete, and vitreous enamels. Silicon also forms covalent compounds, notably the volatile halides and hydrides. Organosilicon oxides are the basis of silicone resins, rubber, and oil. The carbide SiC is an abrasive.

Phosphorus is essential to living protoplasm, and must be incorporated in plant nutritives. The element has several allotropic forms, the best known being the highly reactive white phosphorus. The hydrides and halides are volatile and acidic. The most important practical compounds are PCl_3, PCl_5, and phosphoric acid, H_3PO_4, all made from phosphorus. The element is obtained by carbon reduction of phosphate rock in an electric furnace.

Sulfur is scattered everywhere, and is an unwanted constituent of coal, petroleum, and the ores of copper, lead, and zinc. Its removal in the interest of cleaner air involves many serious problems of chemistry, engineering and economics. Sulfur and the sulfides are the sources of sulfuric acid, which is the workhorse of industry and therefore an indicator of the industrial development of a country.

Chlorine is produced in enormous quantities by the electrolysis of molten NaCl and of salt brine. It is used to purify water, to bleach cotton and paper pulp, and to make plastics, dry-cleaning solvents, and spray-can propellants. There are many important compounds, including NaCl, Al_2Cl_6, $SiCl_4$, PCl_3, NaOCl (household bleach), HCl, and $HClO_4$.

GLOSSARY

adsorb: to gather a dissolved substance from a surrounding gas or liquid and retain it on a surface as a condensed layer, usually by chemical forces. Adsorption differs from absorption in being a surface phenomenon, not a volume effect.

allotrope: one of two or more existing forms of an element, as the diamond or graphite form of carbon.

amphoteric: capable of acting either as an acid or as a base, depending upon the chemical conditions.

autooxidation: a chemical process in which one substance acts as both oxidizing agent and reducing agent.

bidentate: capable of chemical bonding at two sites within the molecule, by simultaneous chemical reaction.

calcine: to convert to an oxide or ash (a calx) by severe heating or roasting.

chelate: a coordination compound or ion formed by a bidentate or polydentate ligand which attaches itself with two or more bonds to a central metal atom or ion.

chlorosilane: a silane (q. v.) in which one or more hydrogen atoms have been replaced by chlorine.

detergent: any cleansing agent, but usually a substance prepared or manufactured expressly to assist in laundering or cleaning.

die-casting: a method of fabricating metals in which hot, partly-melted metal ("slush") is forced into molds under pressure and is allowed to solidify there.

disproportionation: a chemical reaction in which one substance changes into two or more dissimilar ones, usually by autooxidation.

dissociation constant: the equilibrium constant for a dissociation reaction, such as the solvolytic dissociation of an acid or base in water.

divalent: having a valence or combining power of two, as the Zn^{++} ion in $ZnCl_2$ or the $—CH_2—$ group in $ClCH_2Cl$.

equilibrium: a steady state brought about by the functioning of two opposing processes whose effects exactly compensate for each other.

faraday (named after Michael Faraday, 1791–1867): the amount of electricity that will deposit or liberate 1 gram equivalent of an element in an electrolysis. One faraday is equal to 96,494 coulombs of electricity, and also to a flow of 6.023×10^{23} electrons. One coulomb results from an electric current of one ampere flowing for one second.

ferrosilicon: an alloy of iron and silicon.

galvanic action: an electrochemical interaction of two dissimilar metals with an electrolyte, resulting in a flow of electric current through the metals and a corresponding migration of ions in the electrolyte (named after Luigi Galvani, 1737–1798).

geode: a hollow nodular stone lined with crystals.

halide: a compound of one of the halogen elements.

halogen: literally, a salt-former; an element from the group F, Cl, Br, I, At.

hydrogenation: reaction with hydrogen, usually brought about by heating under pressure of H_2 in the presence of a catalyst.

kraft paper: a strong brown paper made from wood chips by the alkaline sulfate process. "Kraft" is the German word for strength.

ligand: an ion or molecule that coordinates to a central atom to form a coordination compound or complex.

matrix: that which surrounds or gives origin or form to a thing, as the intercellular material in a biological tissue, or the polymer that surrounds fibers or particles in a reinforced plastic.

monovalent: having a valence or combining power of one, as an Na^+ ion or a $—CH_3$ group.

opacifier: an agent that renders a material opaque.

oxyacid: an acid containing hydrogen, oxygen, and at least one other element. The other element may be a nonmetal, as in H_2SO_4, or a metal, as in $HMnO_4$, permanganic acid.

oxidation: reaction with oxygen, or any other reaction which results in loss of electrons.

oxidizing agent: a substance which causes oxidation, and is in turn reduced.

polymer: a substance, often of very high molecular weight, consisting of several or many identical structural units joined together by chemical bonds.

reducing agent: that which reduces; a substance capable of furnishing electrons to another substance (the oxidizing agent) in an electron-transfer reaction. Reducing agents become oxidized in such a reaction.

reduction: a gain of electrons; any reaction in which a substance receives electrons from a reagent capable of releasing them, as in the reduction of metals from oxide or sulfide ores.

silane: a binary compound of silicon and hydrogen, such as SiH_4 or Si_2H_6; a derivative of such a hydride, considered to be made by replacing one or more of the hydrogen atoms.

silica: the ancient name for any of the many forms of silicon dioxide.

silicone: an organosiloxane; a polymer consisting of a framework or backbone of alternate silicon and oxygen atoms, with methyl or other hydrocarbon groups attached to the silicon atoms by Si—C bonds.

smelting: the process of obtaining metals from their ores by means of heat, usually with the aid of a reducing agent.

viscosity: the "thickness" of a fluid; the property of a fluid that resists a force tending to make it flow.

vitreous: glass-like.

volatile: capable of evaporating; having a tendency to vaporize.

voltaic cell: an electrochemical cell consisting of two electrodes and an electrolyte in a physical arrangement suitable for generating a potential and causing a flow of electricity in an external circuit (named after Count Allessandro Volta, 1745–1827, who first made such cells).

EXAMINATION QUESTIONS

1. List the elements of the third period in the order of their terrestrial abundance, and estimate the percentage of each.

2. List at least five important products derived from common salt, and indicate by balanced equations (showing all necessary conditions) how each is obtained.

3. What solid substances, often misnamed "chlorine," are added to swimming pools to control the bacterial level? How do they operate to accomplish this?

4. Using equations wherever they would be helpful, compare the chemical properties of the hydrides of Na, Si, P, S, and Cl.

5. Give an example of a Grignard reagent, and show how it is obtained and how it may be used in synthesis.

6. Epsom salt is a hydrated form of magnesium sulfate. One molecule of water is coordinated to the sulfate ion, the rest to the magnesium ion. Write the expected empirical formula.

7. Write an electron-dot formula for the hypothetical $AlCl_3$, and explain from the formula why this is an unlikely compound. What is the actual molecular formula of aluminum chloride, and what is its structure?

8. What is a. "papermaker's alum," and b. a true alum? c. How and why is "papermaker's alum" used?

9. Why is pure aluminum not used for aircraft frames or automobile engines? In what form and composition is aluminum usually used in these applications?

10. Explain the difference between electrical conductivity in a metal and a metalloid like silicon, making use of the band theory.

11. How is elementary silicon obtained, and for what is it used?

12. What is ferrosilicon, and how is it used?

13. What is the chemical composition and structure of a silicone polymer, say methylsilicone?

14. In what ways is a silicone polymer superior to a purely organic polymer made from petroleum, in terms of properties and conservation of resources?

15. By means of equations, indicate how the following chlorides hydrolyze: CCl_4, Al_2Cl_6, $SiCl_4$, PCl_5, and SCl_4.

16. How is sulfuric acid made, and how much is made this way? Give equations and details.

17. What problems are associated with the removal of SO_2 and SO_3 from flue gas by scrubbing the gas with a suspension of lime?

18. How does the acid strength change within the series of compounds H_4SiO_4, H_3PO_4, H_2SO_4, and $HClO_4$? Why does it change in this manner?

19. Why is phosphorus important to agriculture? What compounds of phosphorus are made and sold to aid crop production?

"THINK" QUESTIONS

A. In 1974, hundreds of people had to be evacuated from their homes in Chicago because a huge tank containing nearly one million gallons of $SiCl_4$ began to leak around the valve, and there was no safety shutoff nor any emergency plan for coping with the tank if it did leak. Choking fumes filled the air, and of course there was great danger of the vapors blanketing the entire city. Now the questions:

a. Why does the vapor of $SiCl_4$ cause such a fierce choking when inhaled?

b. Is this dangerous in the sense of causing permanent injury, or just inconvenient?

c. What was the effect of $SiCl_4$ vapor on nearby automobiles, elevators, and other machinery?

d. Given the heat of hydrolysis of $SiCl_4$ (about 75 kcal/mole), and its heat of vaporization (6.86 kcal/mole), was there real reason for expecting the vapor of that $SiCl_4$ to blanket the entire city if the entire million gallons flowed out of the tank and met some water?

e. Was it wise to store that much $SiCl_4$ in one tank in the first place? What safety precautions would you impose on makers and users of silicon tetrachloride?

B. Silicon disulfide, SiS_2, crystallizes in long needles instead of the blunt hexagonal prisms of SiO_2. Moreover, when SiS_2 meets water or moist air there is a strong smell of hydrogen sulfide. How would you explain these two facts?

C. It has been proposed that magnesium oxide be used to remove sulfur dioxide from flue gas in factories and power plants. The MgO may be employed as a solid or as an aqueous suspension, it is said, and an equation is given:

$$MgO(s) + SO_2(g) + 1/2O_2(g) \rightarrow MgSO_4(s)$$

a. Assuming complete removal of SO_2 according to the equation, how much MgO would be required to combine with the 20 million tons of sulfur present in the coal now burned each year in electric power stations?

b. How much magnesite, $MgCO_3$, would have to be mined, calcined, and pulverized to provide this much MgO to do the job?

c. How much $MgSO_4$ would be produced by the process, assuming it is completely successful in capturing all the sulfur present in the coal?

D. In 1972 electric power stations in the U.S. consumed 494 million barrels (67.4 million tons) of residual oil as fuel, along with 386 million tons of coal and 3978 million cubic feet of natural gas. Figures for the sulfur content of the natural gas are lacking, but the oil contains an average of 1.5% S and the coal contains an average of 5% S.

a. If all of this sulfur were to be captured and converted to $MgSO_4$ according to the equation shown in Question 3, how much $MgSO_4$ would have been produced in this way that year?

b. What would you propose be done with all that $MgSO_4$? (Think in terms of its disposal by land or sea, knowing that its solubility in water is 350 g/liter at 20° C, or think of what might be made from it, and the material consequences of what you propose.)

E. Write equations for the production (by a practicable process) of a. NaOH; b. Na_2CO_3; c. $NaBH_4$; d. Mg_2Si; e. silicone rubber of the composition $(CH_3)_2SiO$; f. sodium metasilicate, Na_2SiO_3 (which is called water glass because it is water-soluble) from Na_2CO_3 and SiO_2; g. P_4 from $Ca_5(PO_4)_3F$; h. CS_2 from sulfur; i. CCl_4 by the chlorination of CS_2; and j. ammonium perchlorate, NH_4ClO_4, which is used as the oxidant (oxidizing agent) in solid fuel rockets.

F. Naturally the demand for elementary chlorine would drop drastically after prohibiting the use of DDT, all other chlorine-containing insecticides, all aerosol spray cans using CCl_2F_2, and all vinyl chloride. What would this do to the supply of NaOH? Why?

PROBLEMS

1. Trisodium phosphate (TSP) is a valued cleaning agent for factories and public buildings, although some consider it too alkaline for household use. How much sodium hydroxide is required to convert all the phosphoric acid from 1 metric ton of phosphorus to trisodium phosphate?

2. A 10-g sample of rock was analyzed for sulfate ions by dissolving it in nitric acid, adding an excess of barium chloride solution, and filtering off the $BaSO_4$ that precipitated. After washing and drying the $BaSO_4$, it weighed 0.1035 g. How many moles of sulfate ion were present in the rock sample, and what is the per cent by weight of sulfate ion?

*3. Chloride ion in fresh water is determined analytically by titrating a sample with a solution of silver nitrate:

$$Cl^-(aq) + Ag^+(aq) \rightarrow AgCl(s)$$

A standard 0.001000 M solution of silver nitrate was used to analyze the following five samples of lake water, with the results shown:

Sample	*Volume of Sample, ml*	*Source of Sample*	*End Point Volume of $AgNO_3$, ml*
1	100	Lake Erie, 1906	24.54
2	25	Lake Erie, 1968	17.34
3	100	Lake Ontario, 1906	21.72
4	25	Lake Ontario, 1968	19.39
5	10	Seneca Lake, N.Y., 1968	50.92

*This question courtesy of Professor T. R. Blackburn, Hobart and William Smith College, Geneva, N.Y.

a. What is the chloride content of each sample, in mg of Cl^- per liter?

b. What seems to be happening to the three lakes involved, with passage of time?

c. How could you explain the changes noted in b?

d. What can be done about it?

4. Seawater is alkaline, having a pH of 8.2. In order to extract bromine from seawater, it must first be acidified by adding sulfuric acid. What weight of H_2SO_4 is required to change the pH of 1 metric ton of seawater from 8.2 to 3.5? (Assume the density of the water is 1.0 g/cm^3).

5. Hydrogen sulfide can be eliminated from a smelly water supply by oxidizing it. This is done by chlorinating the water and then passing it through a charcoal filter to remove the suspended sulfur:

$$H_2S(aq) + Cl_2(aq) \rightarrow 2H^+ + 2Cl^- + S(s)$$

If the H_2S content of the raw water is 14 parts per million, how much chlorine is required to remove all the H_2S from one million metric tons of water, which is the daily requirement of a fair-sized city?

6. The density of magnesium metal is 1.74, and that of ordinary mild steel is 7.85. Since some alloys of magnesium (of similarly low density) have about the same transverse strength (yield point) as ordinary steel, one lb of magnesium may take the place of 4.5 lb of steel for most uses.

Let us consider a hypothetical country, Amnesia, which requires 5 million tons of steel annually for buildings, transportation, and consumer goods. Amnesian iron ore is rapidly disappearing, and substitution by magnesium from seawater is being considered.

a. Outline an economical process for obtaining pure magnesium chloride from the seawater, giving balanced equations for the reactions involved.

b. How many tons of Mg are required annually to substitute for all the steel on a strength-equivalent basis?

c. How many tons of $MgCl_2$ does this represent?

d. If seawater contains 0.13% Mg, how many tons of it would have to be processed annually?

e. What part of the total hydrosphere (1.47×10^{18} tons of seawater) is thus robbed of all its Mg each year?

5 ALL ABOUT THE MAIN GROUPS

Having considered the number and organization of the chemical elements in Chapter 1 (the names and numbers of all the players, so to speak), and having surveyed the representative light elements in the next three chapters, it is time now to complete our study of the nontransition elements by considering the main groups in their entirety. This will be done for each group by summarizing the physical and chemical properties of its elements in a comprehensive table, followed by a discussion of general trends and some pertinent points about each important element not previously discussed.

GROUP I, THE ALKALI METALS

Table 5–1 gives a detailed listing of the physical properties of the alkali metals, together with some chemical aspects useful in a comparison. For completeness, data for all the elements are included in Table 5–1 even though some of the figures therein have been used in earlier chapters for other purposes. The remarks in Chapter 1 about trends in melting and boiling points and atomic volumes still hold, of course. As in every group, but to a lesser extent than usual, the individuality of each element asserts itself above the general trends.

All the elements in Group I are soft, silvery metals with body-centered cubic crystal structures, although cesium melts at so low a temperature (and supercools so readily) that samples of it usually are liquid at room temperature. All have but a single valence electron, an s^1 configuration following the previous noble-gas core of filled orbitals, so all are capable of but one oxidation state, +1. The single electron is lost easily, so all are excellent reducing agents and have large negative oxidation potentials. The Pauling electronegativities (derived from selected thermochemical data) are all at the low end of the scale, decreasing progressively from 1.0 to 0.7. The ratios of effective nuclear charge to square of the ionic radius, a physical measure of electronegativity, fall in an even narrower band at the low end of the scale.

TABLE 5–1 GROUP I: THE s^1 ELEMENTS (ALKALI METALS)

	Lithium	*Sodium*	*Potassium*	*Rubidium*	*Cesium*	*Francium*
Symbol	Li	Na	K	Rb	Cs	Fr
Atomic number	3	11	19	37	55	87
Atomic weight	6.939	22.99	39.10	85.47	132.9	(223)*
Valence e^-	$2s^1$	$3s^1$	$4s^1$	$5s^1$	$6s^1$	$7s^1$
mp, °C	186	97.5	63.65	38.89	28.5	27
bp, °C	1326	889	774	688	690	677
d, g/cm³	0.534	0.971	0.862	1.53	1.87	—
Atomic volume	13.1	23.7	45.3	55.9	70.0	—
Atomic radius, Å	1.52	1.86	2.31	2.44	2.62	—
Ion radius, Å	0.60	0.95	1.33	1.48	1.69	—
Pauling EN	1.0	0.9	0.8	0.8	0.7	0.7
Z_{eff}/r^2	0.97	1.01	0.91	0.89	0.86	0.86
Standard potential	−3.05	−2.71	−2.92	−2.93	−2.92	—
Oxidation states	+1	+1	+1	+1	+1	—
Ionization energy*	124	119	100	96	90	—
Heat of vaporization	35.4	25.9	19.4	18.1	19.4	—
Discoverer	Arfvedson	Davy	Davy	Bunsen and Kirchhoff	Bunsen and Kirchhoff	Perey
Date of discovery	1817	1807	1807	1861	1860	1939
rpw* pure O_2	Li_2O	Na_2O, Na_2O_2	K_2O_2, KO_2	RbO_2	CsO_2	—
rpw H_2O	LiOH	NaOH	KOH	RbOH	CsOH	—
rpw N_2	Li_3N	none	none	none	none	—
rpw halogens	LiX	NaX	KX	RbX	CsX	—
rpw H_2	Li^+H^-	Na^+H^-	K^+H^-	Rb^+H^-	Cs^+H^-	—
Flame color	bright red	yellow	violet	purple	blue	—
Mohs hardness	0.6	0.4	0.5	0.3	ul*	—
Crystal structure	cubic bc*	cubic bc	cubic bc	cubic bc	cubic bc	—

*Atomic weights in parentheses are those of the most stable isotope. All energies and heats are in kcal per mole. The letters rpw stand for "reaction product with"; bc = body-centered; ul = usually liquid. For abundances, see Tables 1–2 and 1–3.

All the alkali metals occur as salts in which their M^+ ions are coupled with halide, sulfate, carbonate, or complex silicate negative ions; they never occur as free metals or as oxides or sulfides. Their abundances vary widely, as shown in Table 1–3. Potassium, rubidium, and francium are radioactive, while the others are not. The ^{40}K isotope has a half-life of 1.28 billion years, decaying to ^{40}Ca (and, to a lesser extent, to ^{40}A):

$$^{40}_{19}K \rightarrow ^{40}_{20}Ca + \beta^-$$

$$^{40}_{19}K \rightarrow ^{40}_{18}A + \beta^+$$

There are about 83 mg of ^{40}K in a human body of average weight, and it accounts for about 13% of the unavoidable background radiation we receive.

Rubidium-87 has a half-life of 5×10^{10} years and decays to strontium-87:

$$^{87}_{37}Rb \rightarrow ^{87}_{38}Sr + \beta^-$$

All the isotopes of francium are radioactive, so it has only a fugitive existence in nature. Its most stable isotope, ^{223}Fr, has a half-life of only 22 minutes, so not enough francium has been accumulated to complete its column in the table.

All the alkali metals can be detected and identified by the characteristic colors they impart to a Bunsen flame, as shown in Table 5–1. Their simple atomic structures lead to simple line spectra in the visible region and the patterns of colored lines can readily be recognized, even in the spectrum of a mixture, by using a simple hand-held spectroscope.

All the alkali metals reduce the hydrogen ions of water, leaving the unchanged hydroxyl ions in excess:

$$2M + 2H^+ \rightarrow 2M^+ + H_2$$

This reaction is rapid with lithium, vigorous and self-igniting with sodium, and explosive with the heavier alkali metals. All the resulting hydroxides are strong bases, but LiOH has limited solubility (12.8 g/liter in water at 20° C), so a saturated solution of it appears to be less alkaline than a concentrated solution of NaOH or KOH.

When ignited in air, lithium forms both an oxide, Li_2O, and an ionic nitride, Li_3N, but it is the only alkali metal to form such a nitride. To all the other elements of the group, air is just a diluted form of oxygen.

SPECIAL POINTS ABOUT POTASSIUM. Of that part of the earth which we can see and sample, sodium constitutes 2.6% and potassium 2.4% (see Table 1–3). On a weight basis, these two alkali metals are equally abundant on land, but in the sea there is a 30-to-1 excess of sodium. Why? If all the minerals in the sea have been washed down from the rocks and minerals of the land by rainwater, and if sodium and potassium are equivalent in their chemical behavior, should not those elements be present in equal proportions in the sea?

Yes, if that were so, but there are two good reasons why it is not. First, potassium salts with bulky negative ions are less soluble than the corresponding sodium salts. For example, $KClO_4$ is only 1/279th as soluble as $NaClO_4$ in cold water, and K_2PtCl_6 is only 1/137th as soluble as Na_2PtCl_6. Complex silicate and aluminosilicate ions certainly qualify as large and bulky negative ions, so the difference in the solubilities of their potassium and sodium salts can be appreciated. Secondly, those potassium ions which are dissolved out of rock are seized and held by plants, while dissolved sodium ions are free to continue on to the sea. Potassium is absolutely essential to plant life, and the growth of wild plants is often limited by the slow extraction or dissolution of K^+ ions from the minerals of the soil. Even centuries ago, farmers found that their crops grew better when they spread wood ashes on the soil, without knowing why. What they were doing was returning in a concentrated form some potassium absorbed from the subsoil by trees. In modern agriculture the necessary potassium is supplied by fertilizer in calculated amounts. The earlier source still shows up in the name "potassium," however, for that is just a latinized form of the old word "potash."

Potassium hydroxide and carbonate can indeed be separated from wood ashes by leaching with hot water and concentrating the solution. Even dried seaweed can be burned to provide an ash for the same purpose, because the seaweed extracts and concentrates potassium despite the huge excess of sodium chloride in the water. Most potassium salts used in plant nutrition come from mines, however. The slow crystallization of salts from dried-up ancient European seas has left deposits of sylvite, KCl, and of mixed salts such as carnallite, $KCl \cdot MgCl_2$, and kainite, $KCl \cdot MgSO_4$, while at Searles Lake in California and Carlsbad in New Mexico polyhalite, $2CaSO_4 \cdot K_2SO_4 \cdot 2H_2O$, is mined along with sylvite.

Metallic potassium is not obtained by electrolysis of the fused chloride because the metal reacts with carbon electrodes. Instead, melted KCl (mp

776° C) is reduced by sodium vapor at 880° C in a steel tower, and potassium vapor is drawn off from the top. The process

$$Na(g) + K^+(l) \rightarrow Na^+(l) + K(g)$$

may seem contrary to the activity series of the metals learned in high school, wherein potassium is listed as a more active metal than sodium and should displace sodium from its salts. In practice, however, thermal considerations win over any difference in activity, because at 880° C an equilibrium is set up and both metals are present. Potassium is the more volatile one (bp 760° C vs. 889° C for Na) so it distills off far more readily, displacing the equilibrium and causing the reaction to proceed. By fractional distillation through a packed tower, potassium of 99.5% purity can be produced. Usually, however, a mixture of sodium and potassium is drawn off without fractionating, for the simple reason that mixtures (alloys) between 15 and 55% Na are liquid at room temperature, and such liquid alloys are much easier to transfer and use than solid blocks of either metal. An alloy of 25% Na and 75% K, familiarly known as NaK, or "nack," is an excellent heat-transfer fluid for nuclear reactors and for glass-molding operations. It also is a more active and more convenient reducing agent than either metal alone, and is used in organic synthesis and in commercial organic chemistry, especially for the production of detergents and the modification of fats.

Would we have expected an alloy of Na and K to be liquid at room temperature? This is a good place to pause to look at a simple phase diagram involving two metals, in order to answer that question. A phase diagram is merely a graph which displays many facts about temperature and physical state in an easy-to-read form. We shall consider a simple situation in which we are concerned only with solid and liquid, and not at all with vapor. We want to know how the melting point of a mixture of sodium and potassium varies as we change the proportions of metals. Someone has to prepare mixtures of Na and K by weighing out the proportions very carefully, and then melt the mixtures in an inert atmosphere such as helium. The melting point of each alloy is then measured. Once these experiments have been performed, a graph of melting point vs. composition is prepared, and the solids are examined under a microscope to see what phases have crystallized from the melt. The complete phase diagram then appears as in Figure 5–1, with the melting point curve (called the liquidus line) predominating and the areas of solid phases showing below that line.

The phase diagram tells us that as sodium is added to pure potassium the melting point drops sharply, as is usual (see effect of impurities on melting point, under Colligative Properties in your textbook). Similarly, the addition of potassium to pure sodium lowers its melting point, as shown at the upper right of the diagram. Since the two liquidus curves slope downward, they must intersect somewhere. The point at which they cross is seen to be at a composition of 22.7% Na and 77.3% K, corresponding to a melting point of −12° C. This is called the *eutectic point*, and since we started with fairly low melting points for Na and K, it falls considerably below room temperature. This much of a drop in melting point is not at all unusual. The fact that freezing occurs only at −12° C is very helpful in any process where NaK is used as a heat-transfer liquid, because the material will not freeze in pipes and pumps during shutdowns.

Phase diagrams such as those in Figure 5–1 are very compact and useful sources of information in all areas of metallurgy, and especially in designing new alloys. Note that potassium and sodium form an intermetallic compound, KNa_2. Such intermetallic compounds are very common, particularly among the transition metals. They are not compounds in the sense of a salt or a covalent liquid,

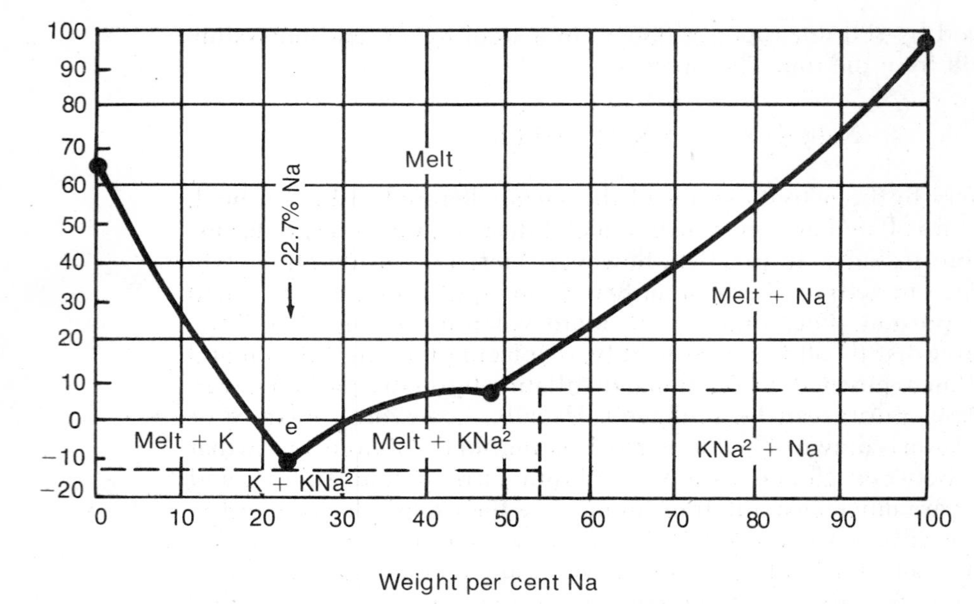

Figure 5–1. Phase diagram for sodium and potassium.

since they exist only in the solid state. They result from considerations of packing within a crystal lattice, more than from chemical attraction, and they are held together by an extension of the concept of metallic bonding.

Metallic potassium is much more active than sodium. It reacts violently with ice even at −100° C, and explosively with acids. When it burns in air it forms some superoxide, KO_2. It detonates upon contact with liquid bromine. It displaces hydrogen from ammonia to form the amide KNH_2, and with carbon monoxide it forms an explosive carbonyl, $K_2(OC{=}CO)$.

Some important potassium compounds are the natural minerals pointed out above, the hydroxide KOH (made by electrolysis of a solution of KCl), the carbonate K_2CO_3, and the nitrate KNO_3. The natural form of potassium nitrate (called nitre) was the oxidizing agent in the original gunpowder, invented by the Chinese several centuries before Christ and still used under the name of black powder. A mixture of 75% KNO_3, 15% charcoal, and 10% sulfur, all finely ground separately, leads upon ignition to the reactions

$$2KNO_3 + 4C \rightarrow K_2CO_3 + 3CO + N_2$$

and

$$2KNO_3 + 2S \rightarrow K_2SO_4 + SO_2 + N_2$$

in which three of the five products are gases, formed in large volume at high temperature. Black powder is not a good commercial explosive for blasting, and so has been replaced by the materials discussed under nitrogen.

RUBIDIUM, CESIUM, AND FRANCIUM. Rubidium is fairly abundant in the crust of the earth, to the extent of 310 parts per million, but it has never had any major use and is usually ignored. It occurs in many complex silicate minerals, often with lithium and cesium. Recent increased production of lithium from lepidolite, a rose-colored mica mineral, has provided a source, but rubidium compounds still cost about $25 per pound. The metal may be obtained from its chloride by electrolysis, or from its carbonate by reduction with magnesium;

the price is about $2,000 per pound. It is a very active metal which ignites when placed in chlorine or bromine, and decomposes water violently. It colors a Bunsen flame purple.

Cesium has interested inorganic chemists ever since its discovery because it is the heaviest alkali metal we can obtain in quantity for study, and hence the element that has the lowest electronegativity and forms the strongest base. It is quite rare but occurs in a distinctive mineral called pollucite, a cesium aluminum silicate, from which it is obtained by calcining with lime and extracting with water. After separating potassium as $KHCO_3$, cesium is separated by gradually adding tin tetrachloride and precipitating Cs_2SnCl_6. Cesium salts with large negative ions are less soluble than the corresponding rubidium and potassium salts, which in turn are less soluble than those of sodium and lithium. The name cesium comes from the Latin word for sky blue, and refers to the characteristic bright blue double line in its spectrum.

Metallic cesium is obtained by reducing Cs_2CO_3 with magnesium, or by electrolyzing the fused cyanide. It is a silvery metal with a melting point so low (28.5° C) that it usually remains liquid at room temperature. It is extremely active, and in moist air it will ignite from the heat of its own reaction with water vapor and oxygen. It burns in dry air to a mixture of Cs_2O, Cs_2O_2, and CsO_2; when the mixture is dissolved in water it evolves oxygen and leaves CsOH. The Cs^+ ion is large enough to crystallize with polyhalide ions, so black CsI_3 and bright orange $CsCl_2I$ are very characteristic.

Metallic cesium is used in the high-pressure arc lights that illuminate streets and parking lots, and in photoelectric cells (where it responds to visible light rather than ultraviolet, and with about the same color sensitivity as the human eye). The radioactive isotope ^{137}Cs is a fission product used in the treatment of cancer.

Francium, at the very bottom of Group I, was unknown until 1939. Various chemists had looked long and hard for it in pollucite and other likely alkali minerals, but it was never found. Then Marguerite Perey at the Curie Institute in Paris discovered that as uranium-235 decays to thorium by way of actinium-227, there is a minor route whereby 1% of the actinium decays by alpha emission to form the 223-isotope of element 87, which she named francium. Larger amounts of it can be made from radium by neutron bombardment. We know only that it behaves as an alkali metal; it has a soluble sulfide and carbonate, and it precipitates with $CsClO_4$ and Cs_2PtCl_6.

GROUP II, THE ALKALINE-EARTH METALS

To the alchemists and the philosophers of the middle ages, an *earth* was any solid substance that did not melt and was not changed by fire into some other substance. They recognized alkalies (a good alchemical name right out of Arabic) by their bitter taste, their astringent effect, and their ability to neutralize acids. Some alkalies, however, melted and reacted with their fireclay containers; KOH (mp 360° C), K_2CO_3 (mp 891° C), and NaOH (mp 318° C) will do this. On the other hand, $CaCO_3$ and $Ca(OH)_2$ did not melt but changed into a white powder in the fire. This powder we recognize as lime, CaO. It melts at 2572° C, a temperature far above any an alchemist could attain. Yet it is a strong alkali, and hence the name *alkaline earth*, which has stuck as a label for the Group II elements to this day.

The characteristics of the alkaline-earth elements are listed in Table 5–2. The metals are all harder, stronger, and higher-melting than the Group I metals

TABLE 5–2 GROUP II: THE s^2 ELEMENTS (ALKALINE-EARTH METALS)

	Beryllium	*Magnesium*	*Calcium*	*Strontium*	*Barium*	*Radium*
Symbol	Be	Mg	Ca	Sr	Ba	Ra
Atomic number	4	12	20	38	56	88
Atomic weight	9.012	24.31	40.08	87.62	137.34	226.05
Valence e^-	$2s^2$	$3s^2$	$4s^2$	$5s^2$	$6s^2$	$7s^2$
mp, °C	1283	650	845	770	725	700
bp, °C	2970	1120	1420	1380	1640	1140
d, g/cm³	1.85	1.74	1.55	2.60	3.51	5
Atomic volume	5.0	14.0	29.9	33.7	39.0	—
Atomic radius, Å	1.11	1.60	1.97	2.15	2.17	2.20
Ion radius, Å	0.31	0.65	0.99	1.13	1.35	—
Pauling EN	1.5	1.2	1.0	1.0	0.9	0.9
Z_{eff}/r^2	1.47	1.23	1.04	0.99	0.97	0.97
Standard potential	−1.85	−2.37	−2.87	−2.89	−2.90	−2.92
Oxidation states	+2	+2	+2	+2	+2	+2
Ionization energy*	215	176	141	131	120	—
Heat of vaporization	73.9	30.6	40.1	36.8	35.7	—
Discoverer*	Vauquelin	Bussy	Berzelius	Davy	Davy	Curie
Date of discovery	1798	1831	1808	1808	1808	1911
rpw* pure O_2	BeO	MgO	CaO	SrO, SrO_2	BaO_2	RaO
rpw H_2O	none	none	$Ca(OH)_2$	$Sr(OH)_2$	$Ba(OH)_2$	$Ra(OH)_2$
rpw N_2	none	Mg_3N_2	Ca_3N_2	Sr_3N_2	Ba_3N_2	Ra_3N_2
rpw halogens	BeX_2	MgX_2	CaX_2	SrX_2	BaX_2	RaX_2
rpw H_2	none	MgH_2	CaH_2	SrH_2	BaH_2	—
Flame color	—	—	brick red	crimson	green	—
Mohs hardness	4	2.0	3	1.8	1.5	—
Crystal structure	hexagonal	hexagonal	cubic fc*	cubic fc	cubic bc*	—

*rpw = reaction product with; fc = face-centered; bc = body-centered. "Discoverer" refers to first isolation; Mg, Ca, and Ba were known to ancients. All energies and heats are in kcal per mole; potentials are in volts.

because twice as many electrons are available for bonding. The atomic volumes are smaller, and much more energy is required to vaporize or to ionize 1 mole of the Group II metal. The electronegativities cover a wider range than in Group I, mostly because the value for Be is so high. The standard electrode potentials *increase* with increasing atomic weight; in Group I they decline.

Chemically, the alkaline-earth elements show a wider range of properties than is found in Group I. We saw that beryllium is seldom ionic in its compounds, and is given over almost entirely to covalency; at the other extreme, barium ranks with sodium and potassium in the base strength of its hydroxide, its electrode potential, and its strongly ionic behavior. Furthermore, beryllium and magnesium are unaffected by water and are customarily used in contact with it, but calcium will decompose boiling water

$$Ca(s) + 2H_2O(l) \rightarrow Ca^{+2}(aq) + 2OH^-(aq) + H_2(g)$$

and barium reacts violently with cold water

$$Ba(s) + 2H_2O(l) \rightarrow Ba^{+2}(aq) + 2OH^-(aq) + H_2(g)$$

All the Group II metals except beryllium react with nitrogen at elevated temperatures to form ionic nitrides, and all except beryllium form hydrides containing H^- ions. All form organometallic derivatives with carbon-metal bonds, but the stabilities of such compounds decline in the order Be-Ba while their reactivities rise in the same order.

Having considered beryllium and magnesium earlier, we shall take a look now at the rest of the family: calcium, strontium, barium, and their radioactive relative, radium.

Calcium is the most common and widespread member of this family, and is the fifth most abundant element on earth. It is everywhere: in the stones as calcium silicate, carbonate, and phosphate; in plants (where it controls water passage through cell membranes); in bones and teeth as $CaCO_3$, $Ca_5(PO_4)_3OH$, and $Ca_5(PO_4)_3F$; and of course in glass and bricks and porcelain, which are mostly calcium aluminum silicates. There is more calcium than sodium or potassium on earth and in the stars and everywhere.

For all this abundance and uniqueness, calcium remains a ho-hum element for most students. We know it geologically as the carbonate in marble, chalk, limestone, and calcite, all of which are deposited in the sea and reformed on land because $CaCO_3$ is so insoluble (0.00014 g/liter at 20° C). It became known centuries ago that all these forms of $CaCO_3$ could be "burned" to quicklime when heated red-hot in a fire, and that the resulting lime made a good mortar when it was mixed with sand and water. In chemical terms, the carbonate was converted to oxide which then hydrated to the strongly alkaline hydroxide, and this slowly picked up carbon dioxide from the atmosphere as the mortar dried and aged, until it reverted at last to something like the original limestone:

$$CaCO_3(s) \xrightarrow{850^\circ C} CaO(s) + CO_2(g)$$

$$CaO(s) + H_2O(l) \rightarrow Ca(OH)_2(s)$$

$$Ca(OH)_2(s) + CO_2(g) \rightarrow CaCO_3(s) + H_2O(g)$$

The Romans made good use of these processes as they built houses and temples and massive aqueducts out of limestone and marble, and of course it was an economical operation because the chips from stonecutting were "burned" on the site to form the mortar for holding the blocks together. Where they did not have limestone or marble for their structures, as in central France, they dug chalk from the ground to make mortar for cementing sandstone blocks, and the enormous caves they dug in the soft chalk to get their $CaCO_3$ remain to this day as the places where champagne is made and aged, millions of bottles in a single cave.

Unfortunately, "burning" $CaCO_3$ to make lime requires a great deal of fuel, and the only fuel available to the Romans was charcoal. It must have taken entire forests to provide the fuel, and denuding the land on such a scale leads to erosion and impoverishment, which lead to poverty and unrest. Later, after the barbarian invasions, the buildings and monuments themselves became a quarry for building materials and mortar to construct houses and walls. By patient investigation, much European history can be traced through the travels of identifiable samples of calcium carbonate.

Metallic calcium is produced in considerable quantity by electrolysis of its chloride (mp 772° C) and is used as a hardening agent in lead and magnesium alloys. It also is used as a reducing agent in the metallurgy of chromium and uranium, and as a deoxidizer and cleanup agent in special steels. In this application calcium combines with both nitrogen and oxygen, both of which weaken the steel if left in it:

$$2Ca + O_2 \rightarrow 2CaO$$

$$3Ca + N_2 \rightarrow Ca_3N_2$$

Calcium nitride and hydride react with water vigorously to give ammonia and hydrogen:

$$Ca_3N_2(s) + 6H_2O(l) \rightarrow 3Ca(OH)_2(s) + 2NH_3(g)$$

$$CaH_2(s) + 2H_2O(l) \rightarrow Ca(OH)_2(s) + 2H_2(g)$$

The carbide, CaC_2, we have already discussed; it is the large-scale precursor of calcium cyanamid for nitrogen fixation, and of acetylene for industrial organic chemistry. The sulfate we have considered under sulfur; its dihydrate gypsum, $CaSO_4(H_2O)_2$, is mined extensively for the production of portland cement and of gypsum wallboard. Moderate heating of gypsum produces plaster of paris, a hemihydrate:

$$2CaSO_4(H_2O)_2(s) \rightarrow (CaSO_4)_2H_2O(s) + 3H_2O(g)$$

When this plaster is mixed with enough water to make a paste, it hardens in a few minutes as it reverts to the dihydrate. The mixture also expands as it hardens, so it forms a sharp impression of anything molded in it. Artists and hobbyists in ceramics are familiar with such plaster molds and their use in the slip-casting of clay.

Strontium never comes into the public eye except as the vaguely ominous strontium-90 isotope, but it has a few chemical lessons to teach. It is a good example of an even-numbered element with three even-numbered stable isotopes, so it is as abundant as the light elements sulfur and chlorine. However, it is so mixed up with calcium and barium that it is spread out thinly in the mineral world. It resembles calcium chemically, and forms a nitride that yields ammonia upon hydrolysis and a carbide that yields acetylene. Most students remember strontium as the element that imparts a spectacular crimson color to a Bunsen flame, and that property still provides strontium with its greatest use—in red flares for highway and railway danger signals, and in fireworks.

Strontium-90 is a major fission product of uranium and plutonium, so it is scattered into the atmosphere whenever nuclear fission weapons are exploded in air. High-altitude winds carry it around the earth, and it settles on pastures and farmland. The danger arises when it gets into the food supply. For example, if ^{90}Sr is ingested by a cow it gets into her milk, and when the milk is drunk the strontium follows the calcium in the milk as it goes into building bones and teeth. Once lodged in the bone structure, ^{90}Sr goes on emitting beta particles because it is intensely radioactive, with a half-life of only 29 years. The radiation damages bone marrow, where red blood cells are made. This situation is precisely like the one encountered years ago by people who worked with radium salts and had radium lodged in their bones; the radioactivity is almost impossible to get rid of, and if severe enough it leads to anemia and cancer. For these reasons a ban on atmospheric testing of nuclear weapons was sought and a test-ban treaty eventually was signed by Britain, the U.S., and the U.S.S.R. However, France and China have not signed the treaty, and they continue to test nuclear weapons in the air, so radioactive debris continues to be put in the atmosphere and some danger still persists. The danger is visited mostly on growing children, whose bones are in the formative stage.

Barium gets its name from the Greek word for heavy, and all its compounds are uncommonly dense. For example, $BaSO_4$ has a density of 4.50 g/cm^3, vs. 2.96 g/cm^3 for $CaSO_4$; barium carbonate has a density of 4.43, compared with 2.71 for $CaCO_3$. Hence barium sulfate and carbonate are mined and put into concrete used to shield nuclear reactors, for the simple reason that this type of shield interposes more mass of material per unit thickness between the radiation source and the outside world, and hence is a more efficient shielder than ordinary concrete. Similarly, finely divided highly purified $BaSO_4$ is administered by a radiologist before making an x-ray examination of a patient's gastrointestinal tract, because the dense barium sulfate absorbs x-rays and enables the radiologist to distinguish the alimentary system over surrounding tissue.

Natural $BaSO_4$ has another use; it is reduced to barium sulfide by heating with carbon, and a filtered solution of the sulfide then reacts with a solution of zinc sulfate (from zinc oxide and sulfuric acid) to precipitate barium sulfate and zinc sulfide:

$$BaSO_4(s) + 4C(s) \xrightarrow{800^\circ C} BaS(s) + 4CO(g)$$

$$\underset{\text{(all ions in water solution)}}{Ba^{+2} + S^{-2} + Zn^{+2} + SO_4^{-2}} \rightarrow \underset{\text{(coprecipitation of two solids)}}{BaSO_4(s) + ZnS(s)}$$

The latter reaction is helped toward completion by the extremely low solubility of both products. Both also are white (insoluble white sulfides are *very* unusual; there is only one other insoluble white sulfide among all the elements, and that is GeS_2) and both are quite dense and refractive, so the combination is a white pigment called lithopone.

Barium oxide reacts so violently with water that lumps of it become red-hot as they absorb steam to form $Ba(OH)_2$. The resulting hydroxide is a very strong base. Barium oxide differs from CaO and SrO in absorbing oxygen when it is hot, forming the peroxide, BaO_2. This liberates hydrogen peroxide when it meets cold water, and thus is a bleaching agent:

$$2BaO(s) + O_2(g) \rightarrow 2BaO_2(s)$$

$$BaO_2(s) + 2H_2O(l) \rightarrow Ba^{+2} + 2OH^- + H_2O_2(aq)$$

Metallic barium emits electrons easily when heated, so alloys of it are used for spark plug electrodes and for cathodes in vacuum tubes. Some barium oxide is used in glassmaking, because it gives a dense and highly refractive (sparkling) glass.

Radium, the heaviest of the Group II metals, was first recognized as an element by Marie and Pierre Curie in 1898. They separated it from a Bohemian uranium ore called pitchblende, where it occurs to the extent of one part for every 3,000,000 parts of uranium. The pure metal can be isolated by electrolyzing a solution of $RaCl_2$ with a negative electrode of mercury. The radium ions discharged at the surface of the mercury dissolve in it, and later evaporation of the mercury in an atmosphere of hydrogen leaves silvery white radium, a very active metal which reacts with water to form a strong base:

$$Ra(s) + 2H_2O(l) \rightarrow Ra^{+2}(aq) + 2OH^-(aq) + H_2(g)$$

It also reacts rapidly with air, forming a black coating of nitride and oxide. All pure radium compounds glow in the dark because their radiations ionize the surrounding air. The same radiations are harmful to living organisms because of the ionization they produce in the tissue. Since tumor tissue is young and embryonic, in a sense, it is more sensitive to radiation damage than older, tougher tissue, and so malignant tumors can sometimes be treated with radium, provided they are not close to vital organs that would be harmed. Slender glass tubes containing radium salts or radon (the gaseous decay product of radium) are implanted in the tumor for a calculated time. Penetrating x-rays or radiation from a gamma ray source can be used similarly in nonsurgical procedures.

Inhaling or ingesting even the tiniest amount of radium is harmful to the human body because Ra^{+2} ions are deposited in the bone structure along with Ca^{+2} ions, as described above in connection with ^{90}Sr damage. Radium is particularly devastating to bone marrow because its principal isotope emits alpha particles of high energy (4.79 meV) that cause a tremendous amount of ionization in their short paths through the marrow tissue. Early experience among the workers who painted luminous figures on watch dials, using radium-activated phosphorescent pigment, showed that radium deposited in bone tissue could lead to cancer even 30 years after exposure. Such experiences, plus all the painful and tragic results of overexposure to x-rays and radioactive materials before the dangers were fully known, have fortunately led to well-established safety regulations for all who handle radioactive tracers or are exposed in any way. If you use radioactive substances of any sort in chemical or biological experiments, be sure to follow the directions and the regulations of your college implicitly. Radiation is implacable and unforgiving.

GROUP III, THE ALUMINUM FAMILY

The properties of the Group III elements are listed in Table 5–3. The elements themselves range from a hard, black, inert metalloid to a soft, reactive metal, with enormous range of melting and boiling points. The electronegativities do not vary monotonically—aluminum is much more electropositive than boron or indium. Although all five elements exhibit +3 oxidation states, boron forms only covalent compounds and its formal oxidation states in the hydrides are variable. The last three elements form +1 compounds with halogens, oxygen, and sulfur, but the compounds of monovalent Ga and In are unstable and difficult to prepare in the pure state. The +1 compounds of Tl, on the other hand, are by far the most stable ones, and trivalent thallium reverts readily to the monovalent state both in solution and in the dry compounds. All five elements are unreactive toward air and water at room temperature, although thallium slowly forms a thick black coating of oxide. When heated in pure oxygen, all react at the temperature given in Table 5–3 to form the oxides shown. Boron and nitrogen combine directly at the high temperatures shown, but the nitrides of gallium and indium have to be made by indirect methods:

$$Ga_2O_3(s) + 2NH_3(g) \xrightarrow{600^\circ\,C} 2GaN(s) + 3H_2O(g)$$

$$(NH_4)_3InF_6(s) \rightarrow InN(s) + 6HF(g) + 3H_2(g) + N_2(g)$$

Similarly, boron does not combine directly with hydrogen, but a number of boron hydrides can be made by indirect methods (see Chapter 3). Under pres-

TABLE 5–3 GROUP III: THE p^1 ELEMENTS (THE ALUMINUM FAMILY)

	Boron	*Aluminum*	*Gallium*	*Indium*	*Thallium*
Symbol	B	Al	Ga	In	Tl
Atomic number	5	13	31	49	81
Atomic weight	10.81	26.98	69.72	114.82	204.37
Valence e⁻	$2s^22p^1$	$3s^23p^1$	$4s^24p^1$	$5s^25p^1$	$6s^26p^1$
mp, °C	2300	660	29.78	156.6	303.5
bp, °C	2550	2327	2403	2000	1457
d, g/cm³	2.34	2.70	5.91	7.31	11.85
Atomic volume	4.6	10.0	11.8	15.7	17.2
Atomic radius, Å	0.88	1.43	1.22	1.62	1.71
Ion radius, Å		0.50	0.62	0.81	0.95
Pauling EN	2.0	1.5	1.6	1.7	1.8
Z_{eff}/r^2	2.01	1.47	1.82	1.49	1.44
Standard potential	−0.90	−1.66	−0.56	−0.34	−0.33†
Oxidation states	covalent 3	+3	+1, +3	+1, +2, +3	+1, +3
Ionization energy*	191	138	138	133	141
Heat of vaporization	128	67.9	68.8	55.4	38.8
Isolated by	Gay-Lussac	Wöhler	Boisbaudran	Reich	Crookes
Date of isolation	1808	1827	1875	1863	1861
rpw* pure O_2	B_2O_3, 1200° C	Al_2O_3, 800° C	Ga_2O_3, 1600° C	In_2O_3, 600° C	Tl_2O, 400°C
rpw H_2O	none	none	none	none	none
rpw N_2	BN, 1200° C	AlN, 740° C	see text	see text	none
rpw halogens	BX_3, 400° C	Al_2X_6, 200° C	Ga_2X_6	In_2X_6	TlX
rpw H_2	see text	see text	see text	see text	see text
Flame color	grass green		violet	blue violet	bright green
Mohs hardness	9.3	2.5	2.0	0.9 B*	2 B
Crystal structure	hexagonal	cubic fc*	cubic	tetrag.	hexagonal

*Energies and heats are in kcal per mole; rpw = reaction product with; B = Brinell hardness, measured by indentation; Mohs hardness by scratching; fc = face-centered.

†Tl potential is for change to Tl^+; from this to Tl^{+3} involves +1.28 volts.

sure and with an organoaluminum catalyst, H_2 is taken up by aluminum powder to form organoaluminum hydrides such as $(CH_3)_2$ AlH, but isolation of a pure hydride is difficult. A polymeric white solid of the composition $(AlH_3)_n$ is formed by the reduction of Al_2Cl_6 by $LiAlH_4$, which in turn is made by the reduction of Al_2Cl_6 by lithium hydride in ether solution:

$$Al_2Cl_6 + 8LiH \xrightarrow{\text{Ether}} 2LiAlH_4 + 6LiCl(s)$$

Other aspects of boron and aluminum have been discussed in previous chapters.

Gallium is promising but unemployed. Its ions are spread all over the mineral world, and it is difficult to concentrate and isolate. The best sources are ores of zinc, germanium, and aluminum, but it costs about $500 per lb of metal to get the gallium out. It is an interesting metal, though; it is liquid at room temperature and all the way up to 2403° C, where it boils. This is the greatest liquid range known among all substances. Furthermore, gallium does not oxidize readily, so it can be exposed to air up to 1600° C before it is affected seriously. Surely someone needs a metal that stays liquid over so wide and so convenient a range, but so far no one has called for it. It was once proposed as a replacement for mercury in thermometers so that they could be used up to 1000° C if the thermometers were made of silica glass, but it turned out that the gallium wet the glass and stuck to it. Anyone who thinks up a good use should talk to the aluminum producers, who might then think about large-scale recovery from bauxite.

Indium is a rare element with a checkered career. It is found in zinc ores, from which it can be separated by a lengthy process ending in electrolysis of a chloride solution. It is a bluish soft metal which forms low-melting alloys with silver and lead. Such alloys conduct heat well, and were used for a time to line the journal bearings of aircraft engines. Chemically, indium is the only element in this group to form isolable divalent compounds; InF_2 is known, and impure $InCl_2$ has been made.

Thallium is a heavy metal that looks and acts like lead, from which it can be separated by electrolysis of its sulfate. In its univalent state thallium behaves like silver in having a white insoluble chloride but a soluble fluoride, and it behaves like the alkali metals in having a soluble and strongly basic hydroxide, TlOH. In the trivalent state it behaves like aluminum in forming colorless Tl^{+3} salts, but it resembles iron in forming a brown oxide, Tl_2O_3.

Thallium trichloride is slowly formed from the elements at 15° C, but when the resulting $TlCl_3$ is warmed to 40° C it decomposes to the more stable monochloride:

$$TlCl_3(s) \rightarrow TlCl(s) + Cl_2$$

In aqueous solution thallium in the +1 state is decidedly more stable than the +3 state, and the reaction

$$Tl^{+3}(aq) + 2e^- \rightarrow Tl^+(aq)$$

has a standard potential of +1.25 volts in 1 M $HClO_4$. That is, Tl^{+3} is an oxidizing agent as strong as manganese dioxide, and almost as strong as chlorine.

Thallium is as toxic* as lead, and its use has been curtailed. All heavy elements, such as mercury, lead, and thallium, are cumulative poisons; they collect in the body because there is no mechanism for excreting them, and within the body their ions concentrate at cell interfaces and cause severe nervous disorders.

GROUP IV, THE CARBON-TO-LEAD GROUP

In Chapter 1 we had a brief look at the properties of the elements of Group IV, which sits in the middle of the two short periods. A more complete listing of the pertinent data for these elements is given in Table 5–4. The properties given there are for the more common forms of the elements; it should be borne in mind that the graphite and diamond forms of carbon are radically different in chemical as well as physical properties, and that gray tin (a loose gray powder) is not at all like the soft, compact, dense metallic form called white tin.

The Group IV elements vary widely. Carbon, a distinct nonmetal with an acid-forming oxide, combines with hydrogen and chlorine in all ratios from 4:1 to 1:1, so its formal oxidation state ranges from −4 to +4. It is ionic in only a few carbides (see Chapter 3); in almost all its millions of compounds carbon forms four covalent bonds, some of them usually to other carbon atoms. Silicon, which is far more electropositive than carbon, forms thousands of tetracovalent compounds and untold numbers of ionic silicates. Divalent silicon occurs only in unstable SiO and in the polymeric dihalides SiF_2 and $SiCl_2$. Germanium com-

*See Agatha Christie, *The Pale Horse*, London: Collins (1961).

TABLE 5-4 GROUP IV: THE p^2 ELEMENTS (THE CARBON GROUP)

	Carbon	*Silicon*	*Germanium*	*Tin*	*Lead*
Symbol	C	Si	Ge	Sn	Pb
Atomic number	6	14	32	50	82
Atomic weight	12.011	28.086	72.59	118.69	207.19
Valence e^-	$2s^2 2p^2$	$3s^2 3p^2$	$4s^2 4p^2$	$5s^2 5p^2$	$6s^2 6p^2$
mp, °C	3570	1414	937	232 a*	328
bp, °C	sublimes	2355	2830	2270	1750
d, g/cm³	2.25 b*	2.33	5.32	7.30 a	11.35
Atomic volume	5.3	12.1	13.6	16.3	18.3
Atomic radius, Å	0.77	1.17	1.22	1.40	1.75
Ion radius (+2), Å			0.73	0.93	1.21
Pauling EN	2.5	1.8	1.8	1.8	1.8
Z_{eff}/r^2	2.50	1.74	2.02	1.72	1.55
Standard potential	+0.39	+0.10	−0.3	−0.15	−0.126
Oxidation states	−4 to +4	−4, +2, +4	−4, +2, +4	−4, +2, +4	+2, +4
Ionization energy*	260	188	187	169	171
Heat of vaporization	171.7	71.0	79.85	61.72	42.06
Isolated by	antiq.	Berzelius	Winkler	antiq.	antiq.
Date of isolation	antiq.	1824	1886	antiq.	antiq.
rpw* pure O_2	CO, CO_2, 600° C	SiO_2, 1200° C	GeO_2, 1000° C	SnO_2, 800° C	PbO, 600° C
rpw H_2O	none	none	none	none	none
rpw N_2	none	Si_3N_4, 1400° C	none	Sn_3N_4, 2000° C	none
rpw halogens	CX_4, 800° C	SiX_4, 400° C	GeX_4, 400° C	SnX_2, 400° C	PbX_2, 400° C
rpw H_2	CH_4, 1000° C	none	none	none	none
Flame color	none	none	none	none	none
Mohs hardness	1 b, 10 d*	7.0	6	3.9 B*	4.2 B
Crystal structure	hexagonal b	cubic d	cubic d	tetrag. a	cubic fc*

*Energies and heats are in kcal/mole; a indicates data for white tin (gray tin has the diamond structure, density 5.75, and is stable only below 18° C); (b) indicates data for graphite form (diamond has density 3.51); B = Brinell hardness by indentation; d = diamond structure; fc = face-centered; potentials are in volts.

pounds are mostly tetracovalent, but under strongly reducing conditions germanium forms a well-established series of Ge(II) compounds such as $GeCl_2$, $GeBr_2$, GeO, and GeS. Ge(IV) remains the more stable oxidation state throughout. Tin divides itself between Sn(II) and Sn(IV) compounds, depending on whether the conditions are oxidizing or reducing. Its chief ore is a +4 compound, SnO_2. Organotin compounds are almost always tetracovalent. Lead shows a marked preference for Pb(II) compounds. Thus $PbCl_2$ is stable, but $PbCl_4$ explodes at 105° C. Lead ores are +2 compounds, PbS and $PbCO_3$. The Pb(IV) state is stable only in the higher oxides and in organolead compounds, such as $(C_2H_5)_4Pb$.

We see that there is a steady shift in stability toward the lower oxidation state among the heavier elements, with the emergence of the +2 state to dominate the chemistry of lead. It is the *p* electrons which are used in the compounds of divalent tin and lead, while the *s* electrons remain idle. This phenomenon is often referred to as the "inert pair effect," and we shall meet it again.

Germanium is a rare element which had a brief but intense popularity when its semiconductor properties were discovered 40 years ago and exploited in the first transistors. It is a brittle metalloid, somewhat more yellowish than silicon and of course more dense. It occurs with zinc and arsenic, and as a complicated sulfide with copper and silver. It is extracted by chlorinating the roasted ore and distilling $GeCl_4$ (bp 83° C) to free it from all the metals with nonvolatile chlorides. Hydrolysis gives GeO_2, which can be reduced by sugar charcoal (a pure form of carbon) under a flux of sodium chloride:

$$\underset{\text{fuming liquid}}{GeCl_4} + 2H_2O(l) \rightarrow \underset{\text{white solid}}{GeO_2} + 4H^+ + 4Cl^-$$

$$GeO_2(s) + 2C(s) \rightarrow \underset{\text{metal}}{Ge(l)} + 2CO(g)$$

Germanium melts at about the same temperature as NaCl (937° C), so it is obtained as fused pellets. These are cast into bars and zone-refined to high purity (see silicon). Germanium transistors function very well, and in fact they opened up the whole field of solid-state electronics, but they are inherently expensive and fail at elevated temperatures. On both counts they were bettered by transistors made from hyperpure silicon, which has taken over.

Just as silicon was seen to be not merely an oversized carbon, so germanium is different from silicon: it is more electro*negative* (where ordinary considerations of group properties would call for more electropositive behavior as the elements get heavier). It occurs with arsenic and copper (not with silicon). Its hydrides are so unreactive toward water that GeH_4 can be prepared in water solution:

$$GeO_2 + BH_4^- + 2H_2O \rightarrow GeH_4(g) + OH^- + H_3BO_3(aq)$$

Notice that GeO_2 is sufficiently soluble in water to be used in this way, whereas the solubility of SiO_2 is exceedingly low.

Tin was one of the nine elements known to the ancients, not because it occurred free in nature, like silver or gold, but because its oxide is readily reduced by the glowing coals of a wood fire:

$$SnO_2 \text{ (the mineral cassiterite)} + 2C(s) \rightarrow Sn(l) + 2CO(g)$$

The low-melting, unreactive metal ran out from the fire and was easily recovered. Tin was needed for bronze, which is an alloy of about 20% tin and 80% copper, and is an enduring metal harder than either of its components. When we think of the bronze age, and then of the tools, household implements, and weapons of Greece and Rome, we should think in terms of the copper that came from the Mediterranean area, the tin that came from Britain, and the brisk trade and cultural exchange that developed between the two. The Phoenician traders carried copper from Cyprus (whence the name copper) to Egypt and to Greece, and they sailed to Britain for tin. Both metals were scarce and expensive then; tin still is, costing five times as much as copper and about one tenth as much as silver. It is valuable enough to be recovered from "tin" cans, which are made from tin-plated steel. Tin is still used in making marine bronze, in solder (a 50-50 alloy of tin and lead), and of course for coating cans. In a recent process for making plate glass, the hot glass is cast directly from the furnace on a pool of molten tin, which always has a mirror surface so smooth that the glass does not need to be ground and polished later.

Pure tin is quite unreactive to air and water, although it will eventually tarnish. The metal itself is nontoxic, and was widely used during the Middle Ages (by those who could afford it) for plates and cups and platters. Many of these can now be seen in museums, where they shine almost as attractively as silver when freshly polished. Colonial pewter is a close relative, made from about 85% tin and the balance lead or a combination of copper, zinc, and antimony. Pewter persisted until the introduction of cheap china during the early nineteenth century, and now is enjoying a revival. The only other use for tin in Western culture is for organ pipes, where it is prized for its tone and its precise tunability.

"Tin disease," which affects tin organ pipes in unheated churches, is actually a phase transformation from the white or metallic form of the element (stable from the melting point, 232° C, down to 13.2° C) to the allotropic gray form, a crumbly powder of much lower density. The "disease" affects only very pure tin, and only if the tin is kept at below-freezing temperatures for long periods of time. Nevertheless, it has been a troublesome thing in expensive and historically famous organs.

The chemistry of tin is divided between the divalent compounds ($SnCl_2$, SnF_2, SnO, SnS, etc.) and the tetravalent compounds ($SnCl_4$, SnO_2, SnS_2, etc.). When tin dissolves in hydrochloric acid the evolved hydrogen keeps it in the reduced divalent state

$$Sn(s) + 2H^+ \rightarrow Sn^{+2} + H_2(g)$$

but when tin dissolves in hot nitric acid or in hot concentrated sulfuric acid it forms tetravalent compounds:

$$3Sn(s) + 4NO_3^- + 16H^+ \rightarrow 3Sn^{+4} + 4NO(g) + 8H_2O$$

$$Sn(s) + 2SO_4^{-2} + 8H^+ \rightarrow Sn^{+4} + 2SO_2(g) + 4H_2O$$

How do the two series of tin compounds differ? In the first place, Sn(II) compounds are obviously reducing agents, because any oxidizing agent or oxidizing atmosphere will convert them to Sn(IV) compounds. Similarly, Sn(IV) compounds are oxidizing agents because they react with reducing agents (such as hydrogen, in the example above) to become Sn(II) compounds. Secondly, Sn(II) compounds are salt-like solids in which tin behaves as a true metal, whereas Sn(IV) compounds are often covalent and resemble the analogous compounds of a metalloid, or even a nonmetal. In other words, tin becomes less metallic in the higher oxidation state. Take the chlorides as examples; $SnCl_2$ is a white, salt-like solid (mp 246° C, bp 623° C, soluble in water), but $SnCl_4$ is a colorless volatile liquid (mp –33° C, bp 114° C, soluble in ether and hydrocarbons) which closely resembles $SiCl_4$ and PCl_3.

We shall meet many other illustrations of similar behavior in other elements, with no exceptions to the trend. In fact, a general principle can be stated: *any multivalent element becomes less metallic and less electropositive in its higher oxidation states, and its compounds become stronger oxidizing agents and form stronger acids.* The reason for this is easily understood—it is progressively more and more difficult to remove successive electrons from an atom. When a tin atom loses two electrons to form an Sn^{+2} ion, the positive charge makes it more difficult to remove two more electrons and form an Sn^{+4} ion. This expectation receives clear quantitative support in the successively higher ionization energies for the electrons of any element.* Greater difficulty of removing electrons and increased tendency to gain electrons are the hallmark of the more electronegative elements at the upper right of the Periodic Table, so any element in its higher oxidation states becomes more like these. This is one of the logical guiding principles of inorganic chemistry. If kept in mind, it will simplify and help to organize a great deal of descriptive chemistry.

Lead is another metal known to the ancients as far back as 3000 B.C. Its chief source is a distinctive heavy black mineral called galena, PbS, which was easy for early people to identify and then easy to reduce to the metal. It was simply roasted in limited air, and then heated with powdered charcoal:

$$2PbS(s) + 3O_2(g) \rightarrow 2PbO(s) + 2SO_2(g)$$

then

$$PbO(s) + C(s) \rightarrow Pb(l) + CO(g)$$

and

$$PbO(s) + CO(g) \rightarrow Pb(l) + CO_2(g)$$

The low-melting metal (mp 328° C) ran out of the mass of charcoal and earthy impurities, and was easily hammered out into sheets which were then cut and joined at the edges by melting. Lead pipes were used for the water supply of Roman villas, and the roof of the Pantheon in Rome was sheathed in lead. Even household vessels and wine storage casks were lined with lead. Many examples of the use of lead in the Middle Ages will be found in old cathedrals, which usually had flashings and gutters made of lead.

Lead is still recovered from its ores in much the same way as formerly, but the plentiful sources are gone and the galena now has to be concentrated by flotation before reduction. Fresh galena is added to the roasted ore to do the reducing, making carbon unnecessary:

$$PbS(s) + 2PbO(s) \rightarrow 3Pb(l) + SO_2(g)$$

The lead so obtained often contains silver, copper, antimony, and bismuth. Most of the copper separates just above the freezing point of the lead, and can be skimmed off. Silver is recovered by stirring the molten lead with molten zinc, which dissolves out the silver and then floats on top as a lighter Zn-Ag alloy which can be separated. The zinc is recovered by distillation in the absence of air (it boils at 907° C, whereas silver boils at 2193° C), and is used over again. Antimony and bismuth can be removed by selective oxidation with air, since they oxidize more readily than lead, but more often extra antimony or bismuth is added in order to make a harder lead alloy.

The largest use of metallic lead is for storage batteries, where it forms the supporting grids for the oxidizing agent (PbO_2) and the reducing agent (spongy lead). Here an alloy of 91% lead and 9% antimony is used. Nearly all the alloy from old storage batteries is recovered and used again, an excellent example of recycling. The second largest use of lead is for making tetraethyl lead, $(C_2H_5)_4Pb$, which is the chief antiknock additive for gasoline. Here lead is melted with sodium to make an alloy which is 90% Pb and 10% Na by weight, and equimolar in Pb and Na. Pellets of this are heated with ethyl chloride, C_2H_5Cl, under pressure:

*See table of ionization energies (for removal of first to eighth electron) for all the elements in B. E. Douglas and D. H. McDaniel, New York: *Concepts and Models of Inorganic Chemistry,* Blaisdell Publishing Co. (1965), pp. 34–36.

$$4NaPb(s) + 4C_2H_5Cl(l) \rightarrow (C_2H_5)_4Pb(l) + 4NaCl(s) + 3Pb(s)$$

Water is then added, and the tetraethyl lead is steam-distilled off. The residual lead is filtered out and reused.

About 3 ml of $(C_2H_5)_4Pb$ added to each gallon of gasoline will raise the octane rating* of the gasoline 10 points, permitting a higher compression ratio to be used in the engine and resulting in a great increase in efficiency. This was a major chemical development of the 1930's, and it promised a 45% saving in gasoline for the same engine horsepower and vehicle weight, thereby saving what some saw clearly to be a rapidly dwindling supply of native petroleum.† Unfortunately, the American public saw fit to take its dividend from this development in the form of increased engine power, and that called for heavier frame and running gear, leading to ever-larger engines and irresponsible waste of gasoline. Now that our resource is disappearing and American production of petroleum is declining, and the costs and dangers of dependence on imported oil are painfully clear, the movement is back toward automotive designs of the 1930's or the small cars of Europe and Japan. Alas, the petroleum that is gone after the horsepower race will never be available to make textiles or plastics, or even the basic foodstuffs that will sorely be needed in the future.

The use of excessive quantities of leaded gasoline for three decades had two other effects—it depleted our lead resources, and it exposed the entire urban population to incipient lead poisoning. After $(C_2H_5)_4$ Pb has done its job of suppressing detonation in an automobile engine, the lead remains as PbO, PbS, and $PbSO_4$. To prevent accumulation of these solids in the combustion chamber, they are converted to the more volatile $PbBr_2$ by the action of dibromoethane, $C_2H_4Br_2$, which is added to the gasoline for that purpose. Most of the $PbBr_2$ is then blown out through the exhaust system as a fine dust, although some remains in the muffler and exhaust pipe to hydrolyze and corrode those parts severely:

$PbBr_2$	+	H_2O	$\rightarrow$	PbO	+	2HBr
from $(C_2H_5)_4Pb$		from combustion of fuel		dust		corrosive acid

Metallic lead is also used for low-melting alloys such as solder (50% Pb) and type metal (10% Sn, 15% Sb, 75% Pb). It also is added to brass and steel to improve machineability. Lead compounds of practical importance, in addition to tetraethyl lead, include "red lead," Pb_3O_4, which is used in corrosion-preventing paints for steel, and "white lead," $Pb_3(CO_3)_2(OH)_2$, formerly the most important white pigment for house paint. Lead oxide is used in glassmaking to increase refractivity and brilliance.

Notice that almost all the stable inorganic compounds of lead, both in natural minerals and in commerce, are compounds of Pb(II): PbS, PbO, $PbSO_4$, $PbCO_3$, and $PbCl_2$. The only exceptions are the oxides PbO_2 and Pb_3O_4. This preference for divalence in the inorganic compounds of lead simply means that the lower

---

*On the meaning and evaluation of octane number of motor fuels, see E. G. Rochow, G. Fleck, and T. H. Blackburn, *Chemistry: Molecules that Matter*, New York: Holt, Rinehart, & Winston (1974), Chapter 14, pp. 417–474, or see the excellent article on "combustion" in the *McGraw-Hill Encyclopedia of Science and Technology*, New York: McGraw-Hill Book Co. (1971).

†For some idea of how far back the energy crisis was clearly predicted, see E. G. Rochow, "Chemistry Tomorrow," *Chemical and Engineering News* 27:1510 (1949).

Cases of lead poisoning reported in a major city during a one-year period. The seasonal pattern of increased incidence during the warmer months is striking.

	Number of Cases Reported
January	25
February	45
March	85
April	100
May	175
June	320
July	375
August	460
September	380
October	250
November	215
December	175

oxidation state becomes more stable in the heaviest element of this group, in accordance with the principle stated above.

NOTE ON THE TOXICITY OF LEAD. With 150,000 tons of lead per year being discharged into the atmosphere of our cities and towns by automobiles, the entire use of tetraethyl lead is being questioned as a health hazard. Fortunately, the mandatory marketing of lead-free gasoline to prevent damage to the catalytic converters of newer cars should start a phase-out of tetraethyl lead as refiners and users see that they can get along without it. Unfortunately, more thought has been given in the past to preventing damage to engines rather than people but now the reintroduction of lead-free gasoline is more a result of concern about the poisoning of people. The symptoms of lead poisoning are well known: colic, anemia, nervousness, headache, and insomnia, followed by tremor, mental disturbance, convulsions, and coma. It also causes sterility in women. Lead may enter the body through the skin (as when ethyl gasoline is improperly used to wash something), or by ingestion (which is why lead-containing paints are now forbidden on children's toys), or by inhalation (the aspect we are concerned with here). Intake of any amount of lead greater than 1 mg per day is considered dangerous. The effect of continued inhalation of amounts less than this is not known, but all agree that it does no good. Perhaps future generations will ascribe the social restlessness of our day to lead poisoning, just as historians believe that lead from cooking pots and lead pipes poisoned the governing class of imperial Rome and caused the birth rate to drop almost to zero.*

*On plumbism in the past, see S. C. Gilfillan, "Lead Poisoning and the Fall of Rome," in the *Journal of Occupational Medicine*, 7(No. 2):53–60 (1965).

TABLE 5–5 GROUP V: THE p^3 ELEMENTS (NITROGEN THROUGH BISMUTH)

	Nitrogen	*Phosphorus*	*Arsenic*	*Antimony*	*Bismuth*
Symbol	N	P	As	Sb	Bi
Atomic number	7	15	33	51	83
Atomic weight	14.007	30.974	74.922	121.75	208.98
Valence e^-	$2s^22p^3$	$3s^23p^3$	$4s^24p^3$	$5s^25p^3$	$6s^26p^3$
mp, °C	−210	44 a*	814 b*	631	271
bp, °C	−196	280 a	sublimes b	1380	1560
d, g/cm³	1.25 g/liter	1.83 a	5.73 b	6.69	9.75
Atomic volume		17	13.1	18.4	21.3
Atomic radius, Å	0.70	1.10	1.21	1.41	1.46
Ion radius (+5), Å	0.11	0.34	0.47	0.62	0.74
Pauling EN	3.0	2.1	2.0	1.9	1.9
Z_{eff}/r^2	3.07	2.06	2.20	1.82	1.67
Standard potential c*	+0.27	−0.06	−0.60	0.51	−0.8
Oxidation states	−3 to +5	−3 to +5	−3 to +5	−3 to +5	
Ionization energy*	335	254	231	199	185
Heat of vaporization	2.04	2.97	7.64	19.60	42.69
Isolated by	Rutherford	Brandt	Albertus	antiq.	Geoffroy
Date of isolation	1772	1669	1250	antiq.	1753
rpw* pure O_2	NO_x, 1200° C	P_4O_6, P_4O_{10}	As_4O_6	Sb_2O_3, 500° C	Bi_2O_3, 700° C
rpw H_2O	none	none	none	none	none
rpw N_2		none	none	none	none
rpw halogens	d*	PX_3, PX_5	AsX_3, AsX_5	SbX_3, SbX_5	BiX_3
rpw H_2	NH_3, 350° C	PH_3, 300° C	d	d	d
flame color	green				
Mohs hardness		0.5	3.5	3.25	2.5
Crystal structure		cubic a	rhomb.*	rhomb.	rhomb.

*Energies and heats are in kcal/mole; a = data on the white form of phosphorus (for other forms see Table 4–4); b = data on gray arsenic, yellow waxy form has density 2.01 g/cm³; c = std. potential for hydride EH_3 to element E, in volts; d indicates indirect preparation not from elements; rhomb. = the rhombohedral crystal system, having two crystal axes at right angles but the third axis at some other angle.

GROUP V, THE NITROGEN GROUP

The properties of the Group V elements are listed in Table 5–5. The first element, nitrogen, is strongly electronegative and its oxides are acid-formers; the second element, phosphorus, is a decided nonmetal and its oxides are still acid-formers. The third element, arsenic, is amphoteric but its oxides are on the acid side; it is sometimes a nonmetal and sometimes a metalloid. The fourth element, antimony, is partly metalloid but mostly downright metal; its oxides are amphoteric but on the basic side. The fifth and heaviest element, bismuth, is a decided metal; its oxides are entirely basic and dissolve only in acids. Even though there are fluctuations in electronegativity, the trend toward more metallic properties with increasing atomic weight is clear throughout the group. As the metallic character increases, the affinity for electropositive elements decreases, so the hydrides become less stable. The oxides become increasingly basic with increasing atomic weight, and the salt-like character of the halides increases also.

As for oxidation states, all the elements exhibit the range −3 to +5, but not necessarily all nine states within that range. Nitrogen is in the +5 state in nitrates and nitric acid; phosphorus is in the +5 state in P_4O_{10}, in phosphates, and in phosphoric acid, although unstable and easily oxidized phosphites and hypophosphites are known; arsenic is amphoteric and combines as both As(III) and As(V), depending on the conditions; antimony is also amphoteric and favors the +3 state but can be forced up to the +5 state; bismuth combines almost exclusively in the +3 state and forms no acids.

Nitrogen was discussed at length in Chapter 3 and phosphorus in Chapter 4, so only the remaining three elements will be considered below.

Arsenic is very widespread, although no more abundant than boron or beryllium. It is a nuisance in the metallurgy of copper and lead because its sulfide, As_2S_3, occurs along with CuS and PbS; it has to be removed during the smelting operations, and this forced recovery gives us more arsenic than we need. Its only use as the element is in lead shot, where 0.2% of it hardens the lead. Its compounds are used as insecticides, weed killers, and rat poisons because of their toxicity. The element is allotropic: yellow arsenic has the structure of white phosphorus, and the more stable gray arsenic (a steel-gray brittle metalloid) has the structure of antimony and bismuth. Either form burns to the oxide As_4O_6 (analogous to P_4O_6), which dissolves in strong acids and in hot concentrated solutions of strong base:

$$As_4O_6(s) + 12HCl(aq) \rightarrow 4AsCl_3(aq) + 6H_2O$$

$$As_4O_6(s) + 6H_2SO_4(aq) \rightarrow 2As_2(SO_4)_3(aq) + 6H_2O$$

$$As_4O_6(s) + 12NaOH(aq) \rightarrow 4Na_3AsO_3 + 6H_2O$$

The salt Na_3AsO_3 is an arsenite. When As_4O_6 is heated with concentrated nitric acid it is oxidized to the higher oxide, As_4O_{10}, which dissolves in water to form arsenic acid, a fairly strong acid with a first dissociation constant of 5×10^{-3}:

$$As_4O_{10}(s) + 6H_2O(l) \rightarrow 4H_3AsO_4(aq)$$

The salts of arsenic acid are arsenates, such as $Pb_3(AsO_4)_2$, lead arsenate, which is an insecticide. The lower oxide, As_4O_6, is soluble only to the extent of 2% in water and gives a very weak acid, H_3AsO_3, but the higher oxide As_4O_{10} is more soluble and gives a much stronger acid. This again illustrates the greater acidity that attends higher oxidation states.

Arsenites and arsenates are reduced by hydrogen to the poisonous and unstable hydride called arsine, AsH_3. This is the basis of the classic Marsh test for arsenic, used in criminal proceedings when arsenic poisoning is suspected. A sample of stomach contents is mixed with hydrochloric acid, and granulated zinc is added:

$$Zn + 2H^+ \rightarrow Zn^{+2} + H_2(g)$$

$$4H_2(g) + H_3AsO_4(aq) \rightarrow AsH_3(g) + 4H_2O$$

The evolved mixture of hydrogen and arsine is conducted through a heated glass tube, where AsH_3 decomposes and forms a metallic mirror on the glass:

$$2AsH_3(g) \rightarrow 2As + 3H_2$$

Other elements will also form volatile hydrides that deposit metallic mirrors, but the deposit can be tested to determine whether it is indeed arsenic. About 0.0005 g of arsenic can be detected in this way.

All arsenic compounds are poisonous. The oxide As_4O_6, called white arsenic, is so toxic that 0.1 g is a lethal dose for man. Arsenic-containing green pigments have long since been banned from commerce, and the use of arsenic insecticides is on the way out as much better and less dangerous materials have been developed. It is interesting to note in connection with arsenic's toxicity, however, that tiny traces of it further the body's uptake of iron and stimulate the formation of red blood cells.

Antimony is a brittle gray metal or metalloid (it seems half of each) which is used chiefly to harden the lead used in storage batteries. It does not react with air or water at ordinary temperatures, but once ignited it burns brilliantly to Sb_4O_6, an amphoteric oxide only very slightly soluble in water. Solution of Sb_4O_6 in strong base gives antimonites

$$Sb_4O_6(s) + 12NaOH(aq) \rightarrow 4Na_3SbO_3 + 6H_2O$$

When finely-powdered antimony is dropped into chlorine, it ignites at once and burns brightly to form white clouds of $SbCl_3$:

$$2Sb(s) + 3Cl_2(g) \rightarrow 2SbCl_3(s)$$

Similarly, all the halogen elements combine directly with antimony to produce the corresponding trihalides. Only two of these, the fluoride and the chloride, can be oxidized further to Sb(V) halides:

$$SbF_3(s) + F_2(g) \rightarrow SbF_5(s)$$

$$SbCl_3(s) + F_2(g) \rightarrow SbCl_3F_2(s)$$

$$SbCl_3(s) + Cl_2(g) \rightarrow SbCl_5(s)$$

Both SbF_5 and $SbCl_3F_2$ can be used as fluorinating agents, as described earlier under fluorine. Any trihalide of antimony hydrolyzes reversibly to the corresponding oxyhalide and hydrohalogen acid:

$$SbBr_3 + H_2O \rightarrow SbBrO + 2HBr$$

In this respect they differ from the trihalides of arsenic, which hydrolyze to arsenious acid:

$$AsBr_3(s) + 3H_2O(l) \rightarrow H_3AsO_3 + 3HBr$$

This difference again emphasizes the greater basic character of antimony.

Bismuth is a yellowish heavy metal which has been known for a long time but has not much use. The solid has a lower density (9.80 g/cm^3) than the liquid (10.03 g/cm^3), so bismuth is one of the few substances that expands on freezing. This convenient property leads to low-melting alloys that expand after being cast, thereby giving a sharp impression of the mold. Several such alloys melt in boiling water; for example, Wood's metal contains 50% bismuth, 25% lead, and 12.5% each of tin and cadmium, and it melts at 65.5° C. A teaspoon cast from it melts in a cup of hot coffee. Low-melting alloys like this also have some serious uses, as in plugs for automatic sprinkler systems and fusible links for fire-alarm systems and oil shutoff valves.

Bismuth burns only at red heat, and gives only the Bi(III) oxide:

$$4Bi(s) + 3O_2(g) \xrightarrow{700^\circ C} 2Bi_2O_3(s)$$

This dissolves in acids to give Bi(III) salts, but does not dissolve in bases:

$$Bi_2O_3(s) + 6HCl(aq) \rightarrow 2BiCl_3(aq) + 3H_2O$$

Similarly, the action of halogens on metallic bismuth gives only the trihalides and, due to the inert-pair effect, the *s* electrons are inactive and compounds of pentavalent bismuth are almost impossible to make. The general picture, therefore, is of a decidedly metallic element which is exclusively basic in its chemistry, and which almost always combines in the Bi(III) oxidation state. We see again that the lower oxidation state is the stable one for the heaviest element.

GROUP VI, THE OXYGEN AND SULFUR GROUP

Although everyone knows about sulfur, the remaining elements of Group VI remain in quiet obscurity. Selenium and tellurium are rare elements that imitate sulfur in part, but also differ from it in being more metallic and less acidic. Polonium we know only in the form of its radioactive isotopes, which show up in

TABLE 5–6 GROUP VI: THE p^4 ELEMENTS (OXYGEN, SULFUR, AND THEIR CONGENERS)

	Oxygen	*Sulfur*	*Selenium*	*Tellurium*	*Polonium*
Symbol	O	S	Se	Te	Po
Atomic number	8	16	34	52	84
Atomic weight	15.999	32.064	78.96	127.60	(210)*
Valence e^-	$2s^22p^4$	$3s^23p^4$	$4s^24p^4$	$5s^25p^4$	$6s^26p^4$
mp, °C	−218	112	217	450	254
bp, °C	−183	444	685	990	962
d, g/cm³	1.43 g/liter	2.07 a*	4.79	6.24	9.32
Atomic volume		15.5	16.5	20.5	22.5
Atomic radius, Å	0.66	1.04	1.21	1.41	1.65
Ion radius (−2), Å	1.40	1.84	1.98	2.21	1.69
Pauling EN	3.5	2.5	2.4	2.1	2.0
Z_{eff}/r^2	3.50	2.44	2.48	2.01	1.76
Standard potential	+1.229	+0.141	−0.40	−0.72	−1.0
Oxidation states	−1, −2	−2 to +6	−2 to +6	−2 to +6	
Ionization energy*	313.9	239	225	208	
Heat of vaporization	3.00	3.01	3.34	11.9	
Isolated by	Priestley	antiq.	Berzelius	Mueller	Curie
Date of isolation	1774	antiq.	1817	1782	1898
rpw* pure O_2		SO_2, SO_3	SeO_2	TeO_2	
rpw H_2O	none	none	none	none	
rpw N_2	NO_x, 1200° C	b*	none	none	
rpw halogens		S_2X_2 to SX_6	SeX_2, SeX_4	TeX_2	
rpw H_2	H_2O, 500° C	H_2S, 400° C	H_2Se, 400° C	b	
Color of element	pale blue	bright yellow	brick red c*	brown c	
Mohs hardness		2	2	2.3	
Crystal structure		orthorh. a*	hexag.	hexag.	

*Energies and heats are in kcal/mole; atomic weight in parentheses is that of the most stable isotope; a = data for orthorhombic S (for properties of other forms see Table 4–4); potentials are for $O_2 + 4H^+ + 4e^- \rightarrow 2H_2O$ and the like; b indicates nitride or hydride formed only by indirect methods; rpw = reaction product with; c = color of nonmetallic form.

all three (uranium, thorium, and actinium) natural decay series. The properties of all five elements are listed in Table 5–6.

The allotropic forms of *selenium* give a clear indication of the oncoming metallic character. While red selenium closely resembles the several forms of sulfur in being composed of Se_8 rings that crystallize in the orthorhombic and monoclinic systems, gray selenium is a metalloid with metallic luster and some electrical conductivity. The conductivity rises when the selenium is exposed to light, increasing with light intensity up to 1000 times the dark current. This interesting aspect led to the first photocells for operating electrical devices by beams of light.

Selenium is very rare, but it clings with sulfur in sulfide minerals and so appears as the solid dioxide in the flue dust of the smelters which produce copper and lead. It is easy to recover the element from SeO_2 because the oxide dissolves in water to form selenous acid, H_2SeO_3, and this is reduced to red selenium by sulfur dioxide:

$$H_2SeO_3(aq) + 2SO_2(g) + H_2O \rightarrow Se(s) + 4H^+ + 2SO_4^{-2}$$

The gray form is obtained by melting red selenium, chilling it to a glassy solid, and then holding this at 75° C until it changes over to the stable gray crystalline solid. Some gray selenium is still used in electrical applications, but by far the largest use for selenium employs the red form to impart a brilliant red color to glass. Selenium compounds are added to the glass furnace near the end of operations, and the yellowish glass is fabricated (signal lights and stoplight lenses are molded; tumblers and stemware are blown). The cooled pieces are then heat-treated at a controlled temperature long enough to allow colloidal red selenium to form within the glass matrix, producing the desired bright or deep red color.

Selenium trioxide is obtained by mixing oxygen with the vapor of SeO_2 and passing the mixture through an ozonizer. Above 180°C SeO_3 decomposes to SeO_2 and oxygen, indicating greater stability of the lower oxidation state. The rest of selenium chemistry follows that of sulfur: the formation of selenite and selenate salts, the formation of volatile halides and oxyhalides, and the formation of many organic derivatives. Special mention should be given to hydrogen selenide, H_2Se, because it is an extremely poisonous gas with a very offensive odor. It can be obtained, if ever it were wanted, by the hydrolysis of aluminum selenide:

$$Al_2Se_3(s) + 6H_2O(l) \rightarrow 2Al(OH)_3(s) + 3H_2Se(g)$$

Tellurium is an extremely rare element that shows up in the mineral world as tellurides of gold, silver, lead, bismuth, and mercury. As expected, its chemistry emphasizes the lower oxidation states; the most stable oxide is TeO_2, but there is even a monoxide TeO. Tellurates and telluric acid can be made, but they are strong oxidizing agents. Hydrogen telluride apparently takes the prize as the foulest-smelling inorganic compound known.

Polonium was discovered by Marie Curie in uranium minerals, where the ^{210}Po isotope exists as the next-to-last decay product of radium. It decays to stable lead-206 by alpha emission, with a half-life of 138 days. Some 29 isotopes of polonium are now known, all of them radioactive. All that is known about the chemistry of polonium has been determined by radioactive tracer techniques. It is chemically similar to tellurium, but more basic and metallic. For example, it forms salts as though it were a metal: $Po(SO_4)_2$, $Po(NO_3)_4$, and $Po(CN)_4$. The only use for polonium is in neutron sources, where its evolved alpha particles react with beryllium nuclei to produce neutrons:

$$^{4}_{2}He + ^{9}_{4}Be \rightarrow ^{12}_{6}C + ^{1}_{0}n$$

GROUP VII, ESPECIALLY BROMINE AND IODINE

The properties of the Group VII elements are listed in Table 5–7. The trends have been examined in Chapter 1, and the chemistry of fluorine and chlorine were discussed in detail in Chapters 3 and 4. We are concerned here principally with bromine and iodine, with a note on astatine.

BROMINE. While chlorine and iodine are named for their colors, and selenium is named for the moon, the word "bromine" comes from the Greek word meaning stench. Its choking odor is not only distinctive, but also destructive; care should be taken not to inhale the vapor or spill any of the liquid on the skin. The appearance is also distinctive, for the liquid is very dark and heavy, and the vapor is brick red. The only other elements which are liquid at room temperature are metals (mercury, gallium, and cesium).

The concentration of bromide ion in seawater is only 0.0066%, yet the sea is

TABLE 5–7 GROUP VII: The p^5 ELEMENTS (THE HALOGEN FAMILY)

	Fluorine	*Chlorine*	*Bromine*	*Iodine*	*Astatine*
Symbol	F	Cl	Br	I	At
Atomic number	9	17	35	53	85
Atomic weight	18.999	35.453	79.909	126.904	(210)*
Valence e^-	$2s^22p^5$	$3s^23p^5$	$4s^24p^5$	$6s^26p^5$	$7s^27p^5$
mp, °C	−220	−101	−7.3	114	
bp, °C	−188	−34	58.8	184	
d, g/cm³	1.81 g/liter	3.21 g/liter	3.12	4.94	
Atomic volume	12.8	14.5	23.5	25.7	
Atomic radius, Å	0.64	0.99	1.14	1.33	
Ion radius (−1), Å	1.36	1.81	1.95	2.16	
Pauling EN	4.0	3.0	2.8	2.5	2.2
Z_{eff}/r^2	4.10	2.83	2.74	2.21	1.96
Standard potential	+2.87	+1.36	+1.07	+0.54	+0.3
Oxidation states	−1	−1 to +7	−1 to +7	−1 to +7	
Ionization energy*	401.7	300.0	273	241	
Heat of vaporization	2.76	2.85	3.58	4.98	
Isolated by	Moisson	Scheele	Balard	Courlois	CMS*
Date of isolation	1886	1774	1826	1811	1940
rpw* pure O_2	O_2F_2 el*	none (i)*	none (i)	none (i)	
rpw H_2O	HF, O_2, O_3	HCl, HOCl	HBr, HOBr	HI, HOI	
rpw N_2	none	none	none	none	
rpw halogens		see section on interhalogen compounds			
rpw H_2	HF	HCl	HBr	HI	
Color of element	pale yellow	yellow-green	dark red	black	
Mohs hardness					
Crystal structure			orthorh.	orthorh.	

*Energies and heats are in kcal per mole; atomic weight in parentheses is that of the most stable isotope; rpw = reaction product with; a = standard electrode potential for reaction $X_2 + 2e^- \rightarrow 2X^-$ (reaction $F_2 + H_2 \rightarrow 2HF$ gives 3.06 V); CMS = Corson, McKenzie, and Segré; (i) indicates oxides, etc., made by indirect methods; el = electric discharge.

a profitable source of bromine. At current prices, the amount in 1 cubic mile of water is worth \$224,000,000 (and there are 320,000,000 cubic miles of seawater available, as we saw in Chapter 2). How can the bromine be gathered up from so dilute a source? First, the normally alkaline seawater is acidified with H_2SO_4 to a pH of 3.5. Then gaseous chlorine is injected, liberating bromine:

$$Cl_2(g) + 2Br^- \rightarrow 2Cl^- + Br_2(g)$$

The acid prevents loss of chlorine by alkaline absorption, and also keeps the bromine as Br_2. A stream of air then sweeps out the vapor of bromine and carries it to a saturated alkaline solution of Na_2CO_3, where it is trapped in concentrated form as sodium bromide and bromate:

$$3Br_2(g) + 3CO_3^{-2} \rightarrow 5Br^- + BrO_3^- + 3CO_2(g)$$

Since an oxidizing agent ($NaBrO_3$) and a reducing agent (NaBr) are already present in the concentrated solution, it is only necessary to add sulfuric acid to liberate clouds of bromine:

$$5Br^- + BrO_3^- + 6H^+ \rightarrow 3H_2O + 3Br_2(g)$$

Some of the bromine is absorbed by ethylene (a refinery by-product) to make dibromoethane for ethyl gasoline

$$H_2C{=}CH_2(g) + Br_2(g) \rightarrow BrH_2C{-}CH_2Br(l)$$

and the rest is used to make a variety of chemical products such as methyl bromide, CH_3Br, an efficient fire extinguisher.

In its relations with oxygen, bromine differs from chlorine and iodine. The usual methods for preparing perchlorates and periodates do not work when applied to bromine, and for many years it was assumed that there could be no perbromates. Then perbromic acid and perbromates were made by fluorine oxidation of bromates, and it was realized that perbromates are just more unstable and are stronger oxidizing agents than perchlorates or periodates. This behavior has something to do with the fact that bromine is the first halogen element to have filled *d* orbitals, with consequent excessive shrinkage of its covalent radius.

The colloquial expression "a bromide," meaning a platitude or a boring play or movie, comes from the use of potassium and sodium bromides as mild sedatives in headache powders. Excessive use of bromides this way leads to mental aberrations which can be corrected only by restoring the proper halide ionic balance in the blood.

IODINE. Named for the violet color of its vapor, iodine was discovered during the burning of dried seaweed. Some plants and some shellfish have the ability to take up iodide ions from water and concentrate the iodine. Everyone needs iodine because it is part of the molecular structure of thyroxine, a growth-regulating hormone produced by the thyroid gland:

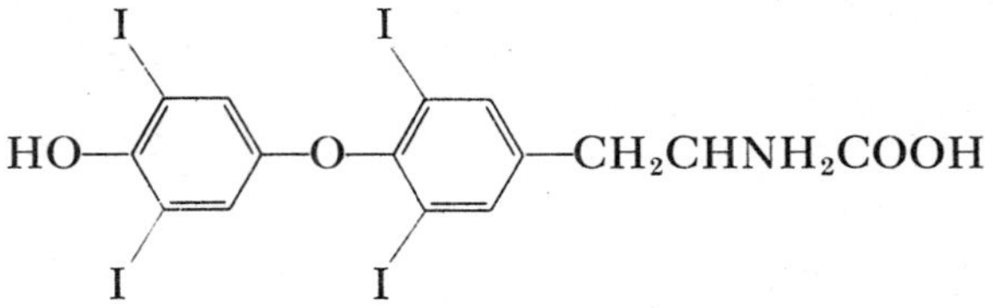

I I
HO—(ring)—O—(ring)—CH_2CHNH_2COOH
I I

Thyroxine

The four iodine atoms necessary to form each molecule of thyroxine can come only from ingested iodides, and they come from sea salt, or iodized salt, or a diet that includes considerable fish and shellfish. If the necessary iodide input is lacking, the thyroid gland enlarges in an attempt to garner more iodine. This is the condition called goiter, which occurs mostly among inland peoples with restricted diet. The addition of 0.01% NaI to table salt ("iodized salt") prevents this condition.

Iodine is more familiar as tincture of iodine (an alcohol solution) than as the black lustrous crystals of the element. Pure iodine dissolves in nonpolar solvents such as CCl_4 and chloroform to impart a distinctive violet color to the solution, a color very close to that of the vapor of iodine. It also dissolves in polar solvents such as water and alcohol, but such solutions are brown. The solubility in water is very low (only 0.29 g/liter); alcohol dissolves ten times as much iodine as water, and still more if some NaI is present. The alcohol solution has long been used as an antiseptic—not only does the iodine kill bacteria by oxidation, but the alcohol itself is a rather potent germicide. However, tincture of iodine stings painfully when applied to a wound, and the free iodine is so potent an oxidizer it destroys human tissue as well as bacteria, thereby delaying healing. For these reasons iodine declined in popularity, but now it is back again without the alcohol, in the form of a water solution of an iodine complex of polyvinylpyrrolidone ("povidone" for short, a polar polymer developed as a blood plasma extender). In this form the iodine is a less severe oxidizer but still a very effective germicide.

Iodine forms ionic and covalent bonds of low energy, so free iodine is liberated by weak oxidizing agents and even by sunlight. A favorite method of testing for oxidizing ability is to add the substance in question to an acid solution of KI. Appearance of a yellow or brown color shows that iodine has been liberated by oxidation:

$$H_2O_2 + 2I^- + 2H^+ \rightarrow 2H_2O + I_2(aq)$$

$$MnO_2(s) + 2I^- + 4H^+ \rightarrow Mn^{+2} + 2H_2O + I_2(aq)$$

If the color is very weak, the test can be made more sensitive by adding some boiled starch solution. This turns dark blue when free iodine is present, as the

Uses of Halogens and Halogen Compounds

I_2	Iodine, a disinfectant
CH_3Br	Methyl bromide, used as a fire extinguisher
NaBr	Sodium bromide, a sedative
NaI	Sodium iodide, used to prevent goiter
CF_4	Carbon tetrafluoride, an aerosol propellant
NaOCl	Sodium hypochlorite, a bleach

OH groups on the starch form an iodine complex. The *amount* of oxidizing agent can be determined by titrating the liberated iodine with a standardized solution of sodium thiosulfate, $Na_2S_2O_3$:

$$I_2 + 2S_2O_3^{-2} \rightarrow S_4O_6^{-2} + 2I^-$$

Disappearance of the iodine color signifies the end point of the reaction as all the iodine is converted to colorless iodide ion. From the volume and molarity of the solution of $Na_2S_2O_3$ the corresponding number of equivalent weights of oxidizing agent may be calculated.

INTERHALOGEN COMPOUNDS. These substances are composed of two different halogen elements united by covalent bonds, formed by reaction of the halogens with each other. There are 1:1 compounds throughout the group, and 1:3 compounds of a larger halogen with a smaller one, as in ClF_3, BrF_3 and ICl_3. Bromine and iodine can form 1:5 compounds such as BrF_5 and IF_5, but only iodine is large enough to form a 1:7 compound, IF_7. These compounds are of interest for two good reasons: they exhibit chemical properties intermediate between those of the halogen elements, and they illustrate very well the possible oxidation states of the halogens and the workings of oxidation number in predicting reactions.

Consider ICl for a start. It is a reddish brown liquid of density 3.18 g/cm^3, compared with 3.12 for Br_2, and it is difficult to tell ICl from liquid Br_2 without determining a melting or boiling point. The only visible difference is that the vapor of ICl is more brownish than that of Br_2. Chemically, ICl is about as strong an oxidizing agent as Br_2. The similarity is so great that the famous nineteenth century chemist Justus von Liebig was chagrined to find he had missed the discovery of bromine. In 1815 von Liebig obtained a dark heavy liquid with reddish vapor by the oxidation of sea salt. He was familiar with iodine monochloride so he labeled the liquid "ICl" and put it aside. Eleven years later Balard discovered and characterized the hitherto unknown element 35, and called it bromine. Upon reading of the discovery, Liebig took his bottle of "ICl" off the shelf and found it matched the properties of the newly announced element exactly.

As for intermediate chemical properties, the compound ClF_3 is only slightly less vigorous a fluorinating agent than F_2 itself. It has the advantage of being a liquid, and so is more convenient to store and handle than fluorine. Experts in fluorine chemistry who use ClF_3 call it "fluorine tricked into being liquid at room temperature."

In each interhalogen compound the oxidation states of the two halogens are readily apparent—in BrF_5 the bromine must be in the +5 state, while fluorine must be in the −1 state, as usual. When such compounds hydrolyze, the oxidation states remain the same. The products are always the hydrohalogen acid of the more negative element, and an oxyacid of the more positive element. The oxidation state of the more positive element tells us which oxyacid to expect, and so we can predict the products of such reactions from that information. In the reaction

$$ClF_3 + 2H_2O \rightarrow 3HF + HClO_2$$

the oxyacid $HClO_2$ is predicted because in it chlorine has the original oxidation state of +3. Similarly, we predict

$$BrF_5 + 3H_2O \rightarrow 5HF + HBrO_3$$

and

$$IF_7 + 4H_2O \rightarrow 7HF + HIO_4$$

ASTATINE. Many searches for the missing element 85 were made from the time of Mendeleev up to 1940, but natural sources yielded nothing. It remained for Corson, McKenzie, and Segré to synthesize element 85 by bombarding natural bismuth with alpha particles accelerated in a cyclotron:

$$^{209}_{83}Bi + ^{4}_{2}He \rightarrow ^{211}_{85}At + 2^{1}_{0}n$$

It was named astatine from the Greek roots *a* (meaning not) and *stat* (meaning stable). Astatine acts like iodine in being slightly soluble both in water and in CCl_4, and it can be oxidized to $HAtO_3$ and H_5AtO_6. The insoluble salts $PbAt_2$, TlAt, and AgAt have been made.

SUMMARY

In this chapter the properties of the main group or "representative" elements have been surveyed. We found that the similarities within any one group are notable, as the elements use their similar complements of *s* and *p* electrons to form chemical compounds. The usage of such electrons is not uniform, however; among the heavier elements of Groups III, IV, and V there is a pronounced tendency to use only the *p* electrons, leaving the two *s* electrons as a socalled inert pair. Thus, all Group III elements form compounds in which they are trivalent, but from gallium onward there is increasing tendency to form univalent compounds as well, so that in thallium compounds the +1 state is dominant. Similarly, in Group IV the group valency is +4, but from germanium onward there is increasing tendency toward +2; so, we find Pb(II) compounds more stable by far than Pb(IV) compounds. In Group V there is a similar trend toward +3 as the more stable state for bismuth. Among the main group elements, then, as *p* electrons become available there is less usage of *s* electrons by the heavier elements, and the lower oxidation states become more stable and prevalent.

Higher oxidation states correspond to more acidic behavior, for the simple reason that the removal of each additional electron from an atom requires more energy than the last one, so that in effect a multivalent element becomes more and more electronegative at successively higher oxidation states. This is the second general principle developed in this chapter, and it finds expression in ionic compounds at lower oxidation states but covalent and acid-forming compounds at higher oxidation states of the same element. For example, Pb(II) is represented by a long series of salts, but Pb(IV) is found only in covalent $Pb(C_2H_5)_4$ and in yellow, oily $PbCl_4$. As for acidity, the higher oxides of arsenic, antimony, sulfur, and selenium form stronger acids than the lower oxides.

For convenient reference, the most important physical properties of all the elements in each of the seven groups are listed in an extensive table for that group. Some chemical characteristics also are listed in these tables, but thorough description of chemical behavior requires individual discussion of each element. Space permitting, the sources of each element are described and at least one method of isolating the element is given. The most important compounds are then considered, together with their applications.

The most important aspects of potassium are (1) its importance to plant nutrition, and hence its place in the food chain, and (2) the radioactivity of the ^{40}K isotope, which contributes a substantial part of the radiation background within our bodies. Rubidium is widespread but almost useless; cesium is rare but useful, especially in photocells and vapor arc lamps.

Calcium is very abundant and is a principal constituent of rocks, shells, limestone, marble, cement, concrete, bricks, tile, and pottery. Lime derived from limestone or shells is the raw material for several large industries, as well as the chief ingredient of builders' mortar. The dihydrate of $CaSO_4$, called gypsum, is the raw material for plasterboard and for plaster of paris. Strontium has no large-scale utility, and barium finds use only in the white pigment lithopone (co-precipitated $BaSO_4$ and ZnS) and in dense concrete for shielding.

Gallium, indium, and thallium have little practical importance, but their compounds illustrate very well the trend toward a lower oxidation state (+1) for the heaviest elements. Thallium(I) compounds are stable and resemble the corresponding compounds of sodium and silver; Tl(III) compounds are strong oxidizing agents and are easily reduced to the +1 state.

In Group IV germanium was found to be a more electronegative element than silicon, and to be unemployed since germanium transistors have been replaced by those of silicon. Tin is important as a metal; it forms the protective coating on sheet steel for cans, and is a major constituent of bronze (Cu + Sn), solder (Sn + Pb), pewter, and low-melting alloys. Tin(II) compounds predominate over tin (IV) compounds, except in organotin compounds. Both tin and lead are easily obtained from their ores, but the ores are now getting scarce. Lead is used in low-melting alloys, in storage batteries, and in tetraethyl lead for gasoline. Lead is a very toxic element, and its use in paints, water conduits, and gasoline is being curtailed.

Arsenic is also toxic, and some of its more poisonous compounds (lead arsenate, for example) were formerly used as insecticides. Today, arsenic has mostly nuisance importance—it is difficult to get rid of it. Antimony is used to harden lead for storage batteries, to harden tin (in pewter), and to make type metal. Bismuth is used in very low-melting alloys for sprinkler systems, and to increase machinability of alloy steels.

Oxygen and sulfur far outshadow the other members of Group VI. Selenium finds some application as a semiconductor and to color glass red, but tellurium and polonium are of interest only in their chemical behavior. Polonium is very rare and always radioactive.

The halogen elements are favorites for pointing out group trends and individual diversity. Their range of electronegativities is extraordinary, and their chemical relations with oxygen vary greatly. For all except fluorine, oxidation states all the way from -1 to $+7$ are found, with accent on the odd-numbered states. Oxyacids of bromine are less stable than those of chlorine and iodine, and perbromates have been made only with the aid of fluorine as oxidizer. All the halogen elements combine with each other in the favorite ratios of 1, 3, 5, and 7, and the oxidation states so defined persist during hydrolysis so that the products of hydrolysis can be predicted.

Bromine is recovered from seawater and from inland brines, and is used to make many organic and inorganic derivatives, among then a lead scavenger for ethyl gasoline. Iodine is now used in a new water-complexed form as an antiseptic, and is essential to nutrition as a source of the growth hormone thyroxine.

GLOSSARY

acetylene: the compound C_2H_2, *ethyne* in systematic nomenclature.
amalgam: a solution of a metal in mercury.
colloidal: composed of, or containing, particles of liquid or solid 10^{-7} to 10^{-5} cm in diameter, too small to be seen by the unaided eye but large enough to be illuminated by a beam of light.
congener: one of the same kind or class.
cubic crystal system: that crystal structure which has three crystal axes of equal length, all at right angles to each other
detonation: a runaway exothermic reaction, in which the reaction proceeds as fast as a shock wave can travel through the material.
fractional distillation: a technique of separating two volatile materials having boiling points close together, in which the mixed vapors from the boiling mixture are washed continuously by a descending stream of condensate in a column of suitable design, thereby returning the less volatile component to the boiler and permitting some of the more volatile component to be recovered in a purified state.
hormone: a regulating substance secreted by an endocrine gland. Examples: insulin, thyroxine, cortisone.
hexagonal crystal system: the structure based on three planar crystal axes at angles of 120° to each other, and all perpendicular to a fourth axis.
index of refraction: see refractive index.
monoclinic: the crystal system in which two of the three crystal axes are perpendicular, but the third is at some angle other than 90°.
orthorhombic: the crystal system in which all three crystal axes are at right angles to each other, but are of different length.
polar: having an unsymmetrical distribution of electric charge, and hence exhibiting a dipole moment.
polyhalide ion: an ion composed of three or more interbonded halogen atoms, such as I_3^- or ICl_2^-.
rectifier: a device that converts alternating electric current to direct current.
refractive index: the ratio of speed of light in vacuum to the speed of light in the substance under consideration, and hence a measure of the degree to which a beam of light will be bent (refracted) as it enters the substance from air or vacuum.
***sp*³ hybrid bonds:** four chemical bonds of equivalent energy formed by molecular orbital combination of an *s* and three *p* atomic orbitals.
tetragonal crystal system: the system in which all three crystal axes are mutually perpendicular but only two are the same length (the third is different).

EXAMINATION QUESTIONS

1. Choose any one of Groups III, IV, V, or VI, and summarize the properties of the elements therein in terms of: a. range of physical properties; b. range of electronegativities and modes of combination; c. oxidation states (with examples); d. sources and extraction of the elements; and e. importance or utility of each element and its compounds.

2. How could you distinguish quickly between: a. RbCl and CsCl; b. LiBr and $BaBr_2$; c. $GaCl_3$ and $TlCl_3$, d. Na_2S and Na_2Se; and e. KBr and KI?

3. By means of chemical equations for major reactions involved, indicate how magnesium and bromine could be recovered from the *same* large sample of seawater.

4. How would you distinguish between minute amounts of arsenic and bismuth?

5. Given a solution containing the chlorides of Li, Na, K, Rb, and Cs, all 1.0 M concentration, how would you go about separating all five alkali-metal elements as completely as possible?

6. Predict the reaction products of the hydrolyses of ICl_3, $BrCl_3$, IBr, ICl, and ICl_5, and write equations for all five reactions.

7. How does lead get into gasoline, and why? Describe the process involved, and the purpose.

8. What happens to the lead in gasoline after combustion of the fuel-air mixture?

9. What are the chief sources of lead poisoning in children, especially in large cities? What are the symptoms? Why is lead toxic? What can be done about the disease?

10. Give evidence for the greater stability of lower oxidation states among the heavier elements of at least three main groups.

"THINK" QUESTIONS

A. Why do odd oxidation states predominate among the halogen elements, whereas even oxidation states predominate in the oxygen-to-polonium group?

B. Assuming that enough of each could be made to see, weigh, and study, what compounds would you expect to be formed between francium and astatine, with or without oxygen? Write formulas for at least three, and predict their colors, general characteristics, stabilities, and oxidizing properties.

C. If tetraethyl lead were outlawed, could triethyl bismuth be prepared to use in its stead? Would it be soluble in gasoline? Would it be any less toxic? Would it be any more desirable to use at the same tonnage level as $Pb(C_2H_5)_4$?

D. Why are the oxides of tin, lead, antimony, and bismuth added to some kinds of glass to be used in lenses, and how do these oxides accomplish that purpose?

E. If tin were only one tenth as costly as it is now, where could much more of it be used to advantage?

F. An old museum piece, a beautiful embossed platter 50 × 70 cm, looks like pewter but may be pure tin. How could you determine its composition without harming the piece?

PROBLEMS

1. Seawater contains 1.125×10^{-6}% of silver and 2.5×10^{-6}% of iodine, by weight. The solubility product of silver iodide is 0.32×10^{-16} at 13° C, and the atomic weights of silver and iodine are 107.868 and 126.905, respectively. How much 1.0 M solution of KI must be added to 1 l of seawater to recover 90% of the silver as silver iodide?

2. Given the following solubility-product constants

$BaCO_3$ 8×10^{-9}	$CaCO_3$ 1.7×10^{-8}
$BaCrO_4$ 2×10^{-10}	$CaCrO_4$ 2×10^{-2}

devise a method for separating Ca^{+2} and Ba^{+2} in a solution wherein both are at a concentration of 0.1 M.

3. What is the oxidation number of each element in the following compounds?

$Na_2S_2O_3$	$CaSiO_3$
$MgClO_4$	Pb_3O_4
Na_3SbO_4	BaO_2

4. Either sodium or magnesium may be used in a certain reduction process. If sodium costs 21 cents per pound and magnesium costs 34 cents per pound, which metal will be less expensive to use? (The atomic weights are Na, 22.99, and Mg, 24.31.)

5. Complete and balance the following equations:

a. $I_2 + H_2S(aq) \rightarrow S(s) + ?$

b. $PbO_2 + Mn^{+2}(aq) + H^+(aq) \rightarrow MnO_4^- + Pb^{+2} + ?$

c. $SnCl_2 + HgCl_2 \rightarrow Hg_2Cl_2 + ?$

d. $H_2S + HNO_3(aq) \rightarrow S(s) + NO(g) + ?$

e. $H_2O + H_2SO_3(aq) + I_2 \rightarrow I^- + H_2SO_4 + ?$

6 THE FIRST SERIES OF TRANSITION METALS

In this chapter we are concerned with the elements from 21 (scandium) through 30 (zinc), which constitute the lightest of three series of transition metals. With the exception of scandium all of these elements are common and useful metals, but six of them are especially plentiful and important: titanium, chromium, iron, nickel, copper, and zinc. It should be no surprise that five of these six have even atomic numbers.

Everyone knows what a metal is: a metal is a hard, shiny solid, cold to the touch, strong but capable of being bent, hammered, and rolled into sheets, and always a good conductor of electricity. So goes the popular view, but students of chemistry can point out immediately that all these criteria are violated by one or more metals. Some of the elemental metals, such as chromium, are brittle and will not bend at all; some (like lead) are soft and weak, three are liquids, and some conduct electricity poorly, with conductivities only 1 or 2% that of copper. In short, it is difficult to define a metal precisely in terms of these physical properties. The most we can say is that all metals are opaque, have a characteristic metallic luster, and conduct electricity to some degree by free flow of electrons, according to Ohm's law.* They are held together in varying degrees by a special kind of bonding wherein fully detached electrons are shared by all the atoms.

Chemically, we can define a metal as an element with many more available orbitals than electrons to fill them. Under this definition we can see that all the transition elements will be metals, for these are the elements in which *d* orbitals are being filled while the next higher *p* orbitals all lie empty.

---

*See Glossary at end of this chapter.

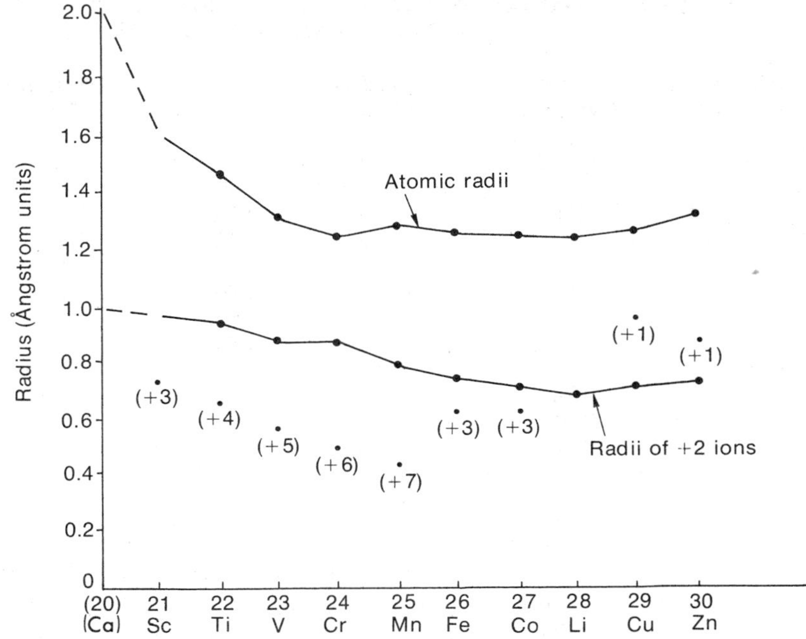

Figure 6–1. Atomic and ionic radii of elements in the first transition series.

TRENDS IN THE FIRST TRANSITION SEQUENCE

ATOMIC AND IONIC RADII. The radii of the elements in the first transition series are plotted against atomic number in Figure 6–1. Filling of the *d* orbitals begins with element 21, scandium, and immediately there is a sharp drop in atomic radius from that of calcium. Since it is a 3*d* orbital that receives this first electron, and since calcium already has filled its 4*s* orbital, the addition may be thought of as an electron going down into an inner space that was already there, so it does not increase the size of the atom. At the same time, the addition of one more positive charge on the nucleus pulls in the entire array of electrons more tightly, sharply reducing the diameter.

As additional *d* electrons are added in the next three elements, the atomic radius decreases further for the same reason. Then it levels off, after the *d* orbitals are all half-filled, and rises slightly at the end of the transition series. The initial drop is called the *d* orbital shrinkage, and is evident in all three transition series.

OXIDATION STATES VS. SIZE. Since a transition element in this first sequence may use its 4*s* electrons and any or all of its 3*d* electrons for chemical bonding, we must expect a variety of oxidation states. For example, titanium has a total of four valence electrons, and so it has oxidation states of +1, +2, +3, and +4 in its compounds. Of these, the +2 and +4 states are the more stable ones, corresponding to removal of the 4*s* pair and then the entire four electrons. The ionic compounds of Ti(IV), such as TiO_2, are colorless because the Ti^{+4} ion consists of the argon core, just like K^+ and Ca^{+2}. On the other hand, compounds of Ti(III) are violet and blue. We shall meet many examples of bright, characteristic colors in this chapter; color is one of the hallmarks of transition elements.

In Figure 6–1 the lower line connects the radii of the +2 ions of the 3*d* elements. The decrease in the radius of Ca^{+2} is much less marked than the de-

TABLE 6–1 OXIDATION STATES AND CORRESPONDING IONIC RADII FOR FIRST TRANSITION SERIES

Oxidation State	Ionic Radius, Å	Oxidation State	Ionic Radius, Å	Oxidation State	Ionic Radius, Å
Sc +3	0.732	Cr +1	0.91	Co +2	0.72
		+2	0.89	+3	0.63
Ti +1	0.96	+3	0.63		
+2	0.94	+6	0.52	Ni +2	0.69
+3	0.76				
+4	0.68	Mn +2	0.80	Cu +1	0.96
		+3	0.66	+2	0.72
		+4	0.60		
V +2	0.88	+7	0.46	Zn +1	0.88
+3	0.74			+2	0.74
+4	0.63	Fe +2	0.74		
+5	0.59	+3	0.64		

crease in atomic radius, because the entire electronic configuration is smaller and more tightly held by surplus positive charges. There is then a gradual decrease in radius to a minimum at Ni^{+2}, followed by a slight rise for Cu^{+2} and Zn^{+2}. The +2 ions are not the most stable for many of these transition elements,

TABLE 6–2 TRANSITION METALS: THE FIRST FIVE 3*d* ELEMENTS

	Scandium	*Titanium*	*Vanadium*	*Chromium*	*Manganese*
Symbol	Sc	Ti	V	Cr	Mn
Atomic number	21	22	23	24	25
Atomic weight	44.956	47.90	50.942	51.996	54.938
Valence e^-	$4s^23d^1$	$4s^23d^2$	$4s^23d^3$	$4s^13d^5$	$4s^23d^5$
mp, °C	1541	1660	1890	1857	1244
bp, °C	2831	3287	3380	2672	1962
d, g/cm³	3.0	4.51	6.11	7.19	7.43
Atomic volume	15.0	10.6	8.35	7.23	7.00
Atomic radius, Å	1.60	1.46	1.31	1.25	1.29
Ion radius (+2), Å		0.94	0.88	0.89	0.80
Pauling EN	1.3	1.5	1.6	1.6	1.5
Z_{eff}/r^2	1.20	1.32	1.45	1.56	1.60
Standard potential	−2.08	−1.63	−1.2	−0.74	−1.03
Oxidation states	+3	+1, +2, +3, +4	+2, +3, +4, +5	+1, +2, +3, +6	+2, +3, +4, +6, +7
Ionization energy*	151	158	156	156	171
Heat of vaporization	81	106.5	106	76.6	53.7
Discovered by	Nilson	Gregor	del Rio	Vauquelin	Gahn
Date of discovery	1879	1791	1801	1797	1774
rpw* pure O_2	Sc_2O_3	TiO, TiO_2	V_2O_3	Cr_2O_3	MnO_2
rpw H_2O	$Sc(OH)_3$	TiO_2	none	none	none
rpw N_2	ScN ns*	TiN ns	VN ns	none	none
rpw halogens	ScX_3	TiX_2, TiX_4	Vx_3, VX_5	CrX_3	MnX_2
rpw H_2	ScH ns	TiH_2 ns	none	none	none
Mohs hardness				9.0	5.0
Crystal structure	hexagonal	hexagonal	cubic bc*	cubic bc	cubic
Modulus of elasticity		14.7 a*	18.5 a		
Compressibility	2.26 c*				0.84c

*Energies and heats are in kcal per mole; a = millions of pounds per square inch; bc = body-centered; c = compressibility in cm²/kg × 10^{-6}; ns means a nonstoichiometric "compound" of variable composition formed when the indicated metals are heated in N_2, NH_3, or H_2; rpw = reaction product with; standard potential is in volts.

but are chosen in order to have a uniform basis for comparison. The radii of the most common oxidation states of the first six transition elements are shown as labeled dots below the line for +2 ions, and above that line the radii of Cu^+ and Zn^+ are also included for comparison. Some further idea of the range of oxidation states, and the corresponding range of ionic radii, can be gained from Table 6–1.

ELECTRONIC STRUCTURE VS. MECHANICAL PROPERTIES. As the first transition series unfolds, the 3*d* orbitals are filled progressively but not always in uniform steps. Table 6–2 lists the usual properties (as in the tables in Chapter 5) for the first five transition elements, and Table 6–3 does the same for the last five elements. Under "Valence e^-" on both tables, notice the distribution of electrons between the 4*s* and 3*d* orbitals. The first three elements gain one *d* electron at a time, but then we find chromium "borrowing" an electron from the 4*s* pair in order to half-fill all of its *d* orbitals. The next element, manganese, then has filled *s* and half-filled *d* orbitals, after which pairing of the *d* electrons begins. The emphasis on completely-filled or half-filled orbitals (which correspond to lower energy states) then leads copper to fill up all five *d* orbitals at the expense of an *s* electron, after which zinc completes the sequence with completely-filled *s* and *d* orbitals.

The structural and mechanical properties of the metals follow suit, as increasing numbers of electrons are available for metallic bonding. The melting

TABLE 6–3 TRANSITION METALS: THE SECOND FIVE 3*d* ELEMENTS

	Iron	*Cobalt*	*Nickel*	*Copper*	*Zinc*
Symbol	Fe	Co	Ni	Cu	Zn
Atomic number	26	27	28	29	30
Atomic weight	55.85	58.93	58.71	63.55	65.37
Valence e^-	$4s^23d^6$	$4s^23d^7$	$4s^23d^8$	$4s^13d^{10}$	$4s^23d^{10}$
mp, °C	1535	1495	1453	1083	420
bp, °C	2750	2870	2732	2567	907
d, g/cm^3	7.87	8.92	9.91	8.94	7.13
Atomic volume	7.1	6.7	6.6	7.1	9.2
Atomic radius, Å	1.26	1.25	1.24	1.28	1.33
Ion Radius (+2), Å	0.74	0.72	0.69	0.72	0.74
Pauling EN	1.8	1.8	1.8	1.9	1.6
Z_{eff}/r^2	1.64	1.70	1.75	1.75	1.66
Standard potential	−0.440	−0.277	−0.250	+0.337	−0.763
Oxidation states	+2, +3	+2, +3	+2	+1, +2	+1, +2
Ionization energy*	182	181	176	178	216
Heat of vaporization	84.6	88.4	91.0	72.8	1.79
Discovered by	antiq.	Brandt	Cronstedt	antiq.	Marggraf
Date of discovery		1735	1751		1746
rpw* pure O_2	Fe_3O_4	CoO	NiO	CuO	ZnO
rpw H_2O	$Fe(OH)_3$	none	none	none	none
rpw N_2	none	none	none	none	Zn_3N_2
rpw halogens	FeX_3	CoX_2	NiX_2	CuX_2	ZnX_2
rpw H_2	none	none	none	†	none
Mohs hardness	4.5			3.0	2.5
Brinell hardness	91	125	75		
Crystal structure	cubic bc*	hexagonal	cubic fc*	cubic fc	hexagonal
Modulus of elasticity	30	30.6 a*	30 a	17 a	
Compressibility	0.60 c*	0.50 c	0.53 c		

*Energies and heats are in kcal per mole; a = millions of pounds per square inch; bc = body centered; c = compressibility in $cm^2/kg \times 10^{-6}$; fc = face centered; potentials are in volts.

†Brick-red powdery CuH can be made by indirect methods, and liberates H_2 with acids.

TABLE 6–4 TENSILE STRENGTHS OF SOME METALS AT 20° C*

Metal	Tensile Strength (lb/in^2)
Aluminum wire	35,000
Brass (Cu + Zn) wire	100,000
Bronze (Cu + Sn) wire	67,000
Copper wire	65,000
Gold wire	20,000
Iron wire	110,000
Magnesium wire	33,000
Monel (Ni + Cu) wire	175,000
Nickel wire	90,000
Platinum wire	50,000
Silver wire	42,000
Sodium wire	250
Stainless steel wire	300,000
Steel (Fe + C) wire	330,000
Tantalum wire	130,000
Tin wire	5,000
Tungsten wire	590,000
Zinc wire	26,000
Cast cobalt	33,000
Cast iron	23,000
Cast tin	4,000
Cast zinc	9,000
Aluminum sheet	32,000
Stainless steel (Fe + Cr + Ni) sheet	112,000
Titanium (pure) sheet	104,000
Titanium alloy (Ti + C + N) sheet	140,000

*Adapted from E. G. Rochow, G. Fleck, and T. H. Blackburn, *Chemistry: Molecules that Matter.* Holt, Rinehart & Winston, New York, 1974. By permission.

points rise to a maximum at V and Cr, then decline a little and level off at about 1500° C during the *d* electron pairing, and finally decrease drastically when all the orbitals are filled and the electrons are less available for metallic bonding. Similarly, the metals become harder at first, reaching a maximum of 9 on the Mohs scale for Cr (the hardness of sapphire), and then declining to the softness of copper and zinc (which are only as hard as gypsum and calcite). Tensile strength follows hardness and incompressibility rather closely, although individual samples vary according to the treatment of the metal. If we concentrate on samples cut from drawn wires, we can compare the tensile strengths given in Table 6–4, which show how much stronger iron and nickel are than copper and zinc. Chromium is too brittle to be drawn into a wire, but the addition of chromium to iron and nickel in stainless steel produces a very strong metal indeed.

The picture that emerges from the tensile strengths of Table 6–4 and the physical properties listed in Tables 6–2 and 6–3 is in full accord with the theory of metallic bonding. Those elements that can furnish only one electron per atom to the "sea of electrons" that pervades the metal and holds it together are very weak, soft, and low-melting; those metals that can contribute more electrons per atom are proportionately stronger. The strongest and hardest metals are found in the middle of the transition series. The strongest metals per unit volume are titanium, chromium, manganese, iron, cobalt, and nickel in the first transition series, and their congeners zirconium, tantalum, molybdenum, and tungsten in the heavier transition series. This comparison of the properties of pure metals is all quite separate from the special properties that can be attained with intermetallic compounds in alloys, about which there will be more later in this chapter.

MAGNETIC PROPERTIES. The magnetic effects of transition-metal compounds are important because they provide a direct link between the theories of chemical bonding and some unambiguous laboratory methods for testing those theories. The resulting interaction between theory and experiment has led to a consistent view of coordination chemistry which now allows us to predict the colors and magnetic properties of most transition-metal coordination compounds before they are made. In a science which has been mostly experimental and empirical for so long, this is a considerable triumph.

We first need to clarify the various kinds of magnetic behavior. The term *paramagnetism* represents a pure interaction of the spin of an unpaired electron

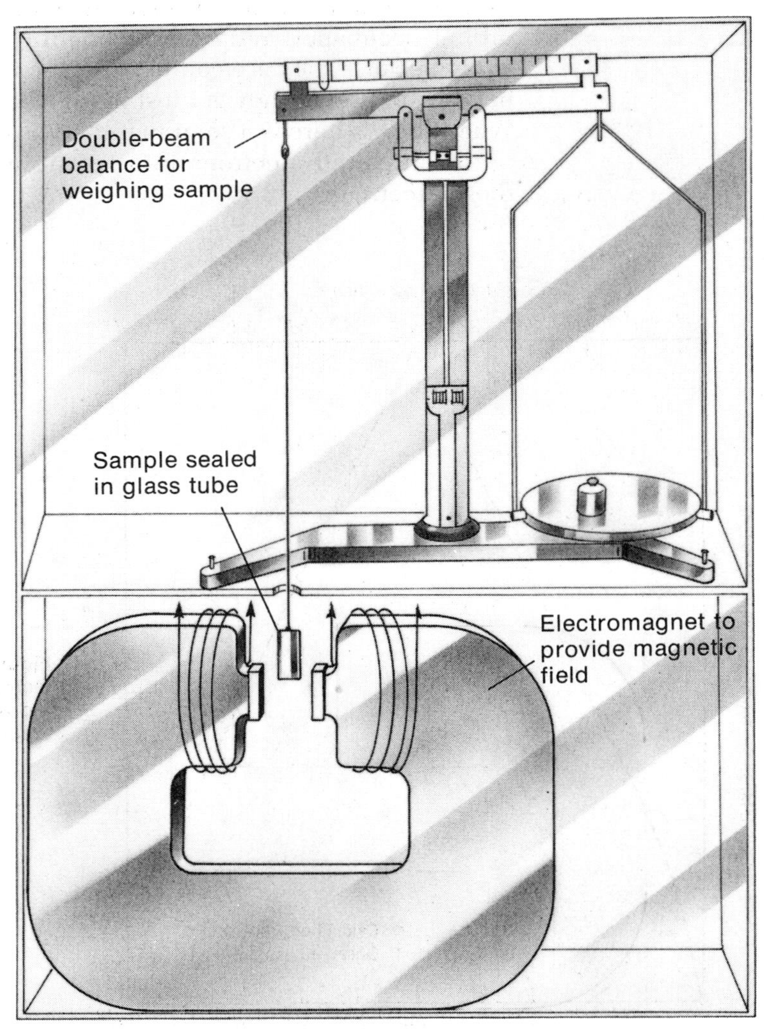

Figure 6–2. Magnetic balance to measure the magnetic properties of a compound. The sample is weighed first with the electromagnet turned off, and then a current is sent through the windings to provide a magnetic field of known intensity. The extent to which the sample is drawn into the field is then measured by the balance.

with an external magnetic field. Isolated atoms or ions containing one or more such unpaired electrons are drawn into the field between the poles of a magnet with a force that can be measured very accurately by a magnetic balance such as that shown in Figure 6–2. What we observe in the behavior of a horseshoe magnet or any other permanent magnet, on the other hand, is *ferromagnetism*, a more complicated phenomenon involving the forced permanent alignment of magnetic domains within a solid. In intensity of magnetic effect, ferromagnetism can far outweigh simple paramagnetism, but of course it cannot apply to gases or liquids. *Antiferromagnetism* is the opposite effect, whereby some solids at low temperatures have their atomic magnets lined up in an *alternating* ordered manner. *Diamagnetism* is a much weaker effect than paramagnetism, and results from the weak repulsion of electron pairs from either pole of a magnet. All substances exhibit diamagnetism to varying degree. We are concerned here only with single-electron effects of paramagnetism, excluding the other kinds of magnetic behavior.

Since paramagnetism is a direct indication of the number of unpaired electrons in an atom or ion, it gives us an instant check on our suppositions about orbital occupancy and type of bonding. Often students are deterred from its study because they are confused by the units used to express the degree of magnetic response, or they get lost in the various temperature effects and anomalies. What follows here is a semiquantitative approach which involves only the number of unpaired electrons in a transition-metal ion, and what we expect from those electrons.

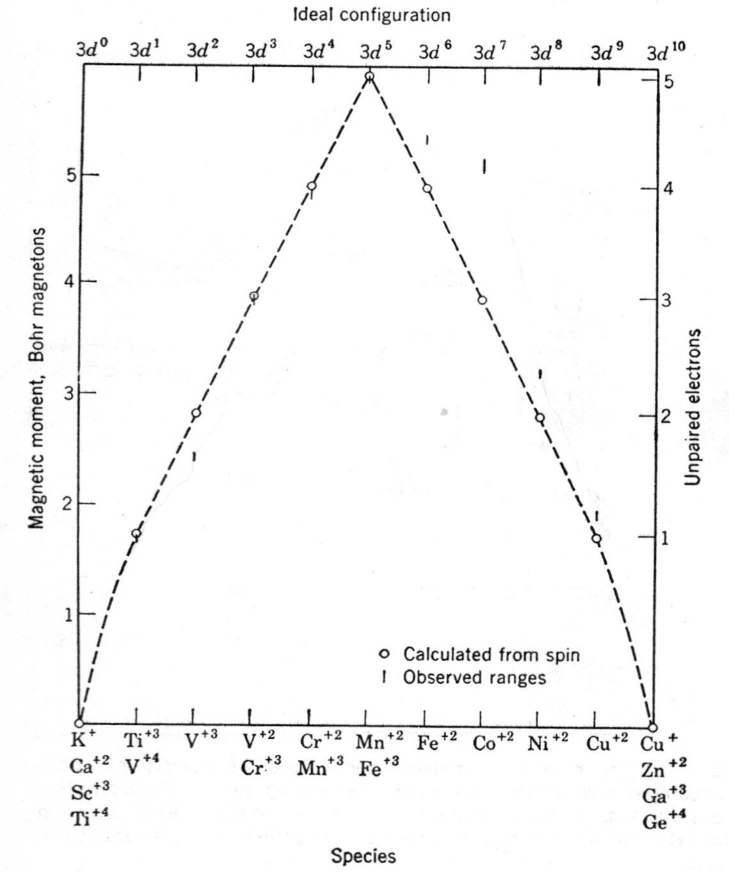

Figure 6–3. Magnetic properties of ions of elements in the first transition series. (From T. Moeller, *Inorganic Chemistry: An Advanced Textbook*, John Wiley & Sons, New York, 1952. By permission.)

In an isolated atom or ion, each unpaired electron acts like a tiny bar magnet and contributes its own magnetic moment to the total additive effect, as shown in Figure 6–3. Here we find the Ca^{+2} ion without any paramagnetic effect because it has no unpaired electrons. In fact, it is stripped down to its argon core of completely filled *1s*, *2s*, *2p*, *3s*, and *3p* orbitals. If scandium loses three electrons, it is in exactly the same condition as Ca^{+2}, so it also has zero magnetic moment. Similarly, Ti^{+4} is stripped down to the argon core and has zero moment. But if titanium loses only *three* of its four electrons, the remaining single electron exerts its magnetic effect, as shown in Figure 6–3. Vanadium, the next element, has a total of five valence electrons, so if it loses four it resembles Ti^{+3} in magnetic effect. If it loses only three, its paramagnetism is found to correspond to two unpaired electrons, as shown. This proves that the two remaining electrons are placed singly in separate orbitals.

In the same manner, V^{+2} and Cr^{+3} correspond to three unpaired electrons, and Cr^{+2} shows that it has four unpaired electrons. The maximum magnetic moment exhibited by any transition metal corresponds to *five unpaired electrons*, one in each of the *d* orbitals. Counting off spaces from the left in the Periodic Table shows that the iron atom has eight valence electrons, and if it loses three of these (the pair of *4s* electrons and one more) the Fe^{+3} ion is left with exactly five electrons, one in each of the *d* orbitals. So Fe^{+3} sits at the maximum of the curve shown in Figure 6–3. Similar counting shows that Mn^{+2} should also be there. The Fe^{+2} ion, however, has one too many electrons to join Fe^{+3} at the maximum; two of its six *d* electrons are paired off, and only four reside singly in separate orbitals, as shown. Successive elements in various oxidation states complete the downward side of the curve, as the electrons pair off and balance each other's spins, until we come at last to Cu^{+}, Zn^{+2}, and the post-transition ions Ga^{+3} and Ge^{+4}, all of which have completely filled *d* orbitals and hence *zero* magnetic moment.

With this clear picture in mind, theoretical discussion of coordination compounds takes on additional meaning and is easier to understand. For example, it helps to look at arrow-in-a-box diagrams of orbital occupancy as though each arrow were indeed a bar magnet. The effect of pairing magnets of opposite alignment when two electrons occupy the same orbital will then be clear, and the effect of introducing additional electrons from donor (ligand) molecules will seem more realistic.

We come now to the effect of coordinated ions or molecules (the so-called "ligands") on the behavior of the transition-metal ion. If this central ion were free to act by itself, all of its compounds would behave in the "high-spin" manner, with the unpaired electrons exerting their full magnetic effect. Some types of ligand, particularly H_2O molecules and F^- ions, have so weak an external electric field that they allow the central transition-metal ion to do just that. When these ligands coordinate, then, they produce high-spin compounds capable of maximum magnetic interaction. On the other hand, ligands with an intense external electric field (such as the CN^- ion, wherein the negative charge intensifies the electron donor capacity of the nitrogen atom) can change the picture, giving rise to "low-spin" compounds.

How this comes about can best be understood by referring to the shapes and positions of the *d* orbitals, as shown in Figure 6–4. Ordinarily all five of the *d* orbitals are equivalent, so that a pair of electrons in any of these orbitals would be at the same energy level as a pair in any other. However, if we think of six potent electron-donors approaching the transition-metal ion along the six axes shown in Figure 6–4, to form an octahedral complex, we see that the intense electric fields of the six donors must surely have an effect. In particular, the six

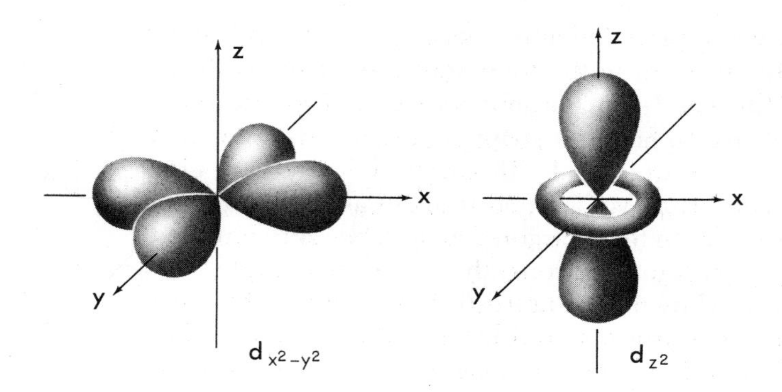

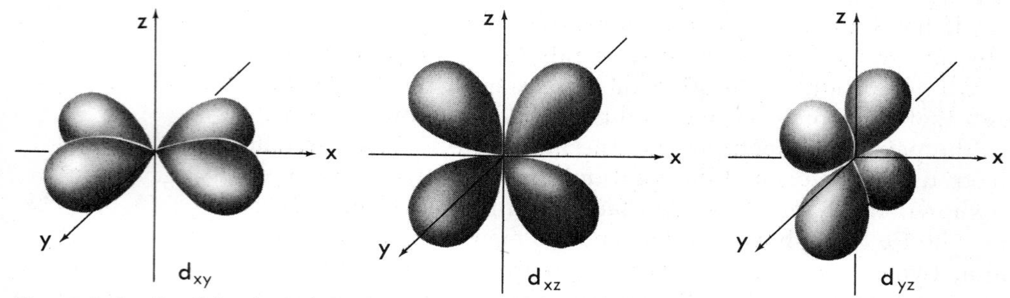

Figure 6–4. Spatial orientation of *d* orbitals. Note that $d_{x^2-y^2}$ and d_{z^2} orbitals are oriented toward ligands approaching corners of an octahedron.

electric fields will interfere markedly with the occupancy of the $d_{x^2-y^2}$ and d_{z^2} orbitals, which extend along these axes. The result can be expressed formally in an energy-level splitting diagram, as shown in Figure 6–5. Here the two orbitals that extend along the axes are assigned a higher energy level than the three interaxial orbitals, which suffer no interference. The amount of splitting depends on the electric field strength of the ligands.

Let us consider now an Fe^{+2} ion about to form a coordination compound (a complex) with six identical ligands. The iron atom began with eight valence electrons ($4s^2$, $3d^6$ as in Table 6–3), and it has lost two (the 4*s* pair) in forming a dipositive ion, so it has six *d* electrons to go into five *d* orbitals. If we draw boxes instead of lines to indicate the split levels shown in Figure 6–5, and then place

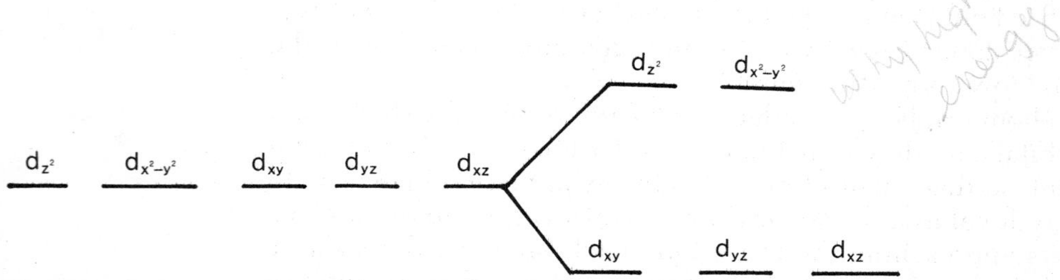

Figure 6–5. Splitting of *d* orbitals by an octahedral crystal field. Two of the orbitals are raised in energy level, while three are lowered.

available electrons in the boxes, we can see what happens. First, suppose six F^- ions approach along the axes. The degree of splitting will be small, because the symmetrical F^- ion has little external electric field:

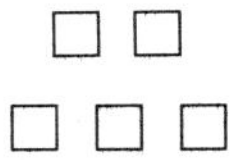

□ □

□ □ □

The energy difference is so small that the six electrons will spread out according to Hund's rule of maximum multiplicity, and we have

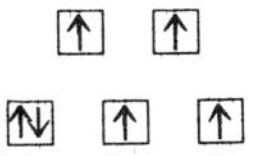

↑ ↑

↑↓ ↑ ↑

for the complex ion FeF_6^{-4}. This will be strongly paramagnetic, at the level of four unpaired electrons (much like the bare Fe^{+2} ion indicated in Figure 6–3). However, suppose we have very strong donors such as CN^- ions, in which the ready availability of an unused pair of electrons on the nitrogen atom at one end of the ion is reinforced by an additional negative charge on the ion. These will exert a strong electrostatic force as they approach an Fe^{+2} ion. Six such CN^- ions, approaching the Fe^{+2} ion symmetrically, will exert so strong a negative electric field as to force the single electrons of iron to pair up within the innermost (lowest level) orbitals of the iron atom, where they find safe refuge, so to speak. The strong electric field of the donors swamps the maximum-multiplicity effect in the bare Fe^{+2} ion, and we have a so-called *low-spin complex:*

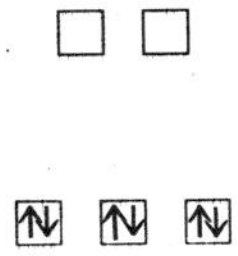

□ □

↑↓ ↑↓ ↑↓

Thus, in the $Fe(CN)^{-4}$ ion, it turns out that all the electrons are paired up, and there is no paramagnetism at all. This is a true low-spin complex.

Not all low-spin complexes have zero paramagnetism. If we start with an Fe^{+3} ion, with only five *d* electrons, and form a cyanide complex, the diagram will be

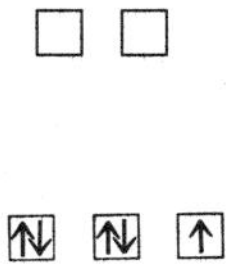

□ □

↑↓ ↑↓ ↑

Hence there will be one unpaired electron left, and a correspondingly small paramagnetic effect.

COLOR OF COMPOUNDS. Obviously we can use the magnetic method as a way to rate ligands according to their field strengths, and this results in a so-called spectrochemical series. In the order of decreasing electric field strength, the common coordinating ions and molecules line up as

$$CN^- > NO_2^- > \text{ethylene diamine} > NH_3 > SCN^- > H_2O > F^- > Cl^-$$

The "spectro-" part of spectrochemical series refers to the optical application of data derived from measurements of paramagnetism. Having determined the degree of splitting of *d* orbitals by a ligand or a combination of ligands, we then know the energy difference between orbitals which are suitable for an optically-observable electron transition. Given Planck's law

$$\Delta E = h\nu$$

where ΔE is the energy difference between the two levels indicated in Figure 6–5, we can calculate the corresponding frequency of the radiation that will be absorbed during the transition under appropriate circumstances. One necessary circumstance is that there must be one unfilled *d* orbital available for an electron to jump to, and of course an electron to do the jumping.

As the donor strength of the ligand decreases, the separation of *d* levels decreases and the energy required to excite an electron decreases, so the color of light absorbed during the transition will move toward the yellow, orange, and red. Since this absorption is subtracted from white light, the residual transmitted light will change in the direction of green to blue to violet. Table 6–5 shows how this works out when frequency is converted to wave length.

Bright colors in transition-metal compounds are not limited to their coordination complexes. Oxygen atoms distributed around manganese in the MnO_4^- ion contribute electrons to the *d* orbitals of that element and give rise to a strong absorption band in the green that can be observed very well with a hand-held direct-vision spectroscope. The resulting intense violet color of permanganates is very impressive and is never forgotten by those who use them. Similarly the chromate ion, CrO_4^{-2}, is intensely yellow, and the dichromate ion, $Cr_2O_7^{-2}$, is orange. Hydrated nickel salts are sea-green, and hydrated copper sulfate is brilliant deep blue, and so on. On the other hand, hydrated zinc sulfate is colorless because there are no empty *d* orbitals on the Zn^{+2} ion, and hence no electron transition is possible.

Some of the major applications of transition-metal compounds stem directly from the colors which arise in the manner just described. The pigment *chrome yellow* is either $PbCrO_4$ or $BaCrO_4$; chrome green is Cr_2O_3. Barn red is hydrated Fe_2O_3, and cobalt blue is a ground-up silicate glass colored deep blue by cobalt oxide. The beautiful colors of stained glass and of pottery glazes and vitreous enamels are all due to transition-metal oxides dissolved in silicate glass. A favorite new blue pigment for automobile finishes is a phthalocyanine complex of copper with the structure

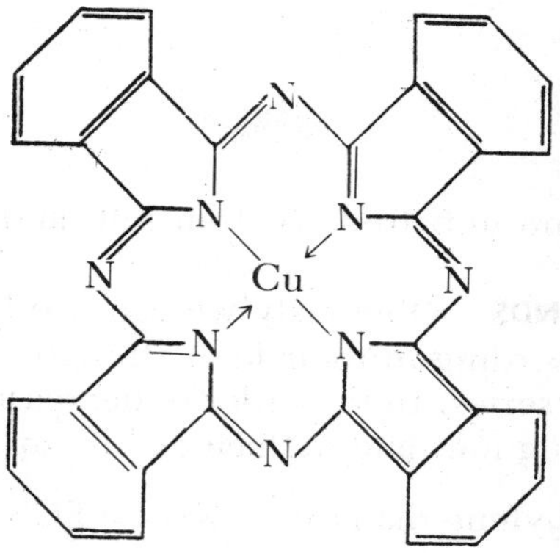

This configuration is much like that of hemoglobin, which gets its red color from the complexed iron atom in its center. There are hundreds of other examples of

TABLE 6–5 COLORS OF COMPLEX IONS OF Co^{3+}*

Complex	Color Observed	Color Absorbed	Approximate Wavelength (Å) Absorbed
$Co(NH_3)_6^{3+}$	yellow	violet	4300
$Co(NH_3)_5SCN^{2+}$	orange	blue	4700
$Co(NH_3)_5H_2O^{3+}$	red	blue-green	5000
$Co(NH_3)_5Cl^{2+}$	purple	yellow-green	5300
trans-$Co(NH_3)_4Cl_2^+$	green	red	6800

*From W. L. Masterton and E. J. Slowinski, *Chemical Principles*, 4th ed., W. B. Saunders Co., Philadelphia, 1977, p. 515. A color photograph of actual samples of these cobalt complexes can be found on Plate 14, *loc. cit.*

color introduced by transition elements, some of which will come up during the discussion of individual elements to follow.

NOTEWORTHY ASPECTS OF SOME COMMON TRANSITION ELEMENTS IN THE FIRST SERIES

Titanium was named for the Titans, the first sons of the earth in mythology. After it was first realized that titanium was an element, in 1791, it took 119 years more to isolate it as a metal. The reason for all this difficulty and elusiveness is that titanium combines avidly with nitrogen, carbon, oxygen, and just about everything else. The ordinary methods of metallurgy do not work when applied to titanium: attempted reduction by carbon results only in a carbide; attempted reduction of the oxide by active metals results in a mixture of oxides and intermetallic compounds; heating the oxide in hydrogen results in an interstitial hydride; electrolysis of an aqueous solution of a titanate gives only oxide. The anhydrous chloride $TiCl_4$ is a volatile covalent nonconducting liquid (bp 136° C), so it cannot be electrolyzed. The only satisfactory way to obtain the metal is by reduction of $TiCl_4$ vapor with red-hot magnesium, followed by separating the $MgCl_2$ and melting the spongy titanium in vacuum.

The principal ore of titanium is ilmenite, $FeTiO_3$, a shiny black mineral which is a constituent of granite and is found mixed with silica sand all along the North Atlantic coast. The paramagnetic Fe^{+2} ions in ilmenite are attracted to a magnetic field, while the diamagnetic silica sand is not, so the ilmenite can be separated from the sand magnetically. It is then mixed with tar or coke, and heated in an atmosphere of chlorine until $TiCl_4$ distills off:

$$FeTiO_3(s) + 3C(s) + 3Cl_2(g) \rightarrow FeCl_2(s) + 3CO(g) + TiCl_4(g)$$

Iron dichloride is nonvolatile and remains behind. The $TiCl_4$ is condensed out of the gas stream, purified, and then vaporized into a vessel filled with hot magnesium shavings:

$$TiCl_4(g) + 2Mg(s) \rightarrow MgCl_2(l) + Ti(s)$$

Most of the liquid $MgCl_2$ (mp 708° C) is drained out, and the rest is washed out with water. The remaining titanium sponge is fused in vacuum in an electric arc and collected in a water-cooled copper pot. It can then be rolled and forged in much the same way as stainless steel, but in a protective atmosphere of argon.

Titanium is our fourth most abundant metallic element, and is so strong and light that it could save an enormous amount of weight and fuel in transportation vehicles. It is only 57% as dense as steel, and has greater strength. It can be alloyed with 5% of Al and 2% each of Fe, Cr, and Mo to minimize the effects of heat. It then retains four fifths of its strength at 400° C, something no aluminum alloy can match. Most titanium produced so far has gone into jet engine parts and airframe construction, but its high corrosion resistance and its high strength/weight ratio recommend it for trains, automobiles, trucks, and bicycles. It is used in chemical plants because it resists all inorganic acids and even withstands dry chlorine. Next to platinum, it has the most resistance to corrosion by seawater.

Titanium dioxide has been mentioned before as a white pigment. It is produced in enormous quantities for that purpose by treating ilmenite with sulfuric acid, filtering the solution, and then partially neutralizing the acid so that hydrated TiO_2 separates:

$$FeTiO_3(s) + 3H_2SO_4(l) \rightarrow FeSO_4(aq) + Ti(SO_4)_2(aq) + 3H_2O$$

$$Ti^{+4}(aq) + 4OH^- \rightarrow TiO_2(s) + 2H_2O$$

When dried and calcined, the hydrated TiO_2 changes to the dense rutile crystalline form of TiO_2. This has a refractive index of 2.7, which is higher than that of diamond (2.42); single crystals of fused TiO_2 actually sparkle and flash more than diamonds, and are sometimes used as gems. With high refractive index comes high dielectric constant, so TiO_2 and various titanates also are used to make capacitors for electronic circuits. Barium titanate, $BaTiO_3$, made by fusing TiO_2 with BaO, has a dielectric constant of 10,000 (compared with 1 for air and 80 for water) and so high capacitance can be achieved in a very small space.*

Compounds of Ti(IV) are colorless for the reasons given above, but Ti(III) salts are blue or violet and Ti(II) compounds are brown or black. Their magnetic properties can be deduced from Figure 6–3. Compounds of Ti(II) and Ti(III) are very strong reducing agents, easily oxidized to the stableTi(IV) state.

Vanadium is not in the public eye very much, but vanadium steel is the toughest ferro-alloy made and is in demand for automobile and truck springs and axles. In another area, vanadium is a nuisance—it is present in Venezuelan crude oil as a residue from bygone animals that used vanadium compounds for their oxygen-transport systems, the way we use iron in hemoglobin. When residual fuel oil containing vanadium is burned, the oxide forms low-melting vanadates which are solvents for the protective oxide coating on stainless steel, and severe corrosion results. Oil refiners must go to some trouble to remove vanadium, while metallurgists are constantly on the lookout for new sources of it.

Vanadium oxide is a catalyst for the oxidation of SO_2 to SO_3 in making sulfuric acid, and various vanadates are used as catalysts in the reductive desulfurization of petroleum. Most of the vanadium produced goes into ferrous alloys, however.

Chromium is a metal everybody knows now, although it did not come into structural or decorative use until the 1930's. Its compounds were known and ap-

---

*A capacitor for storing electric charge has a capacitance directly proportional to dielectric constant, so for the same capacity an air capacitor would have to be 10,000 times as large as a barium titanate capacitor.

preciated a hundred years before that. It was named for the bright colors of its compounds: chromates are bright yellow, dichromates are bright orange, Cr_2O_3 is deep green, $CrCl_3$ is violet, chrome alum is purple, and chromyl chloride, CrO_2Cl_2, is dark red. Chromium resembles iron chemically, and occurs in various iron ores such as chromite, $FeCr_2O_4$. This is reduced by carbon in an electric furnace to obtain ferrochrome, an alloy of iron and chromium which can be used directly to make a variety of stainless steels. To extract chromium itself, the ore is heated in air with sodium carbonate and then leached with water to obtain sodium chromate, Na_2CrO_4. Addition of sulfuric acid gives bright orange crystals of CrO_3, the anhydride of chromic acid, H_2CrO_4. Chromium plating is done electrolytically from a bath of hot chromic and sulfuric acids.

With six electrons available for chemical combination, chromium can assume all oxidation states from +2 to +6. As an element it is decidedly metallic and forms salts such as $CrCl_2$ and $CrSO_4$. In its higher oxidation states it is increasingly acidic, so it appears in the *negative* ion of some Cr(III) compounds, such as sodium chromite, $Na_2Cr_2O_4$, and almost all Cr(VI) compounds, such as sodium chromate, Na_2CrO_4. At the same time chromium leans toward covalency in its higher oxidation states, so CrO_2Cl_2 is a liquid that freezes at −96° C. Chromates and dichromates are both Cr(VI) compounds; they differ only in that dichromates are stable in acids and chromates in bases:

$$\underset{\text{yellow}}{2CrO_4^{-2}} + 2H^+ \rightarrow \underset{\text{orange}}{Cr_2O_7^{-2}} + H_2O$$

Chromates and dichromates are both strong oxidizing agents, and ammonium dichromate has enough oxygen to burn its own ammonium ions to water and nitrogen:

$$(NH_4)_2Cr_2O_7(s) \rightarrow N_2(g) + 4H_2O(g) + Cr_2O_3(s)$$

Hence an ignited pile of bright orange ammonium dichromate crystals looks like a miniature volcano as it burns with a steady blast of flame and sparks, giving off a dust of dark green oxide.

Chromates and Cr_2O_3 are used as pigments, and sodium chromate is used to inhibit corrosion in boiler and radiator systems. Chromyl chloride, on the other hand, is a highly corrosive liquid as well as a strong oxidizing agent; in chloroform solution it is used to oxidize aromatic hydrocarbons to aldehydes and ketones.

Manganese is found in the mineral pyrolusite, which is impure MnO_2 mixed with MnO and Fe_2O_3. It is reduced by carbon in a blast furnace similar to that used for iron, but smaller. The reduction product is ferromanganese, an alloy containing 80% Mn, 14% Fe, and 6% C. Low-carbon ferromanganese is produced by reduction of MnO_2 with silicon in an arc furnace. About 10,000,000 tons of ferromanganese are made per year for use in the steel industry; it takes about 14 lb of manganese to combine with the residual oxygen and sulfur in one ton of steel, making the resulting steel easier to roll and forge when hot.

Pure manganese is an exceedingly tough, hard metal which oxidizes in air to form a brown coating. The metal is not used as such, but a special tough steel containing 12% Mn and 1% C is used for naval armor plate, bulldozer blades, and dredger buckets because it withstands shock and resists abrasion so well. Manganese is also used to strengthen brass and to harden alloys of aluminum and magnesium. Most of our manganese for these purposes must be imported, since very little is found within the U.S.A.

Chemically, manganese is a rather active metal with an electronegativity and a reduction potential similar to those of aluminum. It dissolves in acids to form Mn^{+2} salts, which are pale pink. When heated in air it forms MnO, Mn_3O_4, and MnO_2, but in pure oxygen it can be oxidized all the way to Mn_2O_7. In the higher oxidation states it is more covalent and its compounds become more acidic; green H_2MnO_4 is a rather strong acid, and deep purple $HMnO_4$ is a decidedly strong acid. What, then, is the "true" electronegativity of manganese? The answer is that it is about 1.5 for metallic manganese, but increases considerably in MnO_2 and must reach about 2.5 in permanganates. This shows how misleading it is to assign single values of electronegativity to multivalent elements such as chromium and manganese; these values can only apply to the metal, and give no hint of the acidic behavior in high oxidation states.

Manganese compounds above the oxidation state of +2 are oxidizing agents. The dioxide oxidizes hydrochloric acid to chlorine and is reduced to Mn(II):

$$MnO_2(s) + 2Cl^- + 4H^+ \rightarrow Mn^{+2} + Cl_2(g) + 2H_2O$$

Manganese dioxide is also the oxidizing agent used in ordinary dry-cell batteries. Permanganates oxidize H_2O_2 to oxygen in acid solution, with the manganese again being reduced all the way to +2:

$$2MnO_4^- + 5H_2O_2 + 6H^+ \rightarrow 2Mn^{+2} + 5O_2(g) + 8H_2O$$

In basic solution, however, the permanganate is reduced only to the +4 state:

$$2MnO_4^- + 3H_2O_2 \rightarrow 2MnO_2(s) + 3O_2(g) + 2OH^- + 2H_2O$$

Although hydrated MnO_2 is brown, and dry MnO_2 is black, a small proportion of MnO_2 dissolved in molten glass imparts a pink color which offsets the normally green tinge that comes from dissolved iron. Bottle glass and window glass normally are decolorized in this way. On long exposure to bright sunshine, say for two centuries in a bull's-eye window pane of an old house in Boston or Philadelphia, the glass acquires a beautiful purple tint due to oxidation of the manganese to Mn(VII) by ultraviolet light, and is much prized by antique collectors.

Iron is the old standby of history, commerce, technology, and chemistry courses. If there is time in the classroom or space in the textbook to consider only one metal, iron is rightfully chosen because our culture has been uniquely dependent upon it ever since the industrial revolution. The corrosion of iron and steel by the atmosphere, the hydrolysis of iron salts, and the formation and structure of coordination compounds of iron (including hemoglobin), are all taken up in most textbooks and need not be elaborated here. A little about the production of iron would be in order, though, so that the source and significance of carbon in iron and steel can be understood.

IRON VS. STEEL. Pure iron is a rather soft silvery metal which very few people have ever seen. It is difficult to obtain it free of carbon, oxygen, nitrogen, phosphorus, sulfur, and myriad metallic impurities, and if so obtained it has no use. Iron alone simply would not have made it down through the ages as the preferred metal for weapons, tools, and construction; we depend upon its alloys to achieve the strength, hardness, and workability needed to do the job. The alloys are of two kinds—those that depend solely on carbon, and those that involve alloying with other metals, with or without carbon.

Historically, carbon is the great modifier of iron that brought it into demand.

The most primitive "iron" implements were actually of carbon steel, because charcoal was used to reduce the red ore of iron, and surplus carbon stayed in the metal. Thousands of years ago, powdered iron ore was mixed with powdered charcoal and heated in clay pots to obtain a spongy unmelted metal that could

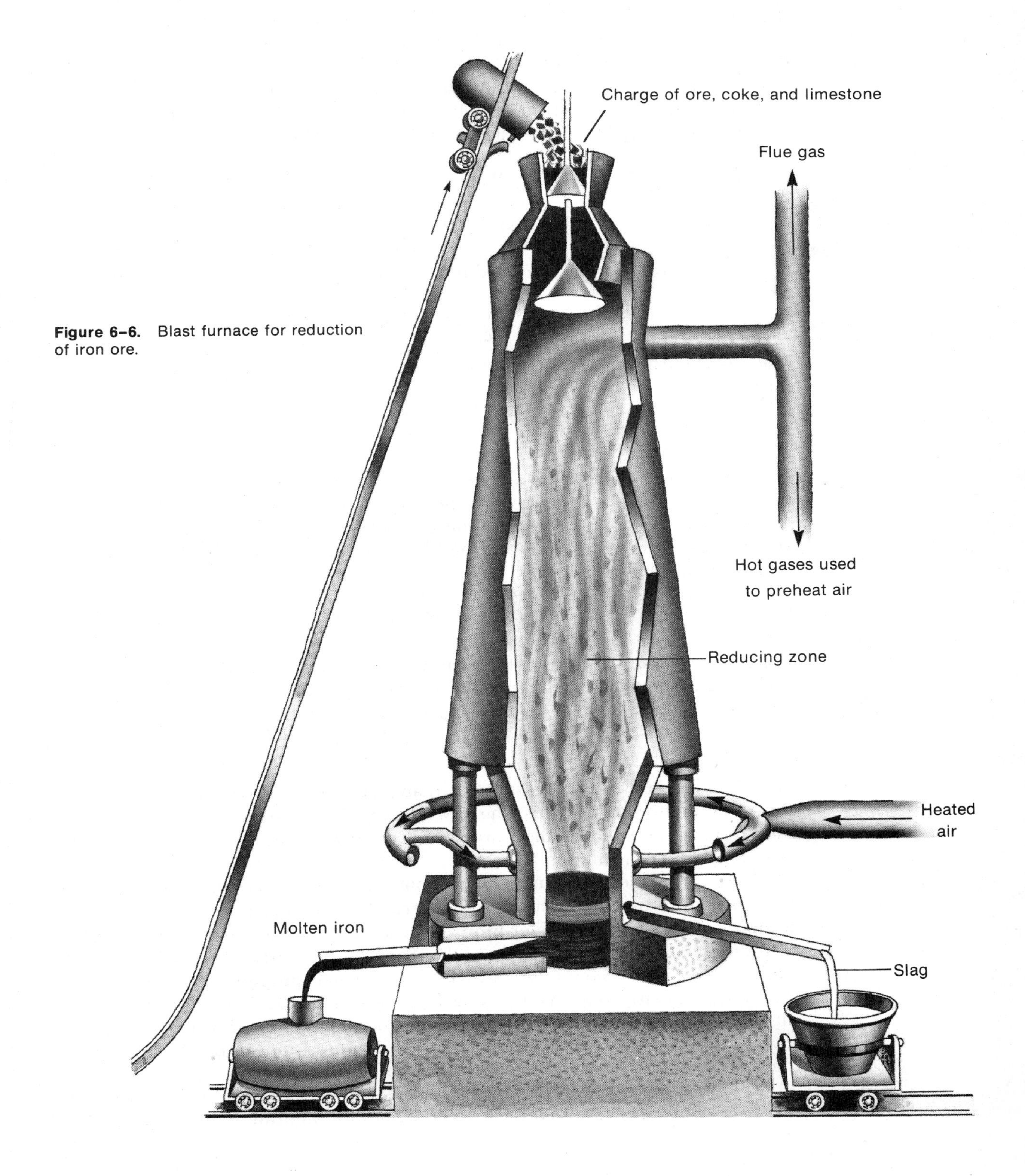

Figure 6–6. Blast furnace for reduction of iron ore.

be reheated and hammered (forged) until a coherent blade or tool was fashioned from it. Later, after invention of the bellows and the blast forge, it became possible to melt the spongy iron in a fireclay crucible under a layer of limestone or chalk, and so obtain an ingot of massive metal that could be hammered into shape while red-hot and plastic. Now coke is mixed with limestone and iron ore, and the mixture is fed into huge blast furnaces. As illustrated in Figure 6–6, the blast furnace is a device for supplying large volumes of heated air to the mixture of ore, limestone, and coke within a shell of refractory material capable of withstanding the high temperature of reaction. The blast of hot air burns the coke to supply the necessary heat, and the carbon monoxide produced at the same time is the principal reducing agent for reaction with the iron oxide:

$$2C(s) + O_2 \rightarrow 2CO$$

$$Fe_2O_3(s) + 3CO \rightarrow 2Fe(l) + 3CO_2$$

Some of the iron oxide is reduced directly by contact with solid carbon:

$$Fe_2O_3(s) + 3C \rightarrow 2Fe(l) + 3CO$$

Much of the carbon dioxide is reduced by contact with the remaining hot solid carbon

$$CO_2 + C \rightarrow 2CO$$

thereby forming more of the active reducing agent. The molten iron runs down through the charge and collects in the bottom of the furnace, from which it is periodically withdrawn by removing a fireclay plug and allowing the stream to run into a ladle or some sand molds. In the meantime, the limestone decomposes to calcium oxide

$$CaCO_3 \rightarrow CaO + CO_2$$

which at the temperature of the reaction can unite with silicon dioxide and silicious gangue to form calcium silicate glass or slag:

$$CaO + SiO_2 \rightarrow CaSiO_3 \text{ (slag)}$$

The viscous, glassy mass is less dense than metallic iron and so collects as a pool above the molten metal. This slag is withdrawn from a separate tap hole, and is used to make cement. A small proportion is blown with air to form rock wool used for thermal insulation.

Few early steel implements have come down to us because the metal had so many impurities it was subject to severe electrolytic corrosion, but we have accounts of its superiority over bronze in war, and we have numerous folk tales relating the legendary prowess of famous steel swords like Excalibur and Nothung. We now know how and why a very small proportion of carbon conferred such greatly improved physical properties on the iron – it was through the formation, solution, and dispersion of iron carbide, Fe_3C, which is 93.31% iron and only 6.69% carbon. The properties of pure iron can be changed quite drastically by the addition of as little as 0.08% carbon.

A look at the iron-carbon phase diagram, shown in Figure 6–7, shows how this is possible. Pure iron melts at 1539° C, but the melting point is reduced to

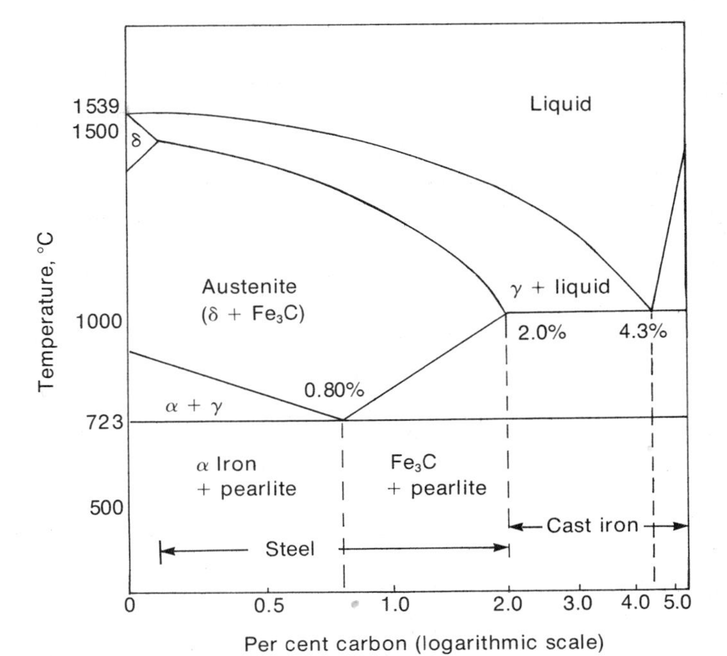

Figure 6–7. The iron-carbon phase diagram.

1015° C (the eutectic temperature) by the addition of 4.3% of carbon. Iron crystallizes in three phases, depending on the temperature: below 910° C it has the body-centered cubic structure called α-iron, from 910° C to 1400° C it has a face-centered cubic structure called γ-iron, and from 1400° C to the melting point it has another body-centered cubic structure called δ-iron. All three phases, α, γ and δ, dissolve carbon and crystallize as solid solutions of Fe_3C in iron in the regions shown. "Pearlite" is just a metallographer's name for the solid phase consisting of alternate platelets or laminations of α-iron and Fe_3C, and "austenite" is a solid solution of Fe_3C in γ-iron.

The solubility of Fe_3C in α-iron (the low-temperature form) is much less than that in γ-iron and δ-iron, so as the alloy cools, Fe_3C must separate. Given enough time, it does so either in the form of pearlite or as crystals of Fe_3C, depending on the concentration. The transformation of tough austenite into malleable pearlite and α-iron depends on the rate of cooling; indeed, the processes can be stopped altogether by rapid chilling. Hence the final properties of a carbon steel depend not only on the percentage of carbon but also on the whole thermal history of the alloy. This explains why the cooling, reheating, tempering, and cold-working of carbon steel is such an art.

Almost any balance of flexibility, hardness, strength, and malleability can be obtained in a carbon steel by adjusting the carbon content and going through the prescribed heat treatment and mechanical working. The cost of carbon steel is minimal, also. There are only two disadvantages: carbon steels corrode easily, and they lose their desired physical properties when hot because they go through the phase changes indicated in Figure 6–7.

Notice that above 2% carbon the situation changes because a different form of iron (the γ modification) crystallizes from the melt. This has an entirely different tolerance for carbon, and a different time-temperature schedule for its separation. When molten iron containing 3% or 4% of carbon is cooled slowly, as by casting, the carbon is not all stable as Fe_3C and some of it has time to crystallize out as graphite. The rest is in the form of Fe_3C dispersed as a coarse precipitate in the iron matrix. In this form the material is a hard, brittle metal called pig iron, with low tensile strength and no flexibility whatever. This is the iron which comes from a blast furnace after reduction of the ore with excess carbon. In order to make steel from it, some of the carbon has to be oxidized by a blast of air until the level is down in the range indicated in Figure 6–7. The oxidation process also gives an opportunity to remove sulfur and phosphorus, two elements which spoil the flexibility and strength of a good steel.

Alloy steels make use of other elements, such as chromium, nickel, molybdenum, and a host of other metals, to change the entire situation in carbon steel. In the first place, the two disadvantages can be alleviated or eliminated: the addition of only 5% of chromium to a carbon steel with 1.5% C will slow down the phase changes so much that thick pieces can be cooled in air instead of quenched in water or oil, and will retain their strength when heated. The corrosion resistance is also vastly improved by the addition of substantial proportions of chromium and nickel. The toughness of steel armor plate is increased by adding manganese, the elasticity of spring steel is improved by adding molybdenum, the durability of keen edges on surgical steel is improved by adding tungsten, and so on. The range of alloy steels is so huge that it will be necessary to limit ourselves here to only a few general classes:

1. *Stainless steels.* The name is given to a great variety of ferrous alloys which have high corrosion resistance. Chromium stainless steels are carbon steels containing 14 to 26% Cr, with the rest iron and carbon. They are capable of being hardened and tempered like any other carbon steel, but are less sensitive to heat changes and corrode very little because of the chemical inactivity of the chromium. Addition of 5% Cr reduces the corrosion rate by 80%, and 16% Cr reduces it to almost zero. Nickel-chromium steels are usually low-carbon steels which are not capable of being hardened or tempered. The most popular is "18-8 stainless," which contains 18 to 20% Cr and 8 to 12% Ni. This is an austenitic steel (face-centered cubic crystals) of excellent corrosion resistance and high strength. Stainless steels of both types achieve tensile strengths of 250,000 to 300,000 lb/sq in, compared with 35,000 for aluminum and 65,000 for copper (see Table 6–4).

2. *Magnetic alloys* may be either permanent magnet types (used in loudspeaker magnets and the like) or nonretentive types used in electric motors, generators, and transformers. Alnico is the general name for the most popular permanent magnet series of high-retention alloys containing Al, Ni, Co, Fe, and sometimes Cu and Ti; Alnico V contains 51% Fe, 14% Ni, 24% Co, 8% Al, and 3% Cu. For AC electrical machinery the object is quite the opposite: to design a magnetic circuit that will permit rapid and efficient magnetization and demagnetization without permanent magnetization. Here 3.5% silicon is dissolved in the steel to increase its electrical resistance seven-fold and to give it the necessary stiffness so that laminations can be stamped out of a rolled sheet. The laminations are insulated from each other by an oxide coating, and are stacked to achieve a solid magnetic core with low losses.

3. *Electrical-resistance alloys* are used for electric heaters, kitchen ranges, clothes dryers, hair dryers, heating pads, and all other electric heating devices. These alloys have a variety of compositions; one contains 72% Fe, 22% Cr, 5% Al, and 1% Co, while others contain more nickel and chromium with less iron. The most popular alloy, Nichrome, contains 80% Ni and 20% Cr, with no iron at all.

4. *Heat-resistant alloys* are special compositions intended to withstand temperatures up to bright red heat (600° C) without destructive oxidation. Two popular compositions are Incoloy, with 46% Fe, 32% Ni, 20% Cr, and 2% Mn

(serviceable to 1100° C), and Inconel, with 74% Ni, 15% Cr, 8.5% Fe, and 2.5% Cu (used in the heating units of electric ranges).

5. *Corrosion-resistant alloys,* beyond the stainless steels considered above, are of special composition used where chemical inertness takes precedence over mechanical properties. Duriron is a cast iron containing 14.5% Si, used for laboratory drains because it resists acids and alkalies. The Hastelloys are super-resistant alloys used in surgical instruments and chemical plant equipment; they contain small amounts of iron and large amounts of nickel, with varying amounts of cobalt, molybdenum, and tungsten.

RECYCLING OF STEEL. In addition to the five types classified above, there are many ferrous alloys that contain special ingredients and are designed to do special jobs. Ours is very much an age of alloy steels, and their variety hinders their recycling. Students of chemistry must understand the difficulties involved in recovering and reusing ferrous metals, so that they can explain these matters to others. For example, why does every town and city (and many a farm, too) have its automobile graveyard, where each car represents a ton or more of wasted metal? Why can't we gather up the cars and remelt the steel for new cars, the way aluminum cans are remelted?

The difficulty is mostly chemical. The cars contain a dozen or more different types of steel, and the alloying agents used in one type ruin the steel for use in another way. Automobile bodies are made from carbon steel containing some lead to improve its deep-drawing properties; obviously if molybdenum spring steel from automobile springs got into this it would make the steel so springy that the body panels could not be drawn or pressed in deep dies. Copper also ruins the deep-drawing properties of steel, and makes it more susceptible to corrosion. Yet the automobile generator and starting motor (plus all the car wiring and all the auxiliary motors) would contribute 20 or 30 pounds of copper if the whole car were remelted, making it impossible to fabricate frames or bodies from that alloy. And so on and on, a litany of troubles due to mixed metals. If all the cars could be completely dismantled and the parts sorted, more of the material could be used. But how shall we get the copper wire out of the generator and starter, when it was wound into the frame by a machine? And how can we afford the labor of dismantling, which is several times the labor of assembly? Obviously we need entirely new designs of cars, in which the recovery of

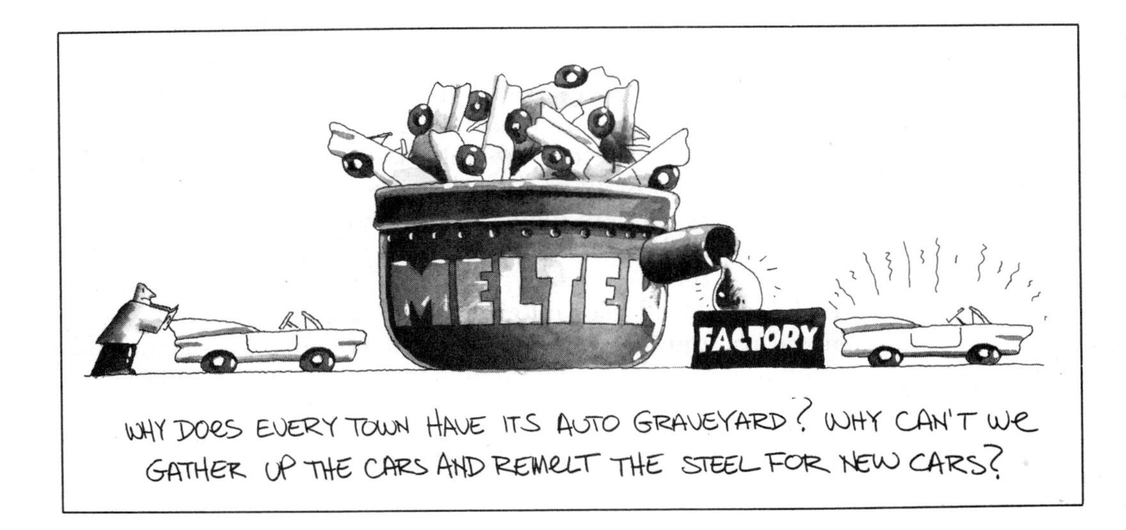

WHY DOES EVERY TOWN HAVE ITS AUTO GRAVEYARD? WHY CAN'T WE GATHER UP THE CARS AND REMELT THE STEEL FOR NEW CARS?

materials is all thought out ahead of time. Or else we need a different mode of transportation, one less wasteful of our resources.

Cobalt is a rather scarce element named after a devil. It occurs along with copper and nickel as the sulfide, from which it is extracted by oxidizing the sulfide, dissolving the residue in sulfuric acid, and fractionally precipitating Co_2O_3 by stepwise addition of $Ca(OH)_2$. The oxide is then reduced by carbon to obtain the steel-gray, hard, tough metal. Cobalt is almost unaffected by air and moisture, and resists nonoxidizing acids but dissolves in 50% HNO_3. Of its two oxidation states, +2 and +3, the lower is the more stable under ordinary conditions: what you find on the reagent shelf are the pink salts $CoCl_2$ and $Co(NO_3)_2$, and the black CoS and CoO. In the presence of ammonia and other strong electron donors, however, stable multicolored complexes of Co(III) form. These are stable in air and resist reduction.

Cobalt is used in many special alloys and metallurgical products, such as Alnico (see ferrous alloys, above) and Carbaloy (a hard metal for the cutting edges of drills and lathe tools, consisting of grains of tungsten carbide embedded in a cobalt matrix). Cobalt oxide is used to color glass and ceramic glazes a beautiful blue. Cobalt is also famous for its colorful octahedral coordination compounds, some of which were historically important in establishing Werner's theory of coordination chemistry 70 years ago. Another of these is of more than just academic interest to all of us—vitamin B_{12}, an essential vitamin, consists of a central cobalt ion coordinated octahedrally to five nitrogen atoms and a CN group. Four of the nitrogen atoms are parts of a planar ring system like that of chlorophyll, while the fifth is part of a very complicated organic phosphate. The entire structure is shown in Figure 6–8. Lack of vitamin B_{12} causes pernicious anemia in man; 1 microgram of it per day will prevent this disease.

Nickel (from the Swedish *kopparnickel,* meaning false copper) occurs with copper and iron in sulfide ores, notably in Ontario and central Africa. The ore is concentrated, roasted to oxide, and then reduced by carbon. Nickel is then separated from the spongy mixture of metals electrolytically, or by warming in an atmosphere of carbon monoxide, whereupon violatile nickel carbonyl (bp 43° C) forms and distills off. The vapors are conducted into a series of tubes heated to 200° C, where the carbonyl decomposes and deposits pure nickel:

$$Ni(s) + 4CO(g) \underset{200^\circ C}{\overset{60^\circ C}{\rightleftarrows}} Ni(CO)_4(g)$$

A natural alloy of nickel containing about 72% Ni, 25% Cu, and 3% Fe, called Monel, is also produced from the same ore merely by reducing and melting the natural mixture of sulfides. This alloy resists oxidation at 800° C, and withstands almost all gases and vapors (including fluorine) at room temperature. Pure nickel itself is highly corrosion-resistant, resembling its congeners palladium and platinum in this respect, so it is easy to understand the key role of nickel in stainless steels and so many other bright-metal alloys. Most of the nickel produced goes into such alloys, the most familiar of which is the coinage metal used in U.S. 5¢, 10¢, 25¢, and $1 coins. This is a malleable alloy of 70% copper and 30% nickel which is rolled into sheets and then explosion-bonded*

*The three sheets of metal are actually bonded or welded together by the impact of an explosion, which forces one metal into the other at the interface.

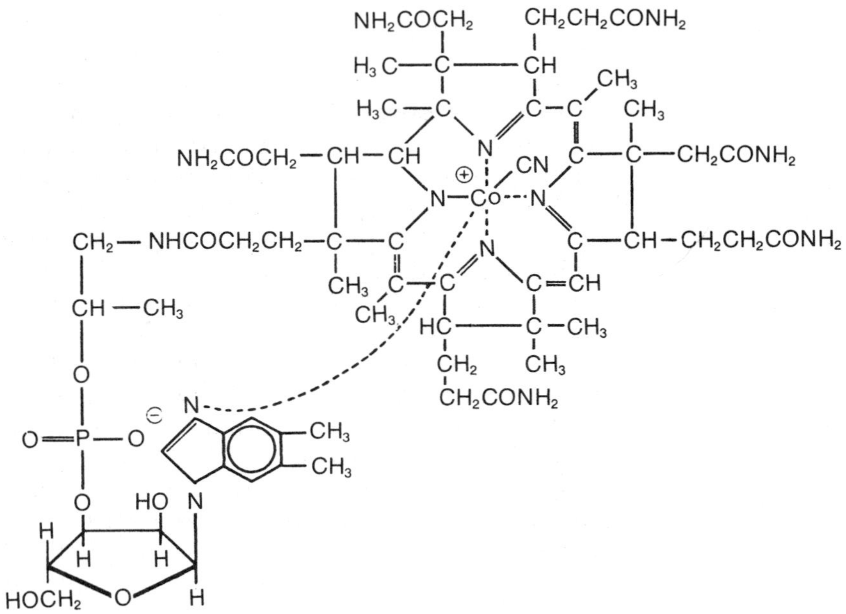

Figure 6–8. Structure of a molecule of vitamin B_{12}. Included in the octahedral coordination sphere around the cobalt ion are four coplanar nitrogen atoms, a cyanide ion, and the nitrogen of an imidazole group. (From E. G. Rochow, G. Fleck, and T. H. Blackburn, *Chemistry: Molecules that Matter,* Holt, Rinehart & Winston, New York, 1974. By permission.)

to a soft copper core. The core reduces the cost and makes the stamping of coins an easier matter, and the cupronickel faces resist wear and corrosion.

While nickel is capable of all oxidation states from zero to four, in most compounds it is present as the Ni^{+2} ion or some coordination complex of it. Solutions of $NiCl_2$ and $Ni(NO_3)_2$ in water are pale sea-green in color due to formation of the $Ni(H_2O)_6^{+2}$ complex. Water can be displaced from this by adding ammonia, precipitating $Ni(OH)_2$ at first but then dissolving the hydroxide as the soluble blue $Ni(NH_3)_6^{+2}$ complex.

A solution of nickel sulfate, $NiSO_4$, is used for electroplating nickel. The dioxide NiO_2 is used as the oxidizing agent in the positive plates of the Edison storage battery, a very durable alkaline-electrolyte battery of indefinite life. (The plates and containers of the battery are made of pure nickel, which is not attacked at all by bases.) Nickel does not form a salt-like hydride, but hydrogen dissolves in the metal and is released by warming. This behavior leads to the use of very finely divided nickel as a hydrogenation catalyst, not only in laboratory work but also in making solid fats out of vegetable oils:

$$(C_{17}H_{33}COO)_3C_3H_5(l) + 3H_2(g) \xrightarrow{Ni} (C_{17}H_{35}COO)_3C_3H_5(s)$$

glycerol trioleate, an edible vegetable oil — glycerol tristearate, an edible solid fat

Copper resembles its congeners silver and gold to the extent of sometimes being found free, as elongated or dendritic crystals of the metal. Very likely this native copper was the first to be found and used by ancient Egyptian and Mediterranean peoples, for it could be hammered into useful shapes and the very act of hammering hardened the metal. Wherever free copper is found brightly colored copper ore is usually also found, so that primitive men were guided to the right minerals and were able to wrest copper from them by roasting and reduction.

Copper introduced Stone Age man to metals, and after copper came its more durable alloys, bronze (Cu + Sn) and then brass (Cu + Zn). Through the years copper and bronze tools, weapons, and coins become scattered and lost, so more and more copper had to be found and refined, and the readily available ores dwindled. Much later, the great demand for bronze church bells, bronze cannon, and huge copper and bronze statues sequestered enormous amounts of copper, as a visit to a museum or old fortress or to the Statue of Liberty or to the Statue of Bavaria or to the Kremlin will show. The result was to force development of a technology to deal with poorer and poorer ores of copper, until today ores containing only 0.25% Cu are considered workable. (By contrast, a titanium ore containing less than 40% Ti is not worth bothering with.) At the same time the price of copper has risen more and more, until now it costs as much as silver did two centuries ago. In the 1960's silver became too scarce and expensive for coins, so the silver coins were withdrawn and replaced by copper. Gold was withdrawn from U.S. coins 30 years before that. So copper has become silver in coins and commerce, and silver has become gold, and gold has gone out of sight. The process will accelerate as 300,000 *extra* people come into the world each day and require articles of copper, so we shall consider ways of recovering and recycling copper later in this section, after considering its metallurgy and chemistry.

The "winning" of copper from its ores begins with prospecting for ore, which is now done with the aid of a microscope and spectrochemical analysis instead of pick and shovel. The principal ores are of two classes—mixed sulfides of iron and copper, such as Cu_3FeS_3 and $CuFeS_2$, and carbonate ores, such as malachite (a mottled green ornamental stone of the composition $Cu_2CO_3(OH)_2$) and azurite (a deep blue or purple basic carbonate, $Cu_3(CO_3)_2(OH)_2$). Once a suitable source is at hand, the steps are:

1. Concentration by flotation, which is described in most textbooks.
2. Roasting, in which the concentrated ore is roasted with enough air to oxidize the iron selectively and leave the less reactive copper as sulfide:

$$2CuFeS_2(s) + 3O_2(g) \rightarrow 2CuS(s) + 2FeO(s) + 2SO_2(g)$$

3. Reduction, in which the roasted ore is mixed with ground limestone, sand, and some fresh concentrated ore, and then heated to 1100° C in a reverberatory furnace (Fig. 6–9). Sulfur in the ore reduces the CuS to Cu_2S, which melts. The lime and silica form a silicate glass which dissolves the iron oxide:

$$CaCO_3(s) + SiO_2(s) \rightarrow CaSiO_3(l) + CO_2(g)$$

$$CaSiO_3(l) + FeO(s) + SiO_2(s) \rightarrow 2(Fe, Ca)SiO_3(l)$$

The silicate slag is less dense than the molten copper(I) sulfide, so it floats thereon as a separate layer and can be drawn off periodically. This is how the iron and copper are separated. The lower layer (called matte) is tapped and run into a Bessemer-type converter where it is blown with air to oxidize the sulfide ions and free the copper:

$$Cu_2S(l) + O_2(g) \rightarrow 2Cu(l) + SO_2(g)$$

4. Refining, in which plates of impure copper are immersed in a solution of $CuSO_4$ and H_2SO_4, alternating with thin sheets of pure copper, and are

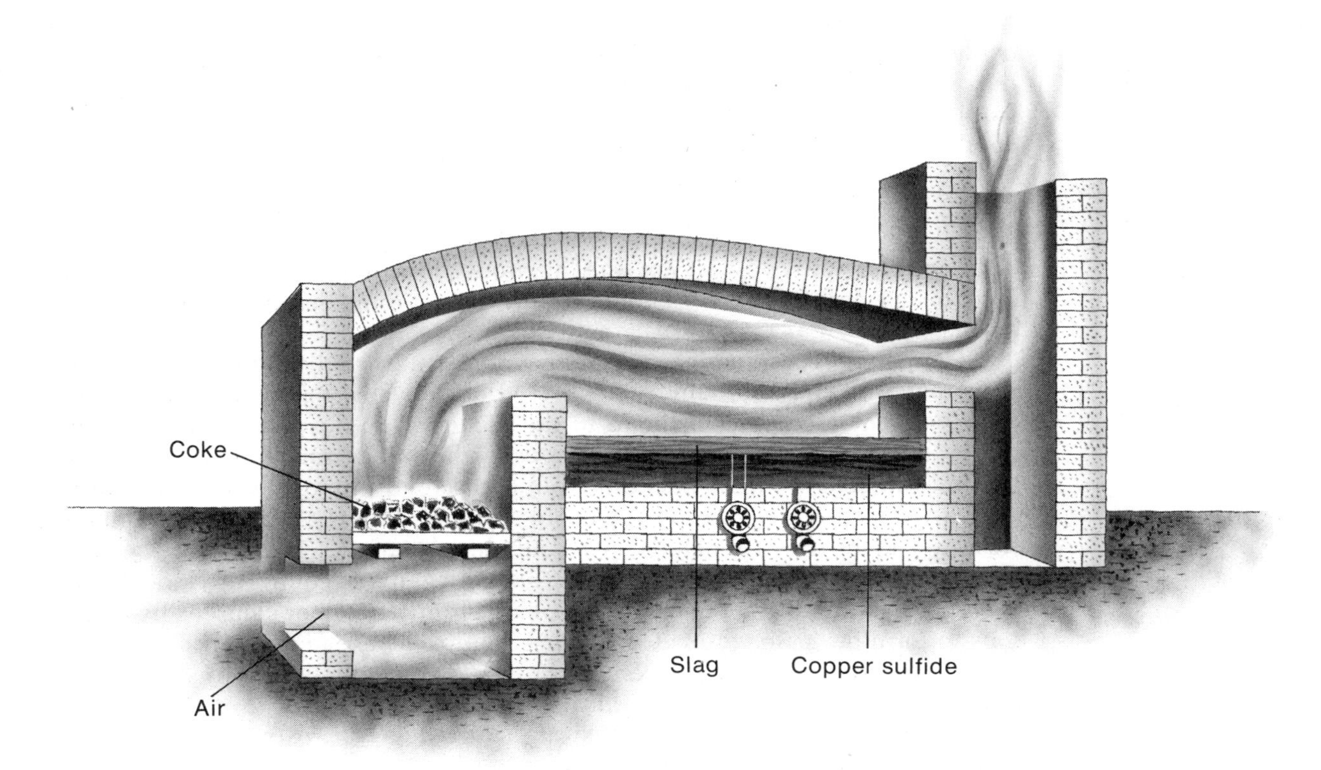

Figure 6–9. Reverberatory furnace for the production of copper matte.

electrolyzed as shown in Figure 6–10. At the positive electrodes the impure copper is oxidized electrolytically and taken into solution; at the negative electrodes pure copper is plated out. As for the undissolved impurities, these fall to the bottom of the tank as a mud or slime which is scraped out occasionally and processed to recover the gold, silver, and platinum. Usually enough precious metal is obtained to pay the entire cost of electrolytic refining.

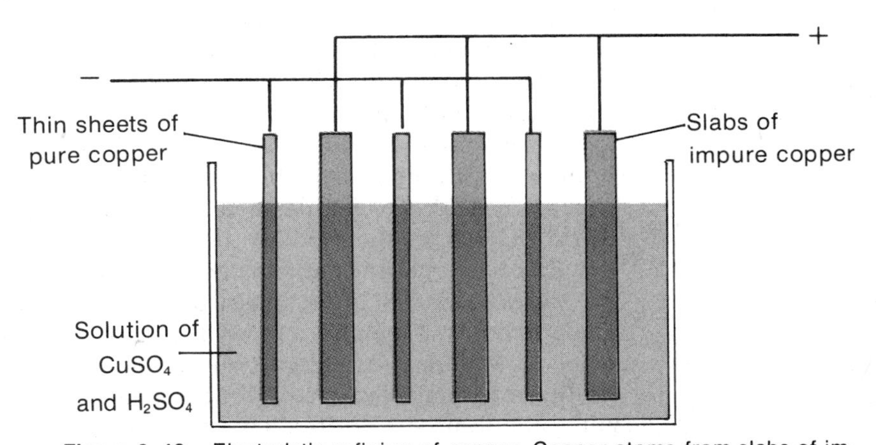

Figure 6–10. Electrolytic refining of copper. Copper atoms from slabs of impure copper (the positive electrodes) are oxidized, passing into solution as Cu^{+2} ions and migrating through the solution to the negative electrodes to be reduced to metal. With careful control of voltage and frequent changing of the solution, very pure copper can be obtained by this method. (From E. G. Rochow, G. Fleck, and T. H. Blackburn, *Chemistry: Molecules that Matter,* Holt, Rinehart & Winston, New York, 1974. By permission.)

Chemically, copper forms two series of compounds, in which it assumes the oxidation states of +1 and +2. Copper(I) salts are diamagnetic, in keeping with the filled *d* orbitals of the Cu^+ ion. They also are unstable and are easily oxidized. In solution Cu(I) salts disproportionate into Cu(II) salts and metallic copper, at a standard potential given by

$$Cu + Cu^{+2} \rightarrow 2Cu^+ \qquad E^0 = -0.37 \text{ volts}$$

The equilibrium constant for the disproportionation is 10^6, so soluble Cu^+ compounds do not survive. The only Cu(I) compounds that persist are insoluble substances such as Cu_2O, Cu_2S, and CuCl.

Copper(II) compounds are paramagnetic and usually highly colored. The principal salt is bright blue $CuSO_4(H_2O)_5$, which is a pentahydrate because four water molecules coordinate to the Cu^{+2} ion in a square planar complex, while a fifth water molecule coordinates to the sulfate ion. In very dilute solution the $Cu(H_2O)_4^{+2}$ ion is pale blue, but addition of ammonia water changes this to the characteristic dark purplish blue of the $Cu(NH_3)_4^{+2}$ ion. The ammonia ligands can in turn be displaced by ethylene diamine, $H_2NC_2H_4NH_2$, which forms a square planar chelate complex that is magenta in color. Ethylene diamine is a strong donor, but it too can be displaced from its copper complex by cyanide ions to form the pale yellow $Cu(CN)_4^{-2}$ complex. These four colorful changes illustrate the successive displacement of ligands from their coordination sites by stronger donors in a spectrochemical series.

All copper compounds are toxic to some degree. An extreme example is copper arsenate, a very potent (but dangerous) fungicide and insecticide. Marine paints containing copper oxide or copper chelate compounds have long been used on ship bottoms to prevent fouling by poisoning marine organisms, but organotin compounds are now preferred for this purpose. The tolerance for copper varies according to the organism; some shellfish actually use copper complexes in their oxygen-transport mechanisms, the way nickel and vanadium were used in now-extinct organisms, and iron is used by man.

RECYCLING OF COPPER. Although copper compounds find other uses, by far the greatest proportion of copper produced goes into copper wire, copper pipe, and copper alloys such as brass, bronze, Duralumin, and coinage metal. There is a market for scrap brass and copper, but unfortunately very little copper wire gets recycled because of its inaccessibility. All that pure copper contained in discarded electric motors and household appliances could easily be remelted and reused, without fear of alloy contamination, if only motors could be designed to be demountable so the wire could be removed. At present everything goes into the city dump or a trash heap if it is pronounced too expensive to repair, and so a great deal of valuable metal is wasted. There is not enough copper in the world to keep up the practice.

Zinc is sometimes associated with copper in sulfide ores, and reduction of the mixed sulfides gives brass, a metal known to the Romans. Zinc was not recognized as a separate element until the end of the eighteenth century. It is twice as abundant as copper, and is found in the form of a carbonate ($ZnCO_3$, smithsonite) and a silicate (Zn_2SiO_4, willemite) as well as the sulfide (ZnS, zinc blende or sphalerite). Two metallurgical routes to the metal are followed. In one, sulfide ore is roasted to the oxide and reduced with carbon:

$$2ZnS(s) + 3O_2(g) \rightarrow 2ZnO(s) + 2SO_2(g)$$

$$ZnO(s) + C(s) \rightarrow Zn(l) + CO(g)$$

In the other an oxide or carbonate ore is treated with sulfuric acid and the resulting $ZnSO_4$ is crystallized, purified, and then electrolyzed in water solution to plate out the zinc. The electrolytic method is more economical when the zinc is to be electroplated anyway, as it is in making electrogalvanized sheet steel. U.S. production of zinc is about half that of copper, and amounts to 500,000 to 800,000 tons per year.

Metallic zinc is a bluish-white lustrous metal that soon turns gray in moist air. It is much lower-melting and more volatile than iron, chromium, and nickel, and it is also much more active chemically. Its chief use is to protect iron and steel from corrosion. Zinc-coated steel ("galvanized iron") can be made by hot-dipping in zinc (mp 419° C), by electroplating, or by spraying with melted zinc. All three methods give about equal protection, because the effect is due more to electrochemical action than to covering up the steel. Sinc zinc has a standard reduction potential of −0.76 volts vs. only −0.44 volts for iron, when the two metals are in contact the zinc will be preferentially oxidized by any corrosive atmosphere or reagent, and the iron will not be affected until all the zinc is consumed.

Zinc itself is not affected much by weather, so it is used in a variety of die-casting alloys for automobile construction. Pure zinc is also used in making dry batteries, where it forms both the container and the negative electrode of each cell. Much zinc is used in making brass, which is the term applied to all Cu-Zn alloys. Red brass contains 75 to 85% copper and is malleable and ductile; yellow brass contains 60 to 70% copper and is harder. Brass is easy to cast, forge, and stamp, so it is used widely for valves, pipes, fittings, and clock parts. Naval brass contains tin as well as copper and zinc, and an alloy called nickel silver or German silver contains copper, zinc, and nickel but no silver at all.

Zinc dissolves readily in dilute acids to form zinc salts and liberate hydrogen, especially if a little copper or iron is present to set up an electrochemical cell. It also dissolves in strong bases to form zincates and liberate hydrogen, so it is amphoteric. The oxidation state of zinc is +2 in both instances:

$$Zn(s) + 2H^+ \rightarrow Zn^{+2} + H_2(g)$$

$$Zn(s) + 2OH^- + 2H_2O \rightarrow Zn(OH)_4^{-2} + H_2(g)$$

Zinc compounds are colorless and diamagnetic. Zinc oxide is an important white pigment and vulcanization activator for rubber, and zinc chloride is used as a Lewis acid catalyst in organic syntheses such as the production of methyl chloride from methanol. Zinc and its compounds do many other diverse jobs with unpublicized effectiveness. Zinc chromate, $ZnCrO_4$, is an effective pigment for anticorrosion paints, and zinc fluoride and chloride are used to treat wood so that it resists termites. Zinc sulfide, which we have already met as a constituent of the white pigment lithopone, fluoresces with a bright green light when struck by x-rays or cathode rays, so it is used on television screens, oscilloscope screens, and x-ray fluoroscopes. Zinc even has biological importance; it is an essential part of the hormone insulin, and so is one of the trace elements necessary for complete nutrition. Usually there is enough in our foods to supply the requirements.

SUMMARY

The elements of the first transition series, elements 21 through 30, differ radically from the representative elements discussed in earlier chapters in at least four ways: most transition metals are exceptionally strong and hard, they can assume a variety of oxidation states in chemical combination, their compounds are usually highly colored and paramagnetic, and the metals and their compounds often exert catalytic effects. These characteristics derive from the filling of the 3*d* orbitals. The filling is not in steady progression, as Tables 6–3 and 6–4 show; variations occur according to the possibility of achieving empty, half-filled, and completely-filled *d* orbitals. Chemical behavior is governed accordingly, as we note oxidation states of +2, +3, and +4 (but no more) for Ti, a maximum of +5 for V, maxima of +6 for Cr and +7 for Mn, and then a decline to +2 for Zn. Similarly, the ions Sc^{+3} and Ti^{+4} at the start of the series and Cu^{+} and Zn^{+2} at the end of the series are diamagnetic, having completely empty or completely filled *d* orbitals. All other bare transition-metal ions are paramagnetic. The ions Mn^{+2} and Fe^{+3}, with *d* orbitals exactly half-filled (single electrons in all five orbitals), exhibit maximum magnetic moment.

All the transition elements are notable for their coordination compounds, where the coordination number is usually six (sp^3d^2 octahedral configuration), but sometimes four (sp^3 tetrahedral or sp^2d square planar configurations). These coordination compounds exhibit isomerism, with attendant color changes. They also change their magnetic behavior according to the electric field exerted by the ligands, giving rise to "high-spin" and "low-spin" complexes.

Another general characteristic of all the transition metals is catalysis of oxidation-reduction reactions. With so many orbitals available, an electron gathered up from an oxidation can be accommodated temporarily and then delivered up to a reduction. In effect, the transition metals swing back and forth between oxidation states. Vanadium is used as catalyst for the oxidation of SO_2 to SO_3 in making H_2SO_4 and activated iron catalysts are the mainstay of ammonia production. Iron chloride carries chlorine as it goes from $FeCl_2$ to $FeCl_3$ and back again in the chlorination of hydrocarbons. Chromium molybdate catalyzes the cyclization of gasoline hydrocarbons. Dissolved coordination compounds of iron, cobalt, nickel, and copper are extremely effective catalysts for the hydrogenation of unsaturated hydrocarbons.

Although all the transition elements are metals, they become more and more electronegative in their higher oxidation states, and may even form acids. The strongest acids (chromic and permanganic) correspond to the highest oxidation states of the elements (+6 for Cr, +7 for Mn). The small sizes and multiple charges of transition-metal ions encourage their changeover to covalent bonding in many halides and carbonyls, and even in anhydrous copper nitrate.

The elements of the first transition series are all relatively abundant and available, and all except scandium are extremely useful structural and mechanical metals. They form a large number of useful alloys with each other. Many of the alloys are within common experience: chromium steel, stainless steel, bronze, brass, Alnico, monel, coinage metal, and nickel "silver." Iron has its own special relationship with carbon, leading to carbon steel with enormous variation in properties. Among various useful compounds of the transition elements we find pigments, catalytic agents, chemicals for treating paper and wood, insecticides and fungicides, termite and barnacle repellents, corrosion inhibitors, oxidizing agents, materials for batteries, and even hemoglobin, vitamin B_{12}, and insulin. No other sequence of elements, either group or period, is as uniformly useful, important, and essential as these light transition metals.

GLOSSARY

aldehyde: a class of organic compounds containing the carbonyl group linked to readily-oxidizable hydrogen; a compound of the type $R{-}\underset{\displaystyle H}{\underset{|}{C}}{=}O$, where R is any hydrocarbon or substituted hydrocarbon group.

alloy: a material composed of two or more metals, or of a metal or metals with a non-metal, intimately mixed by melting together.

anhydrous: devoid of water; completely dried, as by chemical removal of water or by rigorous exclusion of water during synthesis.

austenitic steel: a strong, tough ferrous alloy in which the delta (body-centered cubic) allotrope of iron contains dissolved and dispersed iron carbide, Fe_3C.

capacitance: the property of being able to collect a charge of electricity. The magnitude of capacitance c can be calculated from the relation

$$c = \frac{\epsilon A}{4\pi d}$$

where A is the area of the metal plates of the capacitor, d is the distance between them, and ϵ is the dielectric constant of the insulating material between the plates.

capacitor: a device for accumulating a charge of electricity, consisting of two conducting plates separated by an insulator (dielectric).

chloroform: the compound $CHCl_3$.

dielectric constant: a measure of the electric storage capacity of a substance, given by the ratio of the charge accumulated by a capacitor containing that substance as dielectric to the charge accumulated by the same capacitor containing only air as dielectric.

forge: v.t., to form by heating and hammering, as a blacksmith makes a horseshoe; n., a forced-draft fireplace (usually equipped with bellows) in which metals are heated before shaping.

high-spin complex: a coordination compound in which unpaired electrons are spread out in the *d* orbitals of the central ion with maximum multiplicity, giving maximum paramagnetic effect.

Hund's Rule: the rule of maximum multiplicity, which states that electrons occupy all the available orbitals *of equivalent energy* singly at first, and become paired only after all those orbitals already contain one electron.

hydrate: a substance containing chemically combined water, such as $Na_2SO_4(H_2O)_{10}$.

ketone: a class of organic compounds in which two moieties (portions of a molecule) are joined by the carbonyl group, C=O; a compound of the type R—CO—R′, where R and R′ may be the same or different hydrocarbon groups.

ligand: an ion or molecule which donates a pair of electrons to a central atom or ion to form a coordination compound.

low-spin complex: a coordination compound in which the *d* electrons are paired within the three lower-energy (interaxial) *d* orbitals as a result of the electric fields exerted by the ligands. A low-spin complex has minimum multiplicity of electrons and hence minimum paramagnetic effect.

metallurgy: the science of separating metals from their ores, or of preparing alloys from those metals.

modulus of elasticity: a coefficient of elasticity of a substance, equal to the ratio of pressure exerted on the substance to its fractional change in volume.

Ohm's law: the relation between the voltage E, the current I, and the resistance R as a current passes through an electrical circuit, given by $I = \frac{E}{R}$.

tensile strength: the maximum force per unit of cross-sectional area required to break a given sample of material by pulling it apart.

EXAMINATION QUESTIONS

1. How many unpaired electrons would be expected in an uncoordinated ion of: a. Ti(III)? b. V(III)? c. Mn(II)? d. Ni(II)? E. Co(III)?

2. How is a mechanical property such as tensile strength related to the electronic structure of a metallic element? Summarize the theoretical argument and give examples to support the theory.

3. What is meant by "high-spin" and "low-spin" coordination complexes and how do they arise? Give an example of each, and tell how they differ in physical properties.

4. What oxidation states are at least theoretically possible for an element of atomic number 23? of atomic number 26? Of these, what oxidation states would you expect to be most stable for the two elements?

5. Summarize the sources, metallurgy, physical properties, chemical properties, and uses of nickel.

6. How does the metallurgy of titanium differ from that of manganese? Summarize the chemical reactions involved in the reduction of both metals from their ores, and explain why the methods differ.

7. Making use of chemical equations wherever possible, summarize the entire process for obtaining pure copper from a sulfide ore such as bornite, Cu_3FeS_3.

8. What alloys of copper have been (and still are) in widespread use? Name at least three, and give their approximate compositions.

9. Using diagrams and chemical equations, outline the entire process of making low-carbon steel from an oxide ore such as hematite, Fe_2O_3.

10. Why is carbon so effective in altering the physical properties of iron? What is the effect of a. 0.3%, b. 1%, and c. 5% of carbon on the properties of the iron-carbon alloy?

11. What is "18-8" stainless steel? How does it differ in composition and properties from other stainless steel?

12. What kind of ferrous alloy is used for: a. laboratory drainpipes; b. kitchen knives; c. heater windings of a toaster; d. dredger buckets or bulldozer blades; and e. loudspeaker magnets? State the approximate composition of each alloy, and tell why the various constituents are employed in the alloy.

13. How does zinc differ from all the other elements of the 3*d* series in electronic structure, physical properties, and chemical properties? What uses of zinc depend on these properties?

"THINK" QUESTIONS

A. Zinc oxide is an amphoteric substance. From information given in this chapter, write balanced equations for the preparation of zinc chloride and potassium zincate from zinc oxide, ZnO.

B. Consider the iron-carbon phase diagram, and review the method of making a medieval knight's armor as described in Chapter 3. Then explain the layered structure of such armor, and point out whatever advantages you can ascertain in that structure. What happens if the surface layers are too thin, or are so thick as to join in the center?

C. In Wagner's opera *Siegfried,* the dwarf Mime is unable to repair the broken magical sword Nothung despite all his skill as a blacksmith. Years before, Brunnhilde had given him the pieces of the sword after it was shattered upon the spear of Wotan, their all-powerful god and father.

a. In terms of the chemistry of iron-carbon alloys, why did the sword shatter in the first place, instead of just bending?

"I am Siegmund!
My witness is this sword . . .
Wälse promised me that in
deepest distress
I should one day find it.
Now I grasp it!
Holiest love's highest need . . .
Nothung, keen blade . . .
Come forth from thy scabbard
to me!"
(With a mightly effort,
Siegmund pulls the sword
from the tree and shows it
to the astonished and
delighted Sieglinde.)

ACT 1, SCENE 3 OF "DIE WALKURE": THE HERO SIEGMUND PULLS THE MAGIC STEEL SWORD FROM THE TRUNK OF AN OAK TREE, WHERE THE GODS OF CREATION THRUST IT LONG AGO.

b. Although Mime did his best to join the pieces together by heating and forging, the sword always broke again in the hands of the youth Siegfried, who admittedly gave it a drastic test by attempting to cut through an anvil. Why would the technique of repeated heating to 500° C and hammering be doomed to failure as a method of repair?

c. The impatient Siegfried, contemptuous of his elders and their methods, filed the pieces of sword to powder, heated the powder with a flux in a crucible until the steel melted, and poured the molten steel into a mold and quenched it in cold water. He then reheated the blade only once, and only enough to hammer it into final shape before plunging it into cold water again. After sharpening the blade and fastening it in its handle, he had a sword so strong that it never broke under his full strength; he even cut Wotan's spear with it, and so destroyed that symbol of parental authority. Again, in terms of the chemistry of iron-carbon alloys, explain Siegfried's success.

D. How might the vanadium which is a nuisance in crude oil be extracted or recovered to be used in making steel, where it is beneficial and highly desirable?

E. Think about plating chromium on an automobile bumper from an electrolyte containing chromic and sulfuric acids.

a. Write balanced equations for the reactions at the positive and negative electrodes during the electrolysis.

b. As which electrode does the bumper serve?

c. If the bumper has an area of 1 square meter, how could you be sure of arriving at a plating which is exactly 0.1 mm thick? Be quantitative.

F. Calculate the percentage of cobalt in vitamin B_{12}, the structure of which is shown in Figure 6–8. If the minimum daily requirement of vitamin B_{12} is 1 microgram (10^{-6} g), how much cobalt does the human body require each day to stay healthy?

G. In terms of the properties of the transition metals considered in this chapter, why are zinc-based alloys used for die-casting automobile radiator grilles and door handles? Why not iron, or stainless steel, or chromium, or nickel?

H. If you were designing a space vehicle, what metal would you choose from the first transition series for the critical structural parts? Why?

PROBLEMS

1. One hundred ml of 0.1 M cobalt nitrate solution was used to prepare the bright purple *cis*-isomer of $Co(NH_3)_4Cl_3$. This was analyzed by dissolving one tenth of the product and titrating the solution with 0.01 m $AgNO_3$ solution. (It should be remembered that only one of the three chloride ions in the given compound is available as Cl^- in solution; the other two are tied up in the coordination sphere of cobalt and are not reactive in solution.) How much of the $AgNO_3$ solution was required to precipitate all the available chloride ion as AgCl?

2. If the U.S. production of steel is about 160 million metric tons annually, and if all of this were to come from pure Fe_2O_3 as ore, how much metallurgical coke is required annually by the U.S. steel industry? Assume that the coke is 100% carbon, and ignore all waste, thermal losses, and residual dissolved carbon in the steel.

3. Ores containing only 0.25% Cu are now being worked to obtain the copper. What weight of such ore would be required to produce the Statue of Liberty today, assuming 100% yield from the ore? The statue is made of copper sheets bolted to an iron frame, and the copper in it weighs 200,000 lb.

4. Ilmenite, $FeTiO_3$, is the usual ore of titanium. How much ilmenite is required to make 1 metric ton of $TiCl_4$, preparatory to reducing this to metal? How much magnesium is required for reduction? How much titanium will be obtained as metal? Assume 100% yield in all cases.

5. A popular kind of stainless steel contains 18% nickel. How much sulfide ore, NiS, is required to produce 1 metric ton of this stainless steel, neglecting all losses?

6. One hundred ml of a solution of hydrogen peroxide is titrated with 0.01 M $KMnO_4$ in acid solution, and 27.5 ml of the permanganate solution are required to reach the end point. What was the concentration of H_2O_2 in the sample?

THE HEAVIER TRANSITION ELEMENTS

7

In the previous chapter we considered the general aspects of transition elements, especially insofar as they differ from the representative elements of the main groups. This general description was followed by some details about each of the elements in the first transition series, the 3*d* metals. Now we come to the 4*d* and 5*d* series, 20 elements which are less well known and less available than iron, nickel, chromium, copper, etc., but which merit some consideration because a few of them are very important.

GENERAL TRENDS IN THE 4*d* SEQUENCE

The second transition series, elements 39 through 48, follow rather closely the pattern of the first transition elements in matters of oxidation states, magnetic behavior, and coordination chemistry. The general conclusions about these matters developed in Chapter 6 therefore apply here in full force. At the same time there are four ways in which the second transition series differs from the first:

1. The second transition series lies in the fifth period, and so has a set of seven *f* orbitals lying just above the 4*d* orbitals which are the active valence orbitals in this series. This is the first time *f* orbitals have come within the range of chemical energies, and so we may expect them to be used occasionally the way silicon and sulfur use their 3*d* orbitals. One result is an expansion of the coordination number to eight in some compounds.

2. The combined effects of the normal decrease in atomic radius noted in the development of Periods 2 and 3, and the 3*d* transition-metal shrinkage noted in Chapter 6, result in smaller atomic radii than would be expected for a series of elements with such complex atomic structures. Because of the unusually small sizes of their atoms, the elements of the 4*d* series are somewhat *less* electropositive than those of the 3*d* series. For example, silver is less electropositive and less reactive than copper, and cadmium is less reactive than zinc. This

behavior is contrary to expectation from a study of the main groups alone, where we find rubidium more electropositive and more reactive than potassium, and iodine more electropositive than bromine. The consequence of lower reactivity among elements of the second transition series is increased corrosion resistance and hence increased utility for protective purposes.

3. A further result of unexpectedly small size is increased covalency. Covalent compounds are indeed more prevalent among the heavier transition elements, and the maximum observed covalency is eight, especially among chelate compounds of the type MA_4 (where A represents the bidentate acetylacetone structure).

4. Because the 4*d* elements are heavier than the 3*d* elements, they are scarcer. Except for a big bulge in the neighborhood of isotopes containing 50 neutrons in their nuclei (50 is a magic number), all the elements in the second transition series are rare. For example, the terrestrial abundance of rhodium is only 0.001 parts per million, and that of palladium is only 0.01 ppm. Even the well-known element silver occurs only to the extent of 0.1 ppm. The exceptions in the neighborhood of 50 neutrons occur at zirconium (220 ppm, chief isotope $^{90}_{40}Zr$), yttrium (28 ppm, single species $^{89}_{39}Y$), niobium $^{93}_{41}Nb$ (24 ppm), and molybdenum (10 ppm, chief isotope $^{98}_{42}Mo$).

THE INDIVIDUAL 4*d* ELEMENTS

SOME INTERESTING POINTS ABOUT ZIRCONIUM. Anyone interested in gems or minerals encounters zircon early in his explorations. Zircon is zirconium orthosilicate, $ZrSiO_4$. Usually it is brown or rust-colored, but occasionally clear colorless crystals are found in stream beds, and these are prized as gems because they have a high refractive index and a very high dispersion (which is change of refractive index with wavelength). In fact, zircon has a higher dispersion than diamond, so it has a more colorful flash, or "fire," as a jeweler would call it. Unfortunately its hardness is only 7.5 on the Mohs scale (diamond = 10), so it does not wear as well as sapphire or diamond.

Zirconium is in considerable demand as a metal and as its compounds, so it is extracted from zircon by a process very similar to that used for titanium. The zircon is heated with carbon in an arc furnace to form a hard refractory carbide (the silicon being driven off as the volatile monoxide), and this carbide is chlorinated to obtain $ZrCl_4$, a volatile chloride which sublimes at 300° C:

$$ZrC(s) + 4Cl_2(g) \rightarrow ZrCl_4(g) + CCl_4(g)$$

The chloride is purified by sublimation and then its vapor is reduced with hot magnesium:

$$ZrCl_4(g) + 2Mg(s) \rightarrow Zr(s) + 2MgCl_2(l)$$

The molten magnesium chloride is drained away, and the spongy metal is washed and then fused in an arc furnace in an argon atmosphere. Oxygen, nitrogen, and carbon all act to make zirconium hard and brittle, so the metal has to be safeguarded from these throughout the process. A simple test for hardness tells whether the final product is satisfactory.

Metallic zirconium has high strength, very low corrosion rate, no toxicity, and an extremely low neutron absorption when pure. It is a favorite for the construction of nuclear reactors, but for this purpose it has to be freed of the one per cent or so of hafnium it always contains in the natural state and retains in the metal. Hafnium absorbs 700 times as many neutrons per cm^3 as zirconium, so the separation is a necessity when the metal is intended for reactor construction.

Zirconium is a bright silvery metal that does not tarnish in air. It is very resistant to alkalies, and also to all acids except HF and hot concentrated H_2SO_4 or H_3PO_4. When red-hot it combines vigorously with oxygen to form ZrO_2, with nitrogen to form ZrN, and with chlorine to form $ZrCl_4$. Steam corrodes the metal slowly, but an alloy with 2% tin and small amounts of nickel and chromium is more corrosion-resistant. At elevated temperatures the metal dissolves hydrogen reversibly to form an interstitial hydride without much change in the crystal lattice.

Zirconium oxide is a white refractory powder (mp 2950° C) which is used as a high-temperature insulator and as a component of refractory ceramics. Crude zircon itself ($ZrSiO_4$, mp 2190° C) is used in high-temperature porcelain and in the sand casting of metals. Zirconium carbide (ZrC, mp 3530° C) and nitride (ZrN, mp 2980° C), both hard black solids, are used as abrasives and refractories.

MOLYBDENUM, THE LIGHTER BROTHER OF TUNGSTEN. Molybdenum is in the same group of the Periodic Table as tungsten, which is used for incandescent lamp filaments. There was early interest in molybdenum before tungsten became established, because its properties are not very different and it is easier to obtain from its ores. Since molybdenum is so useful a metal for refractory alloys and for the construction of x-ray tubes, radio transmitting tubes, and electric furnace windings, it is still produced in the thousands of tons. The chief ore is black molybdenite, MoS_2, which has a greasy feel like graphite. This is roasted at 500° C to obtain MoO_3:

$$2MoS_2(s) + 7O_2(g) \rightarrow 2MoO_3(s) + 4SO_2(g)$$

When molybdenum is to be used in steel (where it confers great elasticity and some corrosion resistance), a mixture of crude MoO_3 and powdered limestone is introduced directly into the molten iron. There the lime forms calcium molybdate, which is easily reduced by the hot iron:

$$CaCO_3(s) + MoO_3(s) \rightarrow CaMoO_4(s) + CO_2(g)$$

$$CaMoO_4(s) + 2Fe(l) \rightarrow Mo(l) + Fe_2O_3(s) + CaO(s)$$

The iron and calcium oxides go into the slag layer, while molybdenum melts with the steel.

Molybdenum compounds are used in catalysts for petroleum refining, in ceramic glazes, and in red and orange pigments. The pure sulfide MoS_2 is used as a high-temperature lubricant.

The neighbors of molybdenum, odd-numbered yttrium (element 39), niobium (element 41), and technetium (element 43), are not common or useful enough to warrant discussion. Technetium, in fact, does not even occur in nature. All its isotopes are unstable, with very short half-lives. It is produced at a great rate in nuclear reactors, though, because it is a favored fission product of ^{235}U. It is considered a totally synthetic element; hence its name.

The two elements that follow technetium, ruthenium (element 44) and rhodium (element 45), are very rare and without marked significance. They give pretty-colored compounds (rhodium is from the Greek word for rose), but find no steady work.

PALLADIUM, THE METAL THAT EXTENDS OUR PLATINUM SUPPLY. Palladium is one of the platinum metals, a rather loose term that embraces those rather inert elements that occur in the free state along with platinum and are its neighbors in the Periodic Table: Ru, Rh, Pd, Re, Os, and Ir. Platinum has so many diverse industrial uses, and is so much in demand for all of them, that much attention has been devoted to possible substitutes or helpers. Palladium, directly above platinum in the Periodic Table and therefore its lighter congener, is twice as abundant as platinum and sells for one fourth as much. It functions very much as platinum does in heterogeneous catalysis: finely divided palladium catalyzes the oxidation of SO_2 to SO_3 in the manufacture of sulfuric acid, catalyzes the oxidation of ammonia to make nitric acid, and catalyzes the oxidation of residual hydrocarbons in automobile exhaust gases to CO_2 and water (the reactions that take place in the "catalytic converter" of 1975 and later cars). Unfortunately, about four times as much palladium is required to do these jobs as well as platinum does.

All the platinum elements are difficult to separate from each other, precisely because they do not undergo reactions readily and are so much alike. The mixed metals are found as grains and dust in stream beds and alluvial gravel, and the most plentiful sources are in the Soviet Union. An initial separation is achieved by adding the dust to molten lead, which dissolves platinum and palladium but leaves ruthenium, osmium, and iridium. The lead is then dissolved in nitric acid, which does not affect platinum but does dissolve palladium:

$$3Pd(s) + 2NO_3^-(aq) + 8H^+(aq) \rightarrow 3Pd^{+2} + 2NO(g) + 4H_2O(l)$$

From the nitrate solution PdI_2 is precipitated by adding NaI:

$$Pd^{+2}(aq) + 2I^-(aq) \rightarrow PdI_2(s)$$

Palladium iodide is then converted to the chloride by treatment with HCl, and the metal is recovered by electrolysis of the soluble $PdCl_2$.

Palladium forms two series of compounds, in which it has the oxidation states +2 and +4. The +2 state is the more common. Some typical compounds are PdO, $Pd(OH)_2$, $PdCl_2(NH_3)_4$, and $(NH_4)_2PdCl_4$. The last compound, ammonium chloropalladate, decomposes thermally to volatile ammonium chloride and leaves spongy metallic palladium, an excellent catalyst:

$$(NH_4)_2PdCl_4(s) \rightarrow 2NH_4Cl(g) + Pd(s) + 2HCl(g)$$

Hydrogen does not combine to form a palladium hydride of any of the usual types, but instead dissolves in the metal and diffuses through it. It is believed that the molecules of hydrogen dissociate while they are held on the surface of palladium by adsorption, and the resulting atoms of hydrogen are small enough to move about between the atoms of palladium in the crystal lattice. Since both the rate of diffusion and the rate of dissociation of H_2 increase with rising temperature, at bright red heat the process becomes so rapid that it can be used as a method for obtaining very pure hydrogen. It is astonishing to see such a purifier in action, because it is so simple, quiet, and effective. It consists merely of a smooth palladium tube the size of a pencil, closed at one end and sealed to a glass tube at the other (Fig. 7–1). When the tube is heated in a gas flame, atoms of hydrogen from the hydrocarbons being burned diffuse right through the solid palladium walls and combine at the inner surface as they are pumped away:

$$H + H \rightarrow H_2$$

The palladium acts as an atomic strainer; nothing else diffuses through the metal because no other atoms are small enough to get through the crystal lattice. Oxygen, nitrogen, and carbon are left outside, and only very pure hydrogen gets through. The effect is not minuscule, either; the same pencil-size tube of palladium, heated to 800° C electrically in a stream of impure hydrogen, will pass several liters of extremely pure hydrogen through it each hour. The molecules of H_2 separate at the outer surface of the tube, at the expense of much energy

$$H_2 \rightarrow H + H \qquad Q = 103.7 \text{ kcal/mole}$$

but the hydrogen atoms then diffuse through the metal and recombine at the inner surface, reversing the above equation and liberating exactly as much heat as was consumed, so the process goes on indefinitely without wear or fuss.

SILVER, EVERYBODY'S FAVORITE METAL. Silver was known and used long before recorded history, and it still has universal appeal. Miners and metallurgists like it because it is easy to separate from base metals and is so valuable (not just because of its scarcity but because it is so useful in industry). Artisans like it because it is easy and satisfying to work with, and takes such a brilliant polish. Consumers like it because of its appearance and its heft, and because it kills bacteria, rather than harboring them. Silver also represents a store of value that is timeless and universally recognized. We can examine some of the facts that have given silver such an enviable reputation.

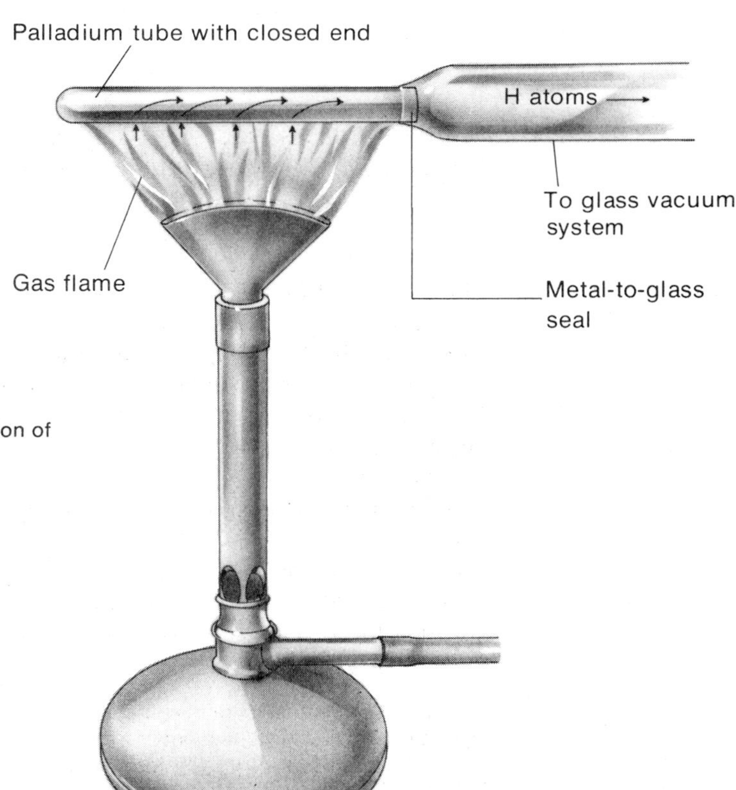

Figure 7–1. Obtaining pure hydrogen by diffusion of hydrogen atoms through hot palladium.

In olden times silver was found in the free state as nuggets or crystals, and sometimes as spongy or mossy silver released from quartzite rock by erosion. Very little of that is left now. We rely on the sulfide ore argentite, Ag_2S, which occurs in very small proportion with the sulfide ores of copper, lead, and zinc. When these base metals are smelted, the silver goes with them and can be recovered during refining. About four fifths of the silver produced is now obtained in this way. World production of new silver is about 7,800,000 kg per year.*

Recovery. The method of recovering silver during the electrolytic refining of copper has already been described in Chapter 6. The recovery from zinc ores is accomplished by cyanide treatment of the reduction residues after the volatile zinc has been distilled away. A dilute solution of sodium cyanide dissolves silver in the presence of air, forming a soluble cyanide complex:

$$4Ag(s) + O_2(g) + 8CN^-(aq) + 2H_2O \rightarrow 4Ag(CN)_2^-(aq) + 4OH^-(aq)$$

Silver ordinarily is not affected by atmospheric oxygen, but the tendency toward cyanide coordination around the small silver ion is so great that in the alkaline cyanide solution any silver ions released in the equilibrium

$$Ag \rightarrow Ag^+ + e^-$$

are immediately snatched up and complexed, so the oxidation continues. The same dissolution in aerated cyanide solution works with silver sulfide and silver chloride ores of silver:

$$Ag_2S(s) + 4CN^-(aq) \rightarrow 2Ag(CN)_2^-(aq) + S^{-2}(aq)$$

$$AgCl(s) + 2CN^-(aq) \rightarrow Ag(CN)_2^-(aq) + Cl^-(aq)$$

In all cases the silver is freed from its cyanide complex by adding zinc dust to the solution, which precipitates silver as a powder:

$$2Ag(CN)_2^-(aq) + Zn(s) \rightarrow Zn(CN)_4^{-2}(aq) + 2Ag(s)$$

The silver is filtered off, washed, and melted.

Silver is recovered from lead by an application of solvent extraction: liquid zinc is insoluble in liquid lead, and is a better solvent for silver than lead is, so when molten zinc is stirred with the lead it dissolves the silver and holds it. The Zn + Ag solution is allowed to rise to the top of the lead and is skimmed off. Simple heating then distills off the zinc to be used over again, leaving the silver.

The Metal. Pure silver is a soft white metal, softer than copper but harder than gold. For most purposes it is hardened by alloying with copper: sterling silver is 92.5% Ag + 7.5% Cu, and the coinage silver formerly used by most countries was 90% Ag + 10% Cu. Silver coins were a favorite medium of exchange for 6000 years, although not without abuses,† but in the mid-1960's world inflation caught up with monetary custom and the silver coins of all the

*The usual unit of weight for silver, gold, and coins made from them is the Troy ounce, which is equal to 31.2 g or 1.1 regular ounces.

†As the cost of government rose, the Romans debased their coinage by replacing silver with tin and copper in the alloy. All governments have been tempted to put less and less actual silver in their coins, and to raise their face value. Too much of this destroys confidence in the money and in the government—it is possible to follow the historical fortunes of a country by studying its coinage.

"hard money" countries became more valuable for their silver content than for their face value. Silver coins were melted and refined to get industrial silver, and were replaced by cupronickel coins.

Silver is a practical household metal because it does not impart any taste of its own, and is self-sterilizing. It forms no oxide in air but does tarnish due to the formation of a black sulfide, Ag_2S. Such tarnishing occurs more rapidly when there is a source of sulfurous vapors nearby, so silver should never be stored near rubber (which is vulcanized with sulfur) or allowed to stand in contact with eggs or mayonnaise (which are rich in sulfur). The sulfide tarnish can be removed by burnishing with an abrasive polish, of course, or it can be reconverted to metallic silver electrolytically by heating the piece in an aluminum pan containing a dilute solution of NaCl. The aluminum, being more active, gives up electrons which are conveyed to the Ag_2S and soon reduce the silver ions:

$$Al \rightarrow Al^{+3} + 3e^-$$

$$3e^- + 3Ag^+ \rightarrow 3Ag$$

The overall equation is

$$3AgS(s) + Al(s) + 3H_2O \rightarrow 3Ag(s) + Al_2O_3(s) + 3H_2S(g)$$

By this method of tarnish removal there is no loss of silver from the piece.*

Silver is easily electroplated on base metals, and silver-plated articles are entirely satisfactory if they are not polished by abrasion. Solid (sterling) silver pieces can be distinguished from plated pieces by their greater density and greater thermal conductivity—the handle of a sterling teaspoon will heat up faster than the handle of a plated spoon if both are put simultaneously into a cup of hot coffee. Pure silver has a heat conductivity 9% greater than that of copper, and an electrical conductivity 6% greater than that of copper.

Since silver does not form an oxide in the heat of an arc, it is preferred to copper for electrical contacts. Metallic silver is also used in brazing alloys, particularly the eutectic alloy with 28% copper which melts at 779° C. When an alloy of silver and tin is mixed with a little mercury, a plastic mass is obtained which is used by dentists to fill teeth. It hardens in several minutes, and although it causes the tooth to darken over the years it also exerts a germicidal effect that prevents further decay. Silver sols (colloidal dispersions of very finely-divided metallic silver in aqueous media) are used in medicine as nonirritating germicides.

Chemical Properties. Silver reacts chemically to form compounds with oxidation states of +1 and +2, just as copper does, but a very strong oxidizing agent is required to force it to the higher state. Fluorine does this, forming AgF_2 (which acts as a convenient fluorinating agent as it releases fluorine and descends to very water-soluble AgF). With the other halogens, white AgCl, creamy AgBr, and yellow AgI are formed, all very insoluble in water. The same three silver halides are formed more conveniently when a solution containing Ag^+ ions is added to a solution of a chloride, bromide, or iodide. All three decompose slowly in daylight as they liberate silver, and all three are used in making photographic materials as explained below.

*Since the aluminum will be darkened, it may be preferable to use a piece of aluminum foil in the bottom of an enameled pan.

The usual source of silver ion for electroplating and for manufacturing photographic film is silver nitrate, obtained by dissolving metallic silver in 50% nitric acid:

$$3Ag(s) + NO_3^- + 4H^+ \rightarrow 3Ag^+ + NO(g) + 2H_2O$$

Solutions of silver nitrate do not darken in light, but contact with room dust or human skin will reduce the silver ions to black metal. Black stains on the skin from silver nitrate or its solutions can be removed by swabbing with tincture of iodine and then removing excess iodine with sodium thiosulfate solution:

$$2Ag + I_2 \rightarrow 2AgI$$

$$2S_2O_3^{-2}(aq) + I_2 \rightarrow S_4O_6^{-2}(aq) + 2I^-(aq)$$

The Role of Silver in the Photographic Process. Photography is a hobby which is appreciated particularly by chemists, because its operations are all delicately chemical. While scattered references are made to these reactions in textbooks, the photographic materials and operations might properly be summarized here because they involve the chemistry of silver. We shall consider: 1. the light-sensitive materials; 2. the initial effect of light on them; 3. the development of the image; 4. the fixing process which renders the final image light-stable; and 5. making positive prints from the negative image.

1. A photographic "emulsion" is prepared by precipitating a silver halide in a solution of gelatin and then coating the "emulsion" on paper or film and drying it, all in complete darkness. Silver chloride is used in the "emulsion" for photographic printing papers because it is the least light-sensitive of the halides. Silver bromide is used for enlarging papers and some films, and silver iodide is used for very fast, supersensitive film. A very small proportion of Ag_2S is co-precipitated with the halide because it sensitizes the silver halide, and various dyes or fluorescing agents may be added to broaden the response to visible light because the halides themselves are especially sensitive to blue and ultraviolet.

2. The camera is merely an enclosure which admits light in the form of an image focused on the film. When a very small amount of light strikes the film, electrons are knocked out of a few bromide or iodide ions in the silver halide lattice, leaving atoms in their place:

$$Br^- + h\nu \text{ (light quantum)} \rightarrow Br + e^-$$

The electrons are mobile and fast-moving in the crystal, but they are trapped eventually at a nearby crystal boundary or at a particle of silver sulfide. There an electron discharges a silver ion and leaves one atom of silver:

$$Ag^+ + e^- \rightarrow Ag$$

The latent image on the film is a collection of such invisible atomic dots of silver. Where the light was strongest the population of silver dots will be greatest; where light was missing they will be absent. The image is "latent" because it is invisible and will not become visible until it is brought out during the development process. Thus the latent-image formation is very different from visible darkening whereby gross exposure to millions of times more light will gradually darken a silver halide without benefit of a developer. Gross exposure to light ruins a latent image long before visible darkening sets in.

3. The film with its latent image, still in total darkness, is transferred to the developing bath. The developer is simply a mild reducing agent. It may be any one of many relatively weak organic reducing agents, usually substances which have two or more hydroxyl or amine groups attached to a benzene ring. Two favorites are hydroquinone and Metol:

OH — benzene ring — OH

Hydroquinone

OH — benzene ring — $HNCH_3$

Metol

The whole trick with developers is to select the proper combination of reducing agent or agents, concentration, temperature, and time such that the reagent *just fails to reduce* unexposed crystals of silver halide. If the reduction potential has been adjusted just right, the unexposed crystals remain unaffected *but* the crystals which contain even a few atoms of silver from Step 2 will be reduced to silver. Note that where silver atoms are trapped in a grain of silver halide, the *entire crystal* is reduced. Thus exposure to a tiny amount of light within the camera has served only to sensitize the silver halide to reduction by the delicately balanced developer. The resulting amplification of the effect of light is over a million to one. It will be seen, furthermore, that large crystals of silver halide in the emulsion lead to fast film, because reduction of a large crystal produces a larger amount of black silver than reduction of a small crystal, even though each has received only a few quanta of light. At the same time, large crystals produce graininess in the picture, so the manufacturer has to balance speed against graininess.

4. Having done its work, the developer must be stopped at once from acting further. This is followed by the step of "fixing" the developed image. Frequently the stopping and fixing are done by the same bath, just by using an acid fixer. A suitable acid is $KHSO_4$, and the usual fixing reagent is sodium thiosulfate, $Na_2S_2O_3$, which complexes the silver by coordinating strongly to the Ag^+ ions:

$$AgBr(s) + 2S_2O_3^{-2}(aq) \rightarrow Ag(S_2O_3)_2^{-3}(aq) + Br^-(aq)$$

Both the thiosulfate complex and the resulting sodium bromide are soluble, so the next step is to wash the film thoroughly in pure water to remove all soluble substances. This can be done in the light. Then at last the film is dried and ready for printing.

5. The finished film bears a negative image, being darkest where the camera image was lightest, and vice versa. In order to make a positive print, the photographer goes through Steps 2, 3, and 4 all over again, this time using a printing paper instead of film. The emulsion used on paper is slower to respond to light, and is sensitive only to blue light so the photographer can work safely in red light. Either the film is placed right on the printing paper and exposed through the negative, or else light is projected through the negative and focused on a faster bromide paper to make an enlarged print. In either case the exposed printing paper must be developed, fixed, and washed in the same way as the original film.

This summary of the photographic processes applies only to black-and-white photography, of course. Color film requires a similar two-step process to create the final positive image, but accomplishes the steps differently. In the case of color print film (Kodacolor, etc.), the sequential steps are carried out separately on the negative and the print, just as with black-and-white film, but dyes are developed to produce the color. For color slide film and movie film (Kodachrome, etc.), development of the negative and reversal to the positive image are done on the same film stock. The process is somewhat complicated, and the reader who wants to do his own color film developing is referred to the special manuals supplied by the manufacturer.

Photography is big business, and the net effect of making and using light-sensitive photographic materials is to convert solid silver bars to great numbers of final photographic images and much waste solution from fixing baths. This is a good example of dispersion of a natural resource, a case of human acceleration

SILVER IS LABORIOUSLY GATHERED TOGETHER FROM MANY SOURCES AT THE EXPENSE OF MUCH EFFORT AND ENERGY, ONLY TO BE SPREAD OVER MILLIONS OF PHOTOGRAPHS, FROM WHICH IT CAN NEVER BE COLLECTED AGAIN. OBVIOUSLY, THIS CANNOT GO ON FOREVER. AFTER SILVER IS USED UP, WHAT THEN?

of the entropy effect. Silver is laboriously gathered together from many sources, at the expense of much effort and energy, only to spread some of the silver over millions of prints (from which it can never be collected again) and disperse the rest as dilute solutions of silver thiosulfate complex poured into the sewers. Obviously this cannot go on forever. The popularity of photography rose so much more rapidly than production of silver that the demand for photographic materials alone created much of the silver scarcity of 1960. After all the silver coins are used up, what then? Silver has so many industrial applications that the demand will force a difficult readjustment. In the meantime, recovery of silver from waste photo-lab solutions and wash water is an essential next step, for only there is the silver concentrated enough to deal with.

CADMIUM, A QUIET RELATIVE OF ZINC. Cadmium is only one thousandth as plentiful as zinc, with which it is found in nature. We get all the cadmium we need from the smelting and refining of zinc ores. Most such ores contain 0.3 to 0.5% of cadmium, which can be removed by adding salt to the sulfide ore during the roasting operation. Volatile $CdCl_2$ forms and sublimes out, collecting in the flues where it can be washed out. The cadmium is recovered from the solution electrolytically, and cast into bars.

Pure cadmium is a soft, bluish white, silvery metal. It has lower melting and boiling points than zinc, and is softer and more malleable. It oxidizes very little in air, but is corroded by acid vapors:

$$Cd(s) + 2H^+(aq) \rightarrow Cd^{+2}(aq) + H_2(g)$$

TABLE 7–1 PROPERTIES OF SOME 4*d* TRANSITION ELEMENTS

	Zirconium	*Molybdenum*	*Palladium*	*Silver*	*Cadmium*
Symbol	Zr	Mo	Pd	Ag	Cd
Atomic number	40	42	46	47	48
Atomic weight	91.22	95.94	106.4	107.87	112.40
Valence e^-	$5s^24d^2$	$5s^14d^5$	$5s^04d^{10}$	$5s^14d^{10}$	$5s^24d^{10}$
mp, °C	1852	2610	1550	961	321
bp, °C	3580	5560	3170	2210	767
Density, g/cm³	6.49	10.22	12.01	10.49	8.65
Atomic volume	14.1	9.4	8.9	10.3	13.1
Atomic radius, Å	1.57	1.36	1.38	1.44	1.49
Ion radius (+1), Å	1.09	0.93	0.80 a*	1.26	0.97 a
Pauling EN	1.4	1.8	2.2	1.9	1.7
Z_{eff}/r^2	1.22	1.30	1.35	1.42	1.46
Standard potential	−1.43	0.0	+0.83	+0.7996	−0.403
Oxidation states	+1, +4	+1, +4, +6	+2, +4	+1, +2	+1, +2
Ionization energy*	160	166	192	175	207
Heat of vaporization	120	112.7	90	60.95	32.15
Discovered by	Klaproth	Scheele	Wollaston	antiq.	Stromeyer
Date of discovery	1789	1778	1803	antiq.	1817
rpw* pure O_2	ZrO_2 e*	MoO_3	none	none	CdO
rpw H_2O	none	none	none	none	$Cd(OH)_2$ e
rpw N_2	ZrN e	none	none	none	Cd_3N_2 e
rpw halogens	ZrX_4	MoX_4	PdX_4	AgX	CdX_2
rpw H_2	i*	i	soluble	none	none
Mohs hardness					2.0
Crystal structure	hexagonal	cubic bc*	cubic fc*	cubic fc	hexagonal
Tensile strength			24,000	25,000	

*Heats and energies are in kcal per mole; rpw = reaction product with; a = radius of +2 ion; bc = body-centered; e = elevated temperatures; fc = face-centered; i = interstitial hydride of variable composition; potentials are in volts, tensile strength in lb/in².

Cadmium and its vapors are considered very toxic, while those of zinc are not. Chemically, cadmium dissolves in acids but not in bases; it is not amphoteric like zinc. Its only stable oxidation state is +2, and its ions are colorless and diamagnetic.

The chief use of cadmium is for electroplating steel to achieve corrosion resistance. The plating is done from baths containing the $Cd(CN)_4^{-2}$ ion, using positive electrodes of pure cadmium. A very thin plating, only 0.0002 inch thick, conveys good protection to steel and is easier to deposit on the steel than zinc. The only other use for cadmium is to make bright yellow cadmium sulfide

$$Cd^{+2}(aq) + H_2S(g) \rightarrow CdS(s) + 2H^+(aq)$$

which is a popular yellow pigment.

TABULATED PROPERTIES OF SELECTED 4*d* ELEMENTS. It is impractical to list the properties of all 20 elements covered in this chapter, simply because some of them are not thoroughly characterized and others are not important in any practical way. To simplify matters, the properties of only the five most useful elements considered above (Zr, Mo, Pd, Ag, and Cd) are listed in Table 7–1. (Similarly, Table 7–4 will list the properties of the four most useful 5*d* elements after they are discussed.)

GENERAL TRENDS IN THE 5*d* SEQUENCE

As we might expect, the metals of the 5*d* transition series (elements 72 through 80) follow the general pattern established for the 3*d* and 4*d* series. At the same time, the individuality of the chemical elements asserts itself clearly and forcibly here, with some surprising changes in the usual group trends. We shall examine the major reason for this behavior, and then consider whatever may be interesting and important about the elements themselves.

Lanthanum, element 57 and rightly the first element in this third transition series, is best considered in connection with the rare-earth or "lanthanide" elements which are the subject of Chapter 8. The rare-earth elements are those in which *f* orbitals are being filled, and while that filling occurs the 5*d* orbital usually stands empty.* We are concerned here with what happens *after* the 4*f* orbitals are filled and the 5*d* development resumes.

The first thing that we note is that hafnium, element 72, has the same atomic radius, 1.57 Å, as its lighter congener, zirconium. Despite 31 intervening elements, hafnium is so much like zirconium that it took 134 years after the discovery of zirconium (and a hundred years after the isolation of zirconium as a metal) to uncover the hafnium that was in it. Up to 1923, when Coster and von Hevesy finally got their clues from x-ray spectroscopy and located the hafnium hiding in natural zircon, every published atomic weight for zirconium was wrong. Furthermore, all the physical constants that were published and relied upon for metallic zirconium actually applied to a natural alloy of zirconium and hafnium. Why was hafnium hidden so long? In brief, it was so much like zirconium in atomic size and in chemical behavior that it was not separated by the usual metallurgical procedures for extracting zirconium.

In previous chapters we saw that the sizes of atoms of the elements decreased from left to right in any period after H and He. Chapter 6 indicated that this contraction or shrinkage was amplified by the filling of the first few 3*d*

*There are three exceptions—when the 5*d* orbital contains one electron while the 4*f* orbital is empty, half-filled, or completely filled. See Chapter 8.

TABLE 7–2 DECREASE IN THE SIZE OF ATOMS AS THE SUCCESSIVE PERIODS DEVELOP

Period	*Range*	*Radius First Element, Å*	*Radius Last Element, Å*	*Decrease*	*% Decrease*
4	K–Ti	2.31	1.46	0.85	36.8
5	Rb–Zr	2.44	1.57	0.87	35.7
6	Cs–Hf	2.62	1.57	1.05	40.1

orbitals as the transition sequence developed. When we come to the elements that undergo still another dose of atomic shrinkage, that which is associated with the filling of the 4*f* orbitals in the lanthanide sequence of elements, the effect is more marked. The facts can be ascertained from the atomic radii given in Table 7–2. In Period 4 the decrease in atomic radius for the first four elements is seen to be 36.8%; in Period 5 the comparable decrease amounts to 35.7% In Period 6, however, with the 14 lanthanide elements intervening between La and Hf, the decrease is greater, amounting to 40.1%. This extra shrinkage due to filling the 4*f* orbitals is called the *lanthanide contraction* and all the elements that follow the lanthanide series have extraordinarily small sizes. Toward the end of the 5*d* series this makes the elements markedly less reactive than their lighter congeners.

SOME SELECTED 5*d* ELEMENTS

TUNGSTEN. Most people think of tungsten solely in connection with the familiar incandescent lamp, but actually only 15% of the 5,000 tons produced annually in the U.S.A. goes into lamp filaments. The rest goes into ferrous alloys for jet-engine parts (40%), nickel-chromium alloys for high-speed cutting tools and surgical instruments (6%), tungsten carbide for dies and lathe tools (38%), and other tungsten compounds. The metal is produced from iron tungstate ore, $FeWO_4$, by fusing the concentrate with Na_2CO_3 and extracting water-soluble sodium tungstate, Na_2WO_4. Addition of H_2SO_4 then precipitates the oxide WO_3, which is reduced with hydrogen at 1000° C to a fine gray powder.

Tungsten is difficult to melt and cast, but can be compacted by the technique of powder metallurgy. Here the powder is pressed into fragile bars which are heated to 1300° C in a hydrogen atmosphere within an electric furnace. The particles of metal adhere to each other and coalesce into a bar strong enough for the next step, where the metal is heated to 3000° C in vacuum by its own electrical resistance. The heat-shrunk bars can then be rolled and swaged into rods, and finally the rods are drawn into wire by pulling the hot metal through diamond dies. The drawn wire has an exceedingly high tensile strength, 590,000 lb/in², and enough ductility so that it can be coiled into lamp filaments.

Tungsten is the most suitable material for incandescent lamps because it has the highest melting point of all the metals. Light emission of an incandescent body (radiation) goes up with the fourth power of the temperature, so even a few degrees higher temperature can be of enormous advantage in getting more light per dollar's worth of electric power. Even so, only 5% of the input energy is radiated as light from an incandescent lamp; the rest is given off as heat. (Fluorescent lamps do much better than this, converting about 30% of the input energy to light.) Those lamps which are purposely designed to operate at maximum tungsten temperature, with consequent short life, are the best buys because they give most light per watt. Consumers should be on guard against incandescent lamps which promise five or ten years of life, for these are simply made to operate at lower tungsten temperature and consequently give much less light per watt. The watts cost much more than the lamp: a 100-W lamp consumes 100 kWh (kilowatt hours) during its 1000-hour life, and at 7.5¢ per kWh this means $7.50 for the power vs. 40¢ for the lamp. A lamp that consumes twice as much power for the same light is indeed a poor bargain, especially if it lasts 2000 or 3000 hours.

All incandescent lamps darken with use, due to sublimation of tungsten from the filament and condensation on the cool glass surface. Tungsten has a very low vapor pressure at 3000° C (it does not boil until 5930° C), but over hundreds of hours of use enough black tungsten will collect on the glass to diminish the light output. The sublimation proceeds fastest in high vacuum, where no gas molecules impede the travel of tungsten atoms from filament to glass. To reduce such evaporation, a small amount of argon (an unreactive gas of Group 0 obtained from liquid air) is put in the lamp after pumping out all the oxygen and nitrogen. The amount is critical; if too much, the gas will carry away heat from the filament and waste power. Because of the light wasted by a blackened lamp, consumers would actually do well to replace lamps as soon as they begin to darken, instead of waiting for them to burn out. Failing that, the most economical lamps to use are the cheap, short-lived ones that must be replaced often simply because they burn out faster.

Pure tungsten resists most reagents, but it is dissolved by the oxidizing action of molten sodium nitrate:

$$W(s) + 2NaNO_3(l) \rightarrow Na_2WO_3(l) + NO(g) + NO_2(g)$$

It resists all acids except a mixture of HF and HNO_3, which dissolves it by oxidizing the surface, dissolving the oxide as oxyfluoride, and then hydrolyzing the fluoride to soluble tungstic acid:

$$W(s) + 2HNO_3(l) \rightarrow WO_3(s) + H_2O(l) + NO_2(g)$$

$$WO_3(s) + 4HF(aq) \rightarrow WOF_4(s) + 2H_2O(l)$$

$$WOF_4(s) + 4H_2O \rightarrow H_4WO_5(aq) + 4HF(aq)$$

Oxidation states of +2, +4, and +6 are known for tungsten (witness the chlorides W_6Cl_{12}, WCl_4, and WCl_6), and of these the +6 state is the most common. As expected, the +6 halides are the most volatile: WCl_6 is a violet solid which melts at 275° C and boils at 347° C, and WF_6 is a gas which condenses at 19° C. Both hydrolyze rapidly by preliminary coordination of water molecules, as predicted by the maximum covalency rule (maximum for this period is 8). As a consequence of its small atomic size, tungsten forms coordination compounds readily, not only with simple ligands like NH_3 and CO but in the form of complicated "heteropoly" acids with other elements. Some examples are green phosphotungstic acid, $H_3PW_{12}O_{40}$, and pale yellow silicotungstic acid, $H_{52}Si_2W_{12}O_{63}$. Despite its enormous molecular weight, the latter compound is soluble in alcohol and ether as well as in water, illustrating once more the covalent tendency of the small, heavy tungsten atoms.

OSMIUM AND ITS HAZARDS. Metallic osmium is rarely seen. It has the highest density of all elements, 22.57 g/cm³; as contrasted with lead, which has a density of 11.36, and uranium, 19.1. However, despite all its rarity and expense

($30 per gram), osmium is well known to biologists and microscopists in the form of its oxide, OsO_4. This substance is easily reduced to black particles of osmium by biological material, and so is used to stain thin sections of organs and tissue before examination under the microscope. Small differences in reducing ability produce marked changes in degree of staining, and hence greatly enhanced contrast of different structures which would otherwise be almost invisible under the microscope. The volatility of covalent osmium tetroxide (mp 40° C, bp 130° C) makes the staining easy because the specimen need only be exposed to the vapor, but the same volatility makes the process dangerous because the vapor of OsO_4 is toxic and can be reduced right on the surface of the human eyeball, leading to a black covering impossible to remove chemically.

PLATINUM: Platinum is encountered early and often by every student of chemistry because it is so effective as a catalyst. It furthers the oxidation of SO_2 to SO_3 in making sulfuric acid, the synthesis of ammonia from nitrogen and hydrogen, the oxidation of ammonia to make nitric acid, and the addition of hydrogen and hydrides to C=C double bonds. True, some of these functions have been taken over by less expensive catalysts in large-scale production, as in the use of V_2O_5 catalyst for oxidizing SO_2 and the use of activated iron catalyst for producing ammonia, but in other areas the superiority of platinum catalysts is so great that nothing else will do. The catalytic converter for oxidizing residual hydrocarbons in automobile exhaust gas is the most recent proof of this.

Platinum is also an admirable alloying ingredient, a protective coating for base metals, a chemically-inert material for laboratory ware, a bright and costly metal for jewelry, a superior nonoxidizing winding for electric furnaces, and the best material for spinnerets (the nozzles from which synthetic fibers are extruded through a multitude of tiny holes drilled in the metal). It is even used for the tools that stir and manipulate molten glass. Platinum does so many things so well that a great deal more of it would be used if it were not so scarce and expensive. Since we have so little of it available, every possible bit is recovered and recycled. It never wears out; crucibles and evaporating dishes that have seen 50 years of laboratory service have scrap value equal to their platinum content at current prices.

The sources of platinum are: 1. Canadian mixed sulfide ores, wherein platinum occurs with nickel and copper; 2. native platinum (alloyed with iridium, osmium, palladium, and other platinum metals) left in Alaskan and Russian placer deposits from the erosion of basic rocks; 3. platinum alloyed with gold in South African deposits; and 4. scrap platinum from a multitude of industrial uses. No method of extraction is applicable to all sources. The recovery of platinum during the electrolytic refining of copper has been described, and also the differentiation of platinum metals by dissolving in lead. Alkaline fusion is effective with some sulfide and arsenide ores, but not with others. One method that always works for the recovery of scrap platinum, and usually works with nuggets and grains of native platinum, is to dissolve the metal in aqua regia, a 3:1 mixture of hydrochloric and nitric acids. The mixture generates chlorine by the oxidation of HCl, and the platinum is chlorinated to orange chloroplatinic acid, H_2PtCl_6:

$$3Pt(s) + 4NO_3^-(aq) + 18Cl^-(aq) + 22H^+(aq) \rightarrow 3H_2PtCl_6(aq) + 4NO(g) + 8H_2O$$

The crude solution is made basic with NaOH, whereupon most impurities precipitate as hydroxides but the platinum remains in solution as sodium chloroplatinate, Na_2PtCl_6. Addition of ammonium chloride to a solution of H_2PtCl_6 or Na_2PtCl_6 then precipitates yellow ammonium chloroplatinate, which can be purified by recrystallization:

$$PtCl_6^{-2}(aq) + 2NH_4^+(aq) \rightarrow (NH_4)_2PtCl_6(s)$$

The nice thing about ammonium chloroplatinate is that it decomposes cleanly when heated, giving off white fumes of ammonium chloride and leaving pure platinum as a gray sponge:

$$3(NH_4)_2PtCl_6 \rightarrow 3Pt(s) + 2NH_4Cl(g) + 2N_2(g) + 16HCl(g)$$

For laboratory use the platinum sponge itself is an effective catalyst; for industrial use the platinum is melted and fabricated. A dilute solution of the same ammonium chloroplatinate can be soaked up by asbestos, or porous aluminum oxide, or any other suitable porous material, and after drying the mass and heating it, finely divided platinum is left spread out on the supporting material in any desired proportion. This is the manner of making a particular platinum catalyst for petroleum refining, one which serves in the cyclization and isomerization of hydrocarbons.

Massive platinum is a heavy, silvery, soft, malleable and ductile metal, more dense than gold and somewhat more resistant to acids than gold. Alkalies attack it somewhat more readily than they do gold, and carbon and sulfur make it brittle. For these reasons, platinum laboratory ware should never be used to fuse alkalies, nor to ignite sulfide precipitates or organic matter. Compounds of lead should also be avoided, because these are easily reduced and the liberated metal may penetrate the platinum to form a low-melting alloy.

Platinum adsorbs hydrogen, oxygen, nitrogen, and carbon monoxide but does not combine with these gases. In the adsorbed state the gases are in a highly condensed or concentrated form, which explains why the surface of platinum is so strongly catalytic in interactions of these substances.

In its halide compounds platinum appears in oxidation states +1, +2, +3, +4, and +6, with +2 and +4 the most common. The Pt^{+2} and Pt^{+4} ions form a host of coordination compounds, and a full account of the chemistry of the organic and inorganic complexes of platinum would fill an entire book of this size. Among its volatile covalent compounds, PtF_6 is noteworthy as the reagent that reacted with xenon and opened up the field of chemical compounds of the noble gases in Group 0, elements which had been considered completely inert until then.

To make it harder and stronger, platinum is usually alloyed with other metals of the platinum group. For laboratory ware 0.5% of iridium is added. For furnace windings, spinnerets, and wire, an alloy containing 10% rhodium is used. For jewelry 5% or so of iridium or ruthenium usually is added to increase hardness. The use of platinum for jewelry puts added strain on the already scarce supply, and it is fortunate for chemistry that such jewelry has gone out of style. Industrial use of platinum alone has pushed the price to around $200 per Troy ounce, about 40% higher than gold. By contrast, 300 years ago, before the industrial revolution, platinum was used to make counterfeit Spanish silver coins!

GOLD, THE MOST NOBLE METAL. Gold has been used as an enduring and valuable metal for at least 8000 years. Since it is the only bright yellow metal, and is "noble" in the sense that it is unaffected in color and integrity through the ages, it was thought to be a part of the sun or a direct representative of the sun god. The Latin name *aurum*, meaning bright dawn, and the alchemical symbol, a circular sunburst, reflect this view. Alchemists and medieval philosophers strove to isolate the divine or essential principle of life and warmth, which they expected would not only heal mankind of its terrible diseases, but would also convert base metals into gold and could be stored there. Failing this, they used gold itself as a cure for various illnesses. The idea persisted for a long time, as old medical books will show.

Some gold is recovered in the refining of copper and silver, but most of the new production comes from two sources: native gold (metal found in the free state, alloyed with 1% to 50% silver) and telluride ore ($AuTe_2$ and $AuAgTe_4$). Native gold may be found either as veins and dust in its original quartzite rock, or as nuggets and dust in alluvial deposits after the weathering and disintegration of such rock. While some spectacular nuggets have been found in the past (one in Australia weighed 600 lb), most of the obvious gold has already been plucked from stream beds and gravel pits, and present production depends upon shaft-and-tunnel mining. In 1970 the world production was 48.2 million Troy ounces, or 1.504 million kg. Of this, South Africa produced by far the most (66.6%), while the rest came from U.S.S.R. (13.7%), Canada (4.97%), U.S.A. (3.73%), and other countries (11.0%).

The methods used to extract gold can be illustrated in microcosm by what happened in Virginia City, Nevada, after gold was discovered there in the middle of the nineteenth century. At first miners and adventurers flocked to the spot and a boom city sprang up, financed by the first easy finds. All that was necessary was to swirl the gravel or crushed rock with water in a flat pan with a practiced hand, and the lighter silicate minerals were washed away, leaving the much heavier gold nuggets and dust behind. Some mechanization followed, but the same separation by gravity was used. Tunnels were dug in the hillsides at likely sites, and the rock and earth were treated the same way. Large piles of discarded gravel and crushed rock ("tailings") accumulated. When this method of recovery no longer paid its way, the miners departed and the mines were abandoned. Then the Chinese cooks and laborers who had been imported to do the menial work proceeded to pick over the leavings themselves, gathering every visible speck of gold. Later some more enlightened people practiced amalgamation recovery, whereby the crushed rock was carried by water down a long riffled chute containing mercury held in place by transverse slats. Heavy particles of gold dust, even if invisible, were dissolved by the mercury as they were carried over its surface. At the end of a day's run the amalgam (solution of metals in mercury) was collected and the mercury was distilled off, leaving the gold as a residue.

When all physical methods had been exhausted, Virginia City became a ghost town. Then, in the early 1900's, an Eastern company moved in with equipment and workers to apply the latest method of recovery, chemical extraction. This was the cyanide method, in which a dilute solution of sodium cyanide trickles through the crushed rock for two or three days while ample air is supplied. Gold is far less reactive than copper and silver, being unaffected by HNO_3 or hot concentrated H_2SO_4, but the gold atom is so small (as a result of the lanthanide contraction) that it has a marked tendency to coordinate. Cyanide ion, with its negative charge and ready electrons, is the strongest coordination ligand commonly known (see the spectrochemical series headed by CN^- ion given in Chapter 6). In the presence of ample NaCN, the electrode potential of metallic gold is reduced from +1.50 volts to −0.60 volts, and even air will accomplish the oxidation. The overall reaction is

$$4Au(s) + 8CN^-(aq) + O_2(g) + 2H_2O \rightarrow 4Au(CN)_2^-(aq) + 4OH^-(aq)$$

The filtered and deaerated solution is then stirred with enough zinc dust to precipitate the gold:

$$2Au(CN)_2^- + Zn(s) \rightarrow 2Au(s) + Zn(CN)_2^{-2}(aq)$$

Silver goes along with gold in the process, so the precipitated powder is added to hot concentrated H_2SO_4, which dissolves silver but not gold:

$$2Ag(s) + SO_4^{-2} + 4H^+ \rightarrow 2Ag^+ + SO_2(g) + 2H_2O$$

The silver is recovered by electrolysis, and the gold is melted, cast into bars, and stamped with the weight and purity. The usual method of expressing purity is fineness, which is simply the decimal fraction of gold present. Most gold coins have a fineness of .900, but for British sovereigns it is .9167 and for Austrian ducats it is .986.

Pure gold is soft, so it is alloyed with copper for coinage and with silver, copper, or nickel for jewelry. The purity of gold for jewelry is given in carats (abbreviation kt), pure gold being 24 kt and a 75% alloy being 18 kt. While pure gold remains untarnished, the alloys do become dulled with age.

Pure gold is so malleable that it can be beaten into foil or "leaf" only 0.000005 in. thick, used for architectural decoration. Thin platings on base metals give excellent protection against corrosion. Gold is an excellent conductor of electricity and a good reflector of infrared rays.

Gold is unaffected by corrosive gases such as SO_2 and H_2S, but chlorine attacks it at elevated temperatures. Alkalis have no effect. By themselves, the acids HCl, H_2SO_4, H_3PO_4, and HNO_3 do not attack it, but aqua regia slowly dissolves gold by converting it to $AuCl_3$ and then chloroauric acid, $HAuCl_4$:

$$Au(s) + NO_3^- + 4Cl^- + 5H^+ \rightarrow HAuCl_4(aq) + NO(g) + 2H_2O$$

Here, under strongly oxidizing conditions, gold is seen to be in the oxidation state of +3, whereas in the cyanide extraction product it has an oxidation state of +1. Both states are common in the chemistry of gold, with the +3 state somewhat more stable. All gold compounds are readily reduced to the metal, for the bonds are weak. The only compounds of practical importance are $NaAuCl_4$ (used in electroplating gold) and $AuCl_3$, a red compound which is the usual source of all other gold compounds.

For the reasons developed earlier in this section, gold has the same atomic radius as silver, 1.44 Å. With 32 more positive charges on its nucleus, and its peripheral electrons at the same distance, gold holds its valence electrons more firmly, and therefore is less reactive and less electropositive than silver. In turn silver is less electropositive than copper, for the reasons outlined in the last chapter. The result is a distinct reversal of the trend which is so clear among all the groups of representative elements. The actual changes in electrode potential, ionization energy, and electronegativity are contrasted in Table 7–3.

MERCURY, THE FLUID BUT INDIFFERENT ONE. Mercury has the simplest metallurgy imaginable—all its natural compounds give metallic mercury when heated moderately in air. For this reason, and because its natural compounds are bright-colored and obvious, mercury was known to the ancients. They considered its liquidity a virtue, and mercury was a favorite healing agent in medieval medicine.

Mercury occurs as bright red cinnabar, HgS, and as the red or yellow oxide, HgO. It is actually rarer than gold, platinum, and uranium, but its sources are so concentrated that it has always been more available. Mercury and its compounds are very important industrially, and the U.S. is dependent upon imported mercury. Three fourths of the world supply (about 7000 tons per year) comes from Spain and Italy.

Metallic mercury is a bright, silvery, dense, monatomic liquid which boils at 356.9° C and freezes at −38.87° C. Its electrical conductivity is only about 1% that of silver, yet the fluidity of mercury is so important to many electrical applications that the high resistance is secondary. Metallic mercury is used in household light switches and as the negative electrode in the electrolysis of salt solution to produce chlorine and sodium hydroxide. A drop of mercury also goes into every fluorescent lamp, where it vaporizes and then emits ultraviolet light at 3650 Å and 3663 Å upon electrical excitation. This ultraviolet light strikes the thin coating of fluorescent powder (made of various silicates and tungstates) on the inner wall of the glass tube, and there excites the coating to fluoresce in the visible region of the spectrum. The output of ultraviolet light represents about 50% of the input energy, and its conversion to visible light in the coating is about 60% efficient, so some 30% of the input electrical energy appears as visible light. This is about six times as efficient as an incandescent lamp, which converts only 5% of the input energy to light, so fluorescent lamps save a great deal of energy and generate less heat.

TABLE 7–3 TRENDS AMONG THE ALKALI METALS VS. THE COINAGE METALS IN PERIODS 4, 5, AND 6

	K	*Rb*	*Cs*	*Cu*	*Ag*	*Au*
Atomic radius, Å	2.31	2.44	2.62	1.28	1.44	1.44
Electronegativity	0.82	0.82	0.79	1.90	1.93	2.54
Standard reduction potential, volts	−2.93	−2.93	−2.92	+0.337	+0.799	+1.50
First ionization energy, kcal/mole	100	96	90	178	175	213

Metallic mercury is also used in thermometers and barometers, of course, and in diffusion-type vacuum pumps and other laboratory equipment. The vapor pressure of mercury is considerable, and even at room temperature it exceeds the safe level for continuous exposure. Laboratory workers are particularly liable to mercury poisoning from the vapor emitted by open pools of mercury, and should be especially careful to avoid spilling any of the liquid. Liberal ventilation of laboratories is essential to carry away the toxic vapor.

The compounds of mercury are mostly covalent, all with small heats of formation and low bond energies. In fact, mercury is not much interested in chemical combination at all. The metal is not oxidized by dry air, but is attacked by sulfur and its compounds, and by chlorine. In the two chlorides Hg_2Cl_2 and $HgCl_2$ mercury has the formal oxidation numbers of +1 and +2, but in both compounds each mercury atom actually forms two covalent bonds:

Cl—Hg—Hg—Cl	Cl—Hg—Cl
Hg_2Cl_2	$HgCl_2$
mercury(I) chloride	mercury(II) chloride

In other mercury (I) compounds the same Hg—Hg bonding appears, and the molecular weight is always found to correspond to Hg_2X_2. The covalent nature of

TABLE 7–4 PROPERTIES OF SOME 5*d* TRANSITION ELEMENTS

	Tungsten	*Platinum*	*Gold*	*Mercury*
Symbol	W	Pt	Au	Hg
Atomic number	74	78	79	80
Atomic weight	183.85	195.09	196.97	200.59
Valence e^-	$5d^4 6s^2$	$5d^9 6s^1$	$5d^{10} 6s^1$	$5d^{10} 6s^2$
mp, °C	3410	1769	1063	−38.87
bp, °C	5930	4530	2966	356.9
Density, g/cm^3	19.3	21.5	19.3	13.5
Atomic volume	9.53	9.10	10.2	14.8
Atomic radius, Å	1.37	1.38	1.44	1.55
Ion radius, Å	0.70(+4)	0.65(+4)	0.85(+3)	1.10(+2)
Pauling EN	1.7	2.2	2.4	1.9
Z_{eff}/r^2	1.40	1.44	1.42	1.44
Standard potential	−0.12 d*	+1.2 b*	+1.42 a*	+0.796 c*
Oxidation states	+4, +6	+2, +4, +6	+1, +3	+1, +2
Ionization energy*	184	207	213	241
Heat of vaporization	36.22	122	81.8	13.0
Discovered by	de Elhuyar	antiq.	antiq.	antiq.
Date of discovery	1783	antiq.	antiq.	antiq.
rpw* pure O_2	WO_3 et*	none	none	HgO et
rpw H_2O	WO_2 et	none	none	none
rpw N_2	none	none	none	none
rpw halogens	WX_4, WX_6	PtX_4, PtF_6	AuX_3	Hg_2X_2, HgX_2
rpw H_2	none	none	none	none
Mohs hardness		4.3	2.8	0
Crystal structure	cubic bc*	cubic fc*	cubic fc	rhombo.
Tensile strength	590,000	20,000	18,000	0

*Heat and energies are in kcal per mole; a = reduction potential for Au^{+3}, Au; b = reduction potential for Pt^{+2}, Pt; c = reduction potential for Hg^{+2}, Hg; d = reduction potential for WO_2, W; bc = body-centered; et = at elevated temperature; fc = face-centered; rpw = reaction product with; potentials are in volts, and tensile strengths are in lb/in^2.

WE RECEIVE MERCURY THROUGH CONTAMINATED DRINKING WATER – ALSO BY EATING FISH WHICH HAVE ABSORBED AND CONCENTRATED THE MERCURY.

$HgCl_2$ is shown by its solubility in alcohol and ether, and its very small dissociation in water. The cyanide $Hg(CN)_2$ has similar covalent properties. Only the nitrate $Hg(NO_3)_2$ and the perchlorate $Hg(ClO_4)_2$ act as true salts, with extensive ionic dissociation and hydrolysis.

Mercury cyanate, $Hg(CNO)_2$, is called fulminate of mercury and is used as a primer in small-arms ammunition because it explodes on impact. Mercury salts are potent germicides because of the general toxicity of mercury, which explains their success in the healing arts hundreds of years before a germ theory of disease was known. Dilute solutions of $HgCl_2$ are still used as disinfectants, and an ointment containing HgO is still used to treat conjunctivitis.

Mercury poisoning, like lead poisoning, is an ever-present hazard heightened by our industrial culture. Mercury and all its compounds are toxic and are easily absorbed but difficult to eliminate. The vapor of the metal is dangerous, as pointed out above. Volatile organic compounds of mercury are particularly hazardous because of their high vapor pressures and their solubility in fatty tissue. Dimethylmercury, which is produced by microorganisms, is an especially bad offender in this respect; it boils at 92° C and is absorbed through lungs and skin. Aqueous solutions of compounds can also be absorbed through the skin; the mad hatter of *Alice in Wonderland* was no whimsical invention, because solutions of $HgCl_2$ were actually used to make felt out of fur, and long exposure produced nervous disorder and insanity as an occupational disease. Today the chlorides of mercury get into drinking water from the waste washed out of industrial electrolytic cells that use mercury electrodes. We may receive mercury not only directly from this source, but also by eating fish which have absorbed and concentrated the mercury. This is why federal control of canned and fresh fish is necessary, and why some supplies have been recalled. Any intake of mercury above the low threshold set by daily elimination leads to cumulative poisoning of a very persistent nature. The progressive symptoms are headache and general malaise, tremors of the hands, loss of teeth and hair, loss of memory and of motor control, and then loss of reason. In its earlier stages the disease can be treated by administering chelating agents that increase elimination of mercury, provided that the sources of intake have been identified and blocked.

SUMMARY

The metals of the 4*d* transition series resemble those of the 3*d* series except for being rarer and somewhat more given to covalency and coordination. They go through the same increase in number of possible oxidation states, to +6 in molybdates and +7 in pertechnitates, before declining to a single +2 state for cadmium. The changes of color and degree of paramagnetism follow the pattern described for the first transition series.

Of the ten 4*d* elements treated in this chapter, only five are useful to the point of economic importance. Zirconium is a strong, corrosion-resistant metal good for constructing nuclear reactors if freed from its persistent impurity, hafnium. Molybdenum makes steel strong and elastic, and is highly prized for this. Palladium is a useful stand-in for scarce and very expensive platinum catalysts, and has the interesting property of passing pure hydrogen through a solid wall of palladium while excluding all other gases. Silver is well known to everyone; its great importance to industry and its universal appeal for jewelry and household articles warranted considerable discussion of its sources and metallurgy, its behavior, and its uses, including its use in photographic materials. Cadmium is lower-melting than zinc and is superior to zinc as an electroplated corrosion-resistant coating on iron and steel, but is decidedly toxic.

The 5*d* transition series begins like the others, but is soon interrupted by the filling of 4*f* orbitals in the rare-earth or lanthanide elements. The increase in nuclear charge each time an electron is added to the inner *f* orbitals results in a drawing-in of the entire orbital pattern, giving rise to sharply reduced atomic radii for the 5*d* elements that follow the last lanthanide. This is called the lanthanide contraction. As a consequence, hafnium is so much like its predecessor zirconium that it remained undiscovered long after zirconium was in use. A further result is more covalent behavior for all the elements of the third transition series, leading to volatile halides for all the elements and even to a volatile oxide for osmium. Still another consequence of the lanthanide contraction is a firmer hold on the valence electrons and a consequent reduction of chemical activity, so that iridium, platinum, and gold are noble metals. This situation produces a characteristic drop in electropositive character and reactivity within all groups at the center of the Periodic Table, in contrast to the rise in electropositive character and increased basic behavior noted throughout the groups of representative elements.

Tungsten is the highest-melting metal we have, so it is the logical choice for the filaments of incandescent lamps. Since the radiation of light from the hot filament goes up with the fourth power of its temperature, incandescent lamps are most efficient when run as brightly as possible, even though they burn out faster. Use of longer-life lamps wastes energy.

Platinum is a very rare and expensive metal, but is much in demand as a catalyst in the chemical and automotive industries. Its utility in catalysis depends upon its ability to adsorb most gases on its surface, where the gases are held as a condensed film at high concentration in a condition favorable for reaction. Platinum also has a very interesting chemical behavior, with thousands of organic and inorganic derivatives and complexes (almost all of them with covalent bonds to Pt). Its most stable oxidation states are +2, +4, and +6, and chloroplatinates such as K_2PtCl_6 are its most useful compounds.

Gold has artistic, emotional, economic, and industrial appeal that far exceeds the scarce supply. The methods of recovery were summarized in terms of their successive uses at Virginia City, Nevada; a new method would be very welcome. Alloys with Cu, Ag, and Ni are used in jewelry, dentistry, and coinage; pure gold leaf only 0.000005 in thick is used architecturally. In its compounds gold has oxidation states of +1 and +3, the latter being more stable. Platinum and gold are not toxic.

Mercury is very useful because it is the only readily available and unreactive metal which is liquid at ordinary temperatures. It shows little interest in chemical combination, and all its chemical bonds are weak. Its compounds are almost all covalent, and often low-melting and volatile. The metal and all of its compounds are toxic. Since the metal has an appreciable vapor pressure at room temperature, there is pronounced danger from inhaling its vapor released from laboratory instruments and from spilled mercury. Solutions of its compounds also are dangerous if handled over extended periods, and volatile organic derivatives such as $(CH_3)_2Hg$ are especially insidious because they can be absorbed so easily through the skin or lungs.

GLOSSARY

alluvial: pertaining to a deposit of sand, gravel, and mud left by flowing water.

braze: to join metals together by heating their cleaned surfaces in contact and melting into the joint an alloy of copper with silver or zinc, often called silver solder.

cyclization: in organic chemistry, a ring-closure reaction, as in the transformation of the straight-chain hydrocarbon hexane, C_6H_{14}, into cyclohexane, C_6H_{12}, or benzene, C_6H_6.

diffusion: an intermingling of atoms or molecules due to random thermal agitation, as in the permeation of a solid by a liquid or a gas.

dispersion, optical: the variation of degree of refraction with changing wavelength of light; the separation of white light into its component colors as a result of such variation.

heterogeneous catalysis: the acceleration of a reaction by a separate, insoluble, solid phase, usually due to greatly increased concentration of the reactants as they are adsorbed and held as a condensed layer on the surface of the solid.

latent: present but not visible or actualized; existing as potential.

Mohs hardness: the scale of relative hardness in terms of talc = 1, rock salt = 2, calcite ($CaCO_3$) = 3, fluorite (CaF_2) = 4, apatite ($Ca_5(PO_4)_3OH$) = 5, feldspar = 6, quartz = 7, topaz = 8, sapphire = 9, and diamond = 10.

placer: in mining, a superficial deposit of gravel containing particles of gold or the like.

powder metallurgy: the technique of preparing alloys or fabricating parts consisting in pressing metallic powders under very high pressure and consolidating the mass by heating hot enough and long enough for diffusion and crystal growth to occur.

sinter: to agglomerate by heating or by any other means that transfers matter from small particles to larger ones by vapor transport.

sol: short for hydrosol, a colloidal dispersion of a solid in a liquid. The analogous colloidal dispersion of a liquid in a liquid is called an emulsion.

solvent extraction: a technique of concentrating a dissolved substance by agitating the solution with an insoluble liquid capable of dissolving more of that substance (per unit volume) than the original solvent. Example: iodine can be extracted from dilute solution in water by shaking the solution with a much smaller volume of CCl_4, whereupon the iodine passes from the poorer solvent (H_2O) to the superior solvent (CCl_4), leaving the water colorless and coloring the CCl_4 purple.

spinneret: the organ by which a spider spins a thread; a thimble-like device with tiny holes through which a solution is pumped in the spinning of synthetic fibers.

swage: to reduce or taper a rod by forging between rotating pairs of grooved steel blocks.

EXAMINATION QUESTIONS

1. Explain why zirconium must be highly purified before it can be used in the construction of nuclear reactors. Why is the purification so difficult?

2. List the five most significant elements of the 4*d* transition series, in your opinion, and summarize for each: a. the usual sources; b. one practicable method for obtaining the pure metal from its source; c. the chief uses of the metal; d. the chief compounds and the oxidation state of the element in each; and e. the toxicity or lack thereof.

3. Summarize the chemical changes that take place within photographic film during exposure, development, and fixing of the film.

4. In what ways does cadmium differ from its lighter congener, zinc, in: a. physical properties; b. chemical behavior; and c. availability?

5. Explain how and why extremely pure hydrogen may be pumped continuously from a closed-end thin-walled tube of metallic palladium while the tube is being heated red-hot in the flame of a burning hydrocarbon.

6. What silver alloys are used for: a. domestic tableware; b. brazing metal, called silver solder; and c. filling teeth? In each instance, explain why those particular alloying agents were used.

7. Which of the following hexafluorides will hydrolyze readily?

   a. MoF_6

   b. SF_6

   c. PtF_6

   d. WF_6

   e. SeF_6

Explain any differences in behavior toward water.

8. What are osmium compounds used for? Which compound or compounds is (are) so used? What dangers are inherent in such use?

9. Explain: a. which compounds of mercury are poisonous; b. why they are poisonous; c. what the symptoms are; and d. what can be done about such poisoning.

10. Contrast the chemical behavior of copper, silver, and gold, touching on: a. their oxidation states; b. their reactions with nitric acid; c. their reactions with oxygen at elevated temperature; and d. their coordination complexes, especially with NH_3 and CN^- as ligands.

"THINK" QUESTIONS

A. Explain why the melting points of the metals in the second transition series rise to a maximum and then decline. Do the boiling points of the metals follow the same sequence as the melting points? Why?

B. Would a prism of lead glass placed in a beam of sunlight produce a spectrum if the refractive index of the glass remained constant at all wavelengths?

C. Would it be wise for an artist to use cadmium yellow and white lead as pigments in the same light yellow paint? Why?

D. Rhodium is sometimes plated over silver in a very thin layer, especially on household silverware. What advantages and disadvantages can you foresee for this procedure?

E. A prestigious international organization of industrial chemists awards a palladium medal to outstanding inventors or organizers in the field of applied chemistry. Why palladium?

F. Cadmium-plated paper clips were once a common item in bookstores and offices, but they have disappeared lately. What advantages would you predict for cadmium-plated paper clips? What reasons can you surmise for their subsequent disappearance?

G. In 3600 B.C. the Egyptian king Menes set the value of silver at two fifths the value of gold, or a gold:silver ratio of 5:2. In the 1870's and 1880's the western silver-producing states of the U.S. wanted the gold:silver ratio fixed at 16:1 by the U.S. Treasury. According to spot-metal prices late in 1975, the ratio of values had risen by then to 31:1; but in mid-1976 the ratio was back down to 24:1. What explanation is there for these changes, in terms of economic and industrial chemistry?

H. Why are even-numbered oxidation states the most stable for hafnium, tungsten, platinum, and mercury, while odd-numbered oxidation states prevail for tantalum, iridium, and gold?

I. Could osmium be used as an effective heterogeneous catalyst for oxidation reactions? If so, what advantages or disadvantages could you predict?

J. Worn-out or malfunctioning mufflers certainly have no worth beyond their value as scrap steel, but could the same be said of an old catalytic converter? Why?

K. Gold-rimmed dinner plates were favored a century ago by housewives who liked fine china, and they can still be found in antique shops. Devise a process for applying a durable gold film to porcelain, if you can. If not, look it up in your chemistry library with the help of the librarian.

L. Could a heated pool of mercury be used as a bath for heating a glass flask in the laboratory? Should it so be used?

PROBLEMS

1. Some countries have recently issued gold coins which are .500 fine. In quantitative terms, how could you distinguish these from older coins that are .900 fine?

2. What will be the molar concentration of Ag^+ ion in a solution prepared by adding to exactly 1 liter of water: a. 1×10^{-6} mole of AgCl; b. 1.05×10^{-5} mole of AgCl; and c. 1×10^{-4} mole of AgCl? The solubility product constant of AgCl is 1×10^{-10}.

3. Complete and balance the following equations:

$$PbO_2 + MnO_3^{-2} + H^+ \rightarrow Pb^{+2} + MnO_4^- + ?$$

$$H^+ + NO_3^- + PbS \rightarrow Pb^{+2} + SO_4^{-2} + NO + ?$$

$$H^+ + NO_3^- + Cd \rightarrow Cd^{+2} + N_2O + ?$$

4. What is the oxidation state of the metal in each of the following compounds?

a. PbO_2

b. Pb_3O_4

c. K_2WO_3

d. $KAu(CN)_4$

e. $(NH_4)_2PtCl_6$

f. Hg_2Cl_2

g. $NaTcO_4$

h. $K_8Ta_6O_{19}$

5. a. Rhenium forms a heptafluoride, ReF_7, but manganese does not, even though the maximum oxidation state is +7 for both these transition elements. How is ReF_7 possible, and what does this indicate that is different about the 5*d* elements?

b. Molybdenum and tungsten form octacyanide ions, $Mo(CN)_8^{4-}$ and $W(CN)_8^{4-}$. Do these facts reinforce your conclusion of a. above, or not?

THE LANTHANIDE RARE EARTHS

Throughout this book the emphasis has been on those chemical elements which are essential to the material of our planet and our own bodies, or are important to the structures we build and the mode of life we strive for. The purpose is to make you familiar with the elements and compounds you will encounter most often, so that you can understand where they came from, how they are used, how they may be reused, and what alternatives may be available in the future. Only with this information at hand can you make intelligent choices in a world of clamoring people and limited resources.

Now we come to a sequence of 14 elements that meet none of the criteria of importance just listed. It seems impossible that sandwiched in between barium and tungsten there should be 14 elements that very few people have ever heard of and that have no major use, but there they are. So far as we know not one of them is essential to plant or animal growth, none is a structural metal or a noble metal, or is of any great importance. The only general uses that have been found for them are to make special absorbing glass for sunglasses, to light the wicks of cigarette lighters, and to sensitize the fluorescent powders used on color television screens, so that each kind of powder emits its own bright color under cathode-ray bombardment.

The 14 elements are those pointed out in the last chapter, commonly called the lanthanide elements or the lanthanide rare earths. They are the elements involved in the development of the first set of seven *f* orbitals, the 4*f* sequence. They are numbered 58 through 71 and are represented in a separate row at the bottom of the Periodic Table. The name *lanthanide elements* seems inappropriate because the elements are not binary compounds of lanthanum; they only follow it in the sixth period. The designation "rare earths" is not entirely appropriate, either, because the 4*f* elements are no more rare than arsenic or tungsten, but they are all reactive basic metals that form refractory oxides ("earths" in alchemical usage). Our purpose in this chapter is to take a brief look at their physical and chemical behavior, just as one aspect of inorganic chemistry that should not be completely neglected.

THE NAMES AND NUMBERS

Lanthanum, element 57, is not a 4*f* element but is the first member of the third transition series. It occurs with cerium, and acts so much like the rare earths that it is best included here. It is an iron-gray metal that burns brilliantly to La_2O_3. Its salts are colorless and diamagnetic.

Cerium, element 58, is the most abundant element in the series (46.1 ppm in the earth's crust). It is found in the monazite sands of North Carolina and Scandinavia, and carries with it lanthanum and the four elements to follow. It and terbium are the only rare-earth elements to form compounds in the +4 oxidation state; these are easily reduced to the +3 state, characteristic of the series. An alloy of cerium with iron and some other rare earths gives off white sparks when filed or struck, and is used as the "flint" in cigarette lighters.

Praseodymium, element 59, named for the two green lines in its absorption spectrum, was first isolated in 1843 from lanthanum as a mixture with its twin, neodymium. The two were considered one element until 1879, when element 60 was separated (hence the name neodymium, the new twin). This feat led to the suspicion that all elements could be separated further if one tried long and hard enough. This haunting idea was eliminated in 1914 by Moseley's determination of actual atomic numbers, which showed that there could be only 14 rare-earth elements.

Neodymium, element 60, was the new twin. Its salts are pink and paramagnetic.

Promethium, element 61, was unknown until isolated from the fission products of uranium. All its isotopes are radioactive, and it does not occur in nature.

Samarium, element 62, is found with cerium. Its compounds are yellow.

Europium, element 63, is extremely rare and has no practical importance. Its +3 compounds are pale pink; its +2 compounds have an f^7 electronic state with half-filled *f* orbitals.

Gadolinium, element 64, stands in the middle of the 4*f* series. Its compounds all contain the Gd^{+3} ion and are always colorless.

Terbium, element 65, forms pale pink +3 salts. Both the metal and its compounds are paramagnetic.

Dysprosium, element 66, is found with the latter rare earths in the dark vein material of granite. Its name means "hard to get at."

Holmium, element 67, forms a pale green oxide, Ho_2O_3, and yellow salts.

Erbium, element 68, occurs with yttrium, element 39, in the second transition series because its atoms are precisely the right size to fit in with those of Yt, an element only half its weight.

Thulium, element 69, is named for *Thule,* the northland. It is the rarest of the rare earths (only 0.2 ppm in the earth's crust), and its properties are not well-established.

Ytterbium, element 70, not to be confused with yttrium, is another rare earth that was eventually resolved into two elements, adding to the fears already described (see section on praseodymium). Its +3 salts are colorless but paramagnetic; its +2 salts have an f^{14} configuration and so are diamagnetic.

Lutetium, element 71, is the last of the rare earths and the twin that was isolated from ytterbium in 1907. Its salts are all colorless and diamagnetic.

OXIDATION STATES: THE SIGNIFICANCE OF f^0, f^7, AND f^{14} CONFIGURATIONS

The rare-earth elements share a space in the Periodic Table with lanthanum, a trivalent element, so we might expect them all to be trivalent. That is, we might expect them all to lose their two 6*s* electrons and one 5*d* electron as they combine chemically, leaving intact whatever electrons they are storing away in their 4*f* orbitals "deep down inside the atom." This is not so—there is a rather jumpy irregular filling of the 4*f* orbitals, sometimes at the expense of the first 5*d* electron. This is not as arbitrary as it seems; it is merely in keeping with the lower-energy condition associated with empty, half-filled, and completely filled *f* orbitals in the neutral atoms. Thus, europium and gadolinium both strive for seven *f* electrons in their atomic structures because that means half-filling of all the *f* orbitals, and europium has to borrow its *d* electron to accomplish that. Hence europium may lose only its 6*s* electrons and form Eu^{+2} ions, leaving its f^7 structure intact. Similarly, ytterbium may lose only its 6*s* electrons, and keep its f^{14} complement intact. Cerium can lose both its 6*s* pair and its 4*f* pair to form Ce^{+4}, leaving it with empty *f* orbitals. These remarks apply only to solid compounds; in solution all the lanthanide rare earths adopt a +3 oxidation state.

MAGNETIC PROPERTIES OF RARE-EARTH COMPOUNDS

In Chapter 6 we saw that the degree of paramagnetism in bare transition-metal ions was a direct function of the number of unpaired electrons. What was thought of as the spin of the electron produced a paramagnetic behavior when not neutralized by an equal and opposite spin of a paired electron in the same orbital, so the overall magnetic effect in the bare ions could be predicted by counting the probable number of unpaired electrons occupying the *d* orbitals.

We might expect the same rule to hold for the ions of rare-earth elements as the *f* orbitals are being filled, but it does not. Figure 8–1 shows the actual situation as a plot of net paramagnetic effect versus atomic number. Instead of a single curve with a central peak at gadolinium, number 64, we find a binodal curve with two unequal peaks. This result was not predicted by any theory. Several explanations after the fact depend upon a combination of orbital angular momentum and spin of the electrons, rather than on the simple spin concept alone.

The magnetic behavior of the rare-earth oxides is important in one practical way: pressed mixtures of the oxides with iron and nickel oxides provide non-metallic magnetic cores for high-frequency currents such as those encountered in radio transmission and reception. Every little transistor radio has a so-called ferrite core for its directional loop antenna, and the properties of such ferrite cores are improved greatly by the addition of rare-earth oxides. The detailed electronic states of transition-metal ions are also important in fixing the color response of fluorescent powders for color TV, so that predictable changes in color can be brought about by manipulation of the rare-earth content.

COLORS AND ABSORPTION SPECTRA OF RARE-EARTH SALTS

The colors of the salts of the 4*f* elements follow a regular pattern. This comes about because the small differences in detailed electronic structure from one rare-earth ion to the next cause distinct small differences in their absorption

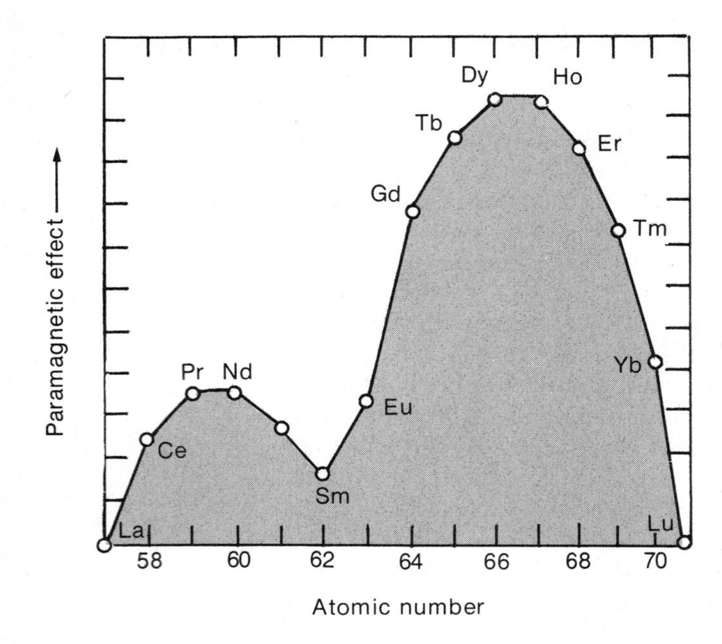

Figure 8–1. Magnetic behavior of tripositive lanthanide ions.

of energy from a beam of white light. The incident light excites one or more electrons within the ion to jump to a higher energy level; the quantized jump involves a characteristic energy absorption, ΔE, which is equal to Planck's constant, *h*, times the frequency of the light absorbed, as explained in Chapter 6 in connection with the transition elements. The extra energy of the excited electron eventually is degraded to heat, so the absorption is continuous and leaves a dark space in the spectrum of the transmitted light. The dark spaces in the absorption spectra of transition-metal salts are broad bands because the electric and magnetic fields of neighboring ions affect the process to varying degree and smear out the effect. The transitions of *f* electrons in rare-earth ions are better shielded from external effects like these, and so the dark spaces in their absorption spectra are lines instead of broad bands. Anyone who wants to know whether a particular solution contains a rare-earth salt or a transition-metal

TABLE 8–1 COLORS OF TRIPOSITIVE LANTHANIDE RARE-EARTH IONS

Atomic Number	*Symbol*	*Color of Salt Solutions*
57	La	colorless
58	Ce	colorless
59	Pr	green
60	Nd	pink or reddish
61	Pm	pale yellow
62	Sm	yellow
63	Eu	pale pink
64	Gd	colorless
65	Tb	pale pink
66	Dy	yellow
67	Ho	pale yellow
68	Er	pink or reddish
69	Tm	green
70	Yb	colorless
71	Lu	colorless

salt need only look at its absorption spectrum to find out—distinct lines mean rare earths, while diffuse bands signify transition elements.

The actual colors of rare-earth salts follow the regularities noted for their oxidation states, and depend similarly on the occupancy of 4*f*, 5*d*, and 6*s* orbitals. In Table 8–1 we find the colors going through a cycle from colorless La^{+3} and Ce^{+3} to colorless Gd^{+3} in the center of the series, and then reversing the order of change all the way back to colorless for the last two ions, Yb^{+3} and Lu^{+3}. Hence there are two elements with green salts, and two elements with yellow salts, and so on. Within any one pair the positions of the lines of the absorption spectra show differences between the two elements, but the overall impression to the eye is the same.

The colors of rare-earth ions are characteristic not only in water solution but also in other solvents. For example, molten silicate glass will dissolve the oxides of transition metals and rare-earth metals, and the glass is then colored by the dissolved ions. We saw that dissolved cobalt colors ordinary soda-lime glass a beautiful deep blue, for example. The oxides of unseparated neodymium and praseodymium color glass a pinkish tan, a shade very helpful in sun glasses because it absorbs bright yellow light in the region where the eye is most sensitive to overexposure, but does not affect the other colors much. It is interesting to note also that such didymium glass absorbs strongly exactly where the sodium D lines occur, and a glassblower who wears such glasses is not blinded by the yellow glare. Furthermore, a room lighted by sodium vapor lamps appears dark from the outside if didymium glass is used in the windows.

SUMMARY

The lanthanide rare earths are the 14 elements in which the 4*f* orbitals are being filled before going on with the development of the 5*d* transition series. They extend the sixth period to a total of 32 elements, but usually the rare earths are consigned to a separate row at the bottom of the Periodic Table to save space.

TABLE 8–2 PHYSICAL PROPERTIES OF THE LANTHANIDE RARE-EARTH ELEMENTS

Atomic Number	*Symbol*	*Atomic Weight*	*Melting Point, °C*	*Boiling Point, °C*	*Density, g/cm³*	*State at Room Temperature*
57	La	138.91	920	3469	6.17	reactive gray metal
58	Ce	140.12	795	3468	6.77	reactive gray metal
59	Pr	140.91	935	3127	6.77	reactive gray metal
60	Nd	144.24	1024	3027	7.00	reactive gray metal
61	Pm	(147)*				reactive gray metal
62	Sm	150.35	1072	1900	7.49	reactive gray metal
63	Eu	151.96	826	1439	5.26	reactive gray metal
64	Gd	157.25	1312	3000	7.86	reactive gray metal
65	Tb	158.92	1356	2800	8.25	reactive gray metal
66	Dy	162.50	1407	2600	8.55	reactive gray metal
67	Ho	164.93	1461	2600	8.79	reactive gray metal
68	Er	167.26	1497	2900	9.05	reactive gray metal
69	Tm	168.93	1356	2800	8.27	reactive gray metal
70	Yb	173.04	824	1427	6.98	reactive gray metal
71	Lu	174.97	1652	3327	9.84	reactive gray metal

*Atomic weight shown in parentheses here is that of the most stable isotope.

In a way this is a pity, for a truly periodic table should show periods of 2, 8, 18, and 32 elements.

Lanthanum is included with the 4*f* rare earths in this chapter because it behaves so much like them. All the rare-earth elements (and La) are reactive hard gray metals with moderate melting points, as shown in Table 8–2. They are widespread elements, although not particularly rare. They occur in the dark low-melting material seen as veins in granite, and in the dense dark sand classified by flowing water after being washed from weathered granite. They can be extracted by acid treatment of the sand, followed by precipitation of the oxides and reduction with calcium to get the mixed metals. This natural alloy of rare-earth metals, often with iron present, is used as the "flint" of cigarette lighters because its filings ignite by friction and burn brilliantly in air. Rare earths are used also to make special absorptive glass for sunglasses, to make nonmetallic magnetic cores for radio circuits, and to activate fluorescent powders to emit colored light from TV picture tubes.

The rare earths are very difficult to separate, but the feat can be accomplished by using an ion-exchange column. The isolated elements show variations in oxidation states, in magnetic behavior, and in colors or absorption spectra. The variations in color and oxidation state are shown to correspond closely to electronic configuration of the atoms and of the ions derived from them, and especially to low-energy states resulting from empty, half-filled, and completely filled *f* orbitals. The variations in paramagnetism follow a more complicated pattern.

GLOSSARY

absorption spectrum: the spectrum of light (or other radiation) that has already passed through the sample under investigation, and therefore consists of the spectrum of the original light source minus those wavelengths or frequencies which have been absorbed by the sample. (See Plates 1 and 2.)

cathode ray: a stream or a narrowly-defined beam of electrons accelerated by electrical means.

didymium: the name given to a natural mixture of praseodymium and neodymium, which are difficult to separate.

earth: in alchemy, any of several oxides difficult to reduce, such as Al_2O_3, CaO, BaO, or La_2O_3.

lanthanide elements: the 4*f* rare-earth elements.

orbital angular momentum: the product of the mass of a body times its velocity in a circular path. Early Bohr theory of atomic structure portrayed electrons as bodies moving in circular orbits around the nucleus, like a miniature solar system.

Prometheus: in Greek mythology, the god who gave the gift of fire to man. Element 61 was named after him because it is a product of nuclear energy, the gift of which is likened in importance to the gift of fire.

rare earth: see **earth.** A rare-earth element is one of the series comprising atomic numbers 58 through 71 (the lanthanide rare earths), or 90 through 103 (the actinide rare earths).

QUESTIONS

1. Anhydrous cobalt nitrate is blue, but hydrated cobalt nitrate is pink because the $Co(H_2O)_6^{2+}$ ion is pink. Various gadgets make use of this change to indicate indoor humidity or show when plants need watering. By adjusting concentrations, it is possible

to prepare a water solution of neodymium nitrate which looks exactly the same color as a solution of cobalt nitrate. How could you tell them apart by: a. a physical method; and b. a chemical method?

2. List the different kinds of colored glass you have encountered so far in this book, and tell precisely what is responsible for each color.

3. Table 8–2 shows that ytterbium has a much lower density, melting point, and boiling point than its neighbors thulium and lutetium. What explanation can you find for this in terms of atomic structure? What does it tell us about the dimensions of the crystal lattice of metallic ytterbium?

4. The standard reduction potentials of some rare-earth elements are:

Ce	−2.48 volts	Gd	−2.40 volts
Pr	−2.47 volts	Ho	−2.32 volts
Eu	−2.41 volts	Lu	−2.25 volts

Would these metals make good protective platings for steel, like chromium or nickel? Would they make good consumable electrodes for practical batteries? Support your answer with facts.

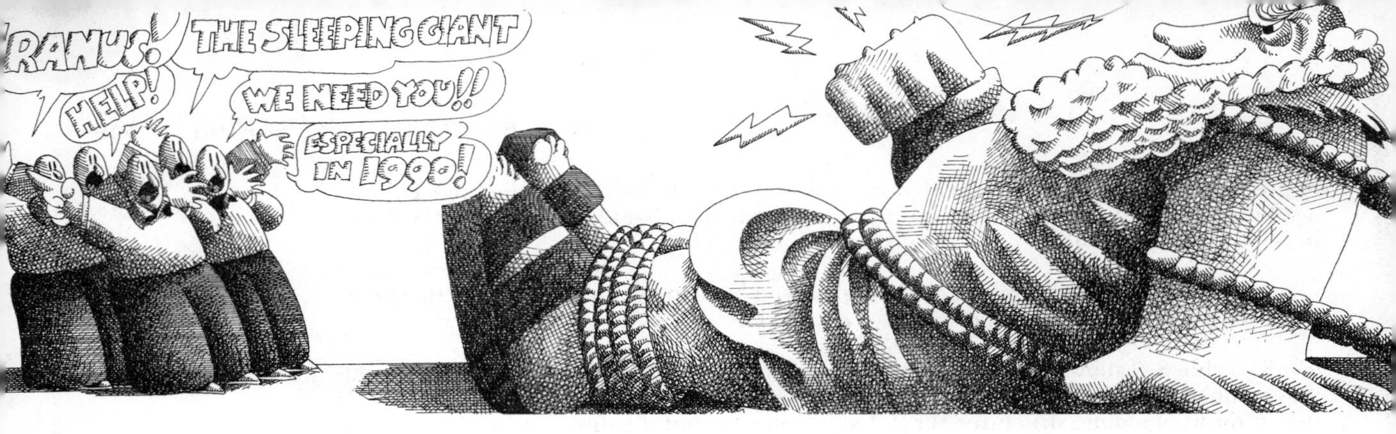

9 THE ACTINIDE RARE EARTHS

In the last period of the Periodic Table there occurs another series of 14 elements in which *f* orbitals are being filled, this time the 5*f* orbitals. The series comprises elements 90 through 103, often called the actinide elements or actinium series because they follow actinium (element 89) in just the way the rare earths follow lanthanum. Only two of these 5*f* elements are found in appreciable amounts in nature, thorium and uranium. The element between thorium and uranium, protactinium, is a temporary decay product of the radioactive disintegration of uranium; all the other elements in the series are synthetic in the sense that they were made from uranium or its products by nuclear reactions. All the elements in the series are radioactive; none has any stable isotopes. We shall consider the three most significant elements of the series—thorium, uranium, and plutonium—and then go on to a brief discussion of all 14 elements as a coherent series.

THORIUM, THE MOST PLENTIFUL ONE

Thorium has been known for 150 years, and for half that time its compounds were used by the public without any regard for their radioactivity because Becquerel had not yet discovered radioactivity. No one saw or felt anything that would suggest danger. Fortunately natural thorium (all ^{232}Th, no isotopes) has such an extremely long half-life, 140 billion years, that its radiation level is very low and its compounds can be handled and used without great danger as long as they are kept outside the body.

Thorium occurs as the silicate $ThSiO_4$, in which it was discovered, and as the oxide in various uranium ores such as pitchblende ($UO_2 + UO_3 + ThO_2 +$ PbO, etc.). It was extracted by acid treatment of the ore and converted to two useful compounds: thorium dioxide or thoria, ThO_2, a true "earth" in the old chemical sense because it is refractory (mp 2800° C) and difficult to reduce, hence a good material for crucibles; and thorium nitrate, $Th(NO_3)_4$, a soluble colorless salt which decomposes to white thoria when heated. During the gas-light era every city dweller in Europe and America made use of these two thorium compounds in the gas mantle, a simple device that was used to light homes and streets and public buildings for nearly half a century. The gas mantle was a little knitted bag of cotton 1 or 2 inches in diameter which was fastened to a porcelain rim and impregnated with a solution of thorium nitrate. It could be attached by anyone to the holder suspended from a gas jet, and the first time the gas was lit the cotton burned away, leaving a white skeleton of ThO_2. A gas flame by itself is dim and yellow because it is made visible only by incandescent carbon particles, but a thoria mantle heated to 1400° C or so by the gas-air reaction glowed with a brilliant white light.* A descendant of such gas mantles is used today in gasoline lanterns used by campers.

Thoria is still used to make crucibles and refractory insulators, but by far the greatest interest in thorium now centers on its use in extending the scarce supply of uranium for nuclear power plants and ship propulsion. It is only the ^{235}U isotope of natural uranium that is fissionable with slowed-down neutrons, and ^{235}U constitutes only 1/142 part of what is mined. If we were to generate even half the electric power we need from nuclear energy via fission reactors, we should soon consume all the available ^{235}U. Are there any substitutes? In order to answer the question we need to know what makes one isotope of uranium fissionable with slow neutrons while another is not.

Fission is not a reaction peculiar to uranium. Experiments conducted in cyclotrons and synchrotrons have shown that all elements above atomic number 80 undergo fission if bombarded at the right energy. Sorting out the results, three conclusions can be drawn:

1. The elements beyond uranium undergo fission spontaneously as one mode of nuclear disintegration. For example, every once in a while a nucleus of ^{254}Cf or ^{256}Fm simply flies apart into two unequal pieces, emitting some spare neutrons in the process. It seems that all nuclei with masses above 250 behave this way, just because they are too large to be stable. No bombardment is needed, but of course bombardment increases the rate of fission.

2. Between mass numbers 225 and 250, nuclei do not undergo fission spontaneously, but will do so if a small amount of additional energy (about 6 meV) is supplied. This activation energy can be supplied by fast-moving electrons, or slower protons or alpha particles or neutrons, or even by gamma rays, but until it is supplied the nucleus stays intact.

3. Below a mass number of 225 a rapidly-increasing amount of activation energy is required. For example, at a mass of 200 about 20 meV of energy per atom (46×10^7 kcal per mole) is required to induce fission.

Nuclear chemistry teaches that most nuclei are capable of absorbing and binding at least one additional neutron, and that the binding energy given off during such an event varies according to the nuclear structure. For nuclei with an odd number of neutrons the energy given off during the binding of an additional neutron is always greater than that given off by a nucleus with an even number of neutrons, due to the pairing effect. It follows that in the region of mass 225 to 250 (situation 2, above) an isotope with an odd number of neutrons or protons can be excited to the fissionable state by a slow neutron more easily than

*The ThO_2 acted as a surface catalyst for the gas-air reaction as it sat between the two gases, and so achieved a much higher temperature than that of the ordinary gas flame.

an isotope with an even number of neutrons and protons, just because it will add its own higher binding energy to the kinetic energy of the bombarding neutron. This is why ^{235}U is more readily fissionable than ^{238}U. And this is precisely where thorium comes into the picture—it can supply another easily fissionable odd-numbered isotope to supplement the meager supply of ^{235}U.

All natural thorium is ^{232}Th, and when this absorbs a neutron it becomes an excited nucleus of ^{233}Th which soon decays by two beta emissions to give ^{233}U:

$$^{1}_{0}n + ^{232}_{90}Th \rightarrow ^{233}_{90}Th^{*} \xrightarrow[25\text{ minutes}]{-\beta} ^{233}_{91}Pa \xrightarrow[27\text{ days}]{-\beta} ^{233}_{92}U$$

This new synthetic even-odd isotope of uranium is an alpha emitter with a half-life of 162,000 years, and is a perfectly good nuclear fuel for fission reactors. Since there is three times as much thorium in the world as uranium, and 426 times as much thorium as ^{235}U, it makes sense to convert some of this thorium to fissionable fuel. That can be done in a breeder reactor, which is devoted principally to "breeding" additional nuclear fuel in this manner, or it can be done as a conservation measure in all existing research reactors and power reactors. We shall see how this is accomplished when we come to the discussion of fission reactors, in the next section.

URANIUM, AN OBSCURE ELEMENT SUDDENLY BECOME PROMINENT

Uranium was known a hundred years before Mendeleev. It occurs to the extent of 4 ppm in crystal rocks, especially granite. The chief minerals are the black mineral uraninite, UO_2, and the black mineral pitchblende, a mixture of U, Th, and Pb oxides mentioned in the previous section. Carnotite, a yellow complex uranite-vanadate with the formula $K_2(UO_2)_2(VO_4)_2(H_2O)_x$, is the ore found in Colorado. Uranium was a by-product of the extraction of thorium for gas mantles during the latter half of the nineteenth century, but it had no use except to color glass a fluorescent greenish yellow. Then in 1896 Becquerel discovered radioactivity while investigating the fluorescence of uranium compounds, and in 1939 Hahn and Meitner discovered nuclear fission in uranium. From then on uranium became more valuable than gold. Its metallurgy was studied exhaustively in a crash effort to prepare hyperpure metal and to make alloys suitable for fuel assemblies, and its chemistry was studied with equal intensity to discover how best to separate the isotopes of uranium and how to separate uranium from its myriad fission products.

METALLURGY. Metallic uranium is obtained from its ores by several methods, the simplest of which is the reduction of UF_4 with calcium. The concentrated ore is leached with a solution of nitric acid, with free access of air, thereby oxidizing the uranium to the +6 state and putting it in a solution as uranyl nitrate, $UO_2(NO_3)_2$:

$$2UO_2(s) + 4H^+(aq) + 4NO_3^-(aq) + O_2(g) \rightarrow 2UO_2(NO_3)_2(aq) + 2H_2O(l)$$

The uranyl nitrate is sufficiently covalent to be soluble in organic solvents, even hydrocarbons, so it can be extracted into such a solvent while the impurities are left behind in the water. Crystalline uranyl nitrate hexahydrate is then obtained by evaporating the solvent, and the crystals are decomposed thermally to uranium trioxide:

$$2UO_2(NO_3)_2(H_2O)_6(s) \xrightarrow{500^\circ C} 2UO_3(s) + 4NO_2(g) + O_2(g) + 6H_2O(g)$$

Without removing it, this oxide is then reduced to UO_2 in a stream of hydrogen, and then converted to UF_4 in a stream of hydrogen fluoride:

$$UO_3(s) + H_2(g) \xrightarrow{700^\circ C} UO_2(s) + H_2O(g)$$

$$UO_2(s) + 4HF(g) \xrightarrow{550^\circ C} UF_4(s) + 2H_2O(g)$$

Finally, the UF_4 ("green salt") is mixed with powdered calcium and ignited, starting an exothermic reaction which leaves both products molten:

$$UF_4(s) + 2Ca(s) \xrightarrow{1300^\circ C} U(l) + 2CaF_2(l)$$

The dense metal settles to the bottom of the crucible and solidifies there.

If the uranium is to be enriched in ^{235}U, the UF_4 is first oxidized to UF_6 with fluorine:

$$UF_4(s) + F_2(g) \xrightarrow{300^\circ C} UF_6(g)$$

The hexafluoride is a volatile covalent compound that sublimes readily at room temperature in vacuum to form large colorless birefringent crystals. These have a vapor pressure of 1 atm at 56.2° C, and melt under 2 atm of their own vapor at 69.2° C. Hence UF_6 is by far the most volatile compound of uranium, and is the one used for separation of the 235 and 238 isotopes by vapor diffusion through porous barriers. This method of separating the isotopes of uranium is still widely used because every American-designed nuclear power reactor uses enriched uranium (with 4% ^{235}U instead of the 0.72% in natural uranium) in its fabricated alloy fuel assemblies.

Pure uranium is a dense, reactive, strongly electropositive, ductile, and malleable gray metal. It crystallizes in at least three different phases, all with different densities, which complicates its metallurgy and often weakens the metal as it transforms from one crystal structure to another. It is attacked by air, water, steam, and acids, so it has to be enclosed in some corrosion-resistant casing before it can be used as a fuel in water-cooled reactors. It reacts vigorously with chlorine to form UCl_3, UCl_4, UCl_5, and UCl_6. Its odd oxidation states are unstable; +4 and +6 states predominate. The most commonly used oxyacid salts are those of the uranyl ion, UO_2^{+2}, such as uranyl nitrate (see above) and uranyl sulfate, UO_2SO_4.

FISSION. The nuclear fission of uranium is not a simple process, so no single equation can be written for it. The products of its fission are spread over half the Periodic Table, and no one can tell in advance which elements will result from the fission of a particular atom of uranium. Hence a statistical approach is necessary, and we must think in terms of the probability that this or that element will appear in the fission product mixture. Furthermore, almost all the products are in the form of highly radioactive isotopes which may change rapidly from one element to another as they decay by beta or positron emission.

Chemical analysis, supplemented by counting (that is, quantitative measurement of the radiation from a separated and identified sample), tells us what elements are present in the product mixture, and how much of each. From this information the fission yield curve of ^{235}U has been worked out, and is re-

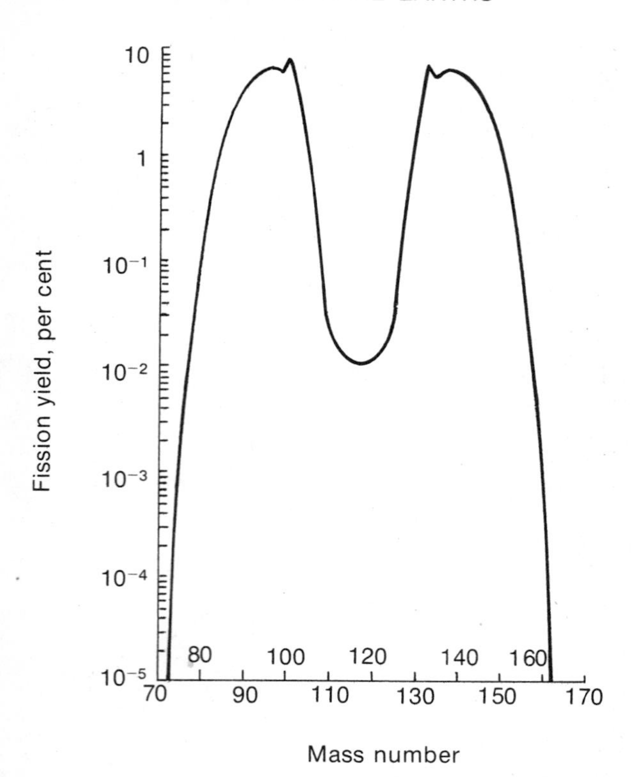

Figure 9–1. Fission yield curve for uranium-235. (From G. Friedlander, J. W. Kennedy, and J. M. Miller, *Nuclear and Radiochemistry,* John Wiley & Sons, New York, 1955. By permission.)

produced in Figure 9–1. We see a binodal curve having two broad peaks centered approximately at mass numbers 93 and 136, with small "blips" that occur where magic numbers of neutrons (50 and 82) appear in the isotopes. So most uranium nuclei split into two unequal parts, with masses of 85 to 100 (in the range of bromine to silver) and 135 to 150 (in the range of the lanthanide rare earths) especially favored. However, no isotope occurs in an amount over 7% of the mixture, and the average yield is only about 0.01%. Since the product nuclei are all much lighter than the uranium nucleus, they must contain a smaller proportion of neutrons (see the stability rules in Chapter 1). The surplus neutrons are liberated, and therein lies the key to a chain reaction. If enough neutrons are left, after absorption and escape take their toll, then further fission can be achieved without an external supply of neutrons.

From this information we can write some sample equations, with the qualification that they are only random examples of a thousand possibilities:

$$^{1}_{0}n + ^{235}_{92}U \rightarrow ^{82}_{35}Br + ^{151}_{61}Pm + 4\ \beta^- + 3\ ^{1}_{0}n$$

$$^{1}_{0}n + ^{235}_{92}U \rightarrow ^{87}_{38}Sr + ^{146}_{57}La + 3\ \beta^- + 3\ ^{1}_{0}n$$

$$^{1}_{0}n + ^{235}_{92}U \rightarrow ^{72}_{30}Zn + ^{160}_{62}Sm + 4\ ^{1}_{0}n$$

The fission product nuclei fly apart from each other at high velocity, due to the mutual repulsion of their large positive charges. The neutrons also are emitted at high velocity, by a recoil effect. The fission products are soon slowed down by massive collisions, converting their kinetic energy to heat. The neutrons must be slowed down by artificial means, so that at least one from each fission event may be absorbed by a ^{235}U nucleus and induce another fission. The slowing down is done by a carefully selected light material of low absorption called a moderator.

Nuclear Reactors. A nuclear reactor is any device in which a nuclear fission reaction is carried out and manipulated for a particular purpose. The discussion here will be limited to research reactors and nuclear power reactors for the generation of electricity, ignoring devices for marine propulsion and nuclear weapons. While the treatment centers on uranium, it applies also to any other nuclear fuel.

This discussion of reactors is intended to be purely factual, and its inclusion does not constitute an endorsement of nuclear energy in preference to other forms of energy. Some pros and cons of nuclear power will be considered later so that the reader may arrive at his or her own conclusion.

WHAT A NUCLEAR POWER REACTOR MUST HAVE TO OPERATE. Any nuclear reactor that utilizes the fission of uranium or plutonium by slow neutrons as its source of energy must have these six components:

1. A *nuclear fuel,* that is, a fissionable isotope present in sufficient quantity to sustain a steady fission reaction, and fabricated in a design suitable for transferring the resulting heat to the coolant.

2. A *moderator* to slow down the fast neutrons that come out of the fission reaction to thermal velocities so that they can be absorbed by the fissionable isotope and cause more fission. A moderator must have very low neutron absorption, lest it waste neutrons, and should preferably contain light nuclei that can reduce the speed of neutrons (by exchange of momentum) in a minimum number of collisions. The best moderators, in order, are:

   a. helium, which has zero absorption but needs much space because it is a gas;

   b. heavy water, D_2O, which has very low absorption and makes an excellent moderator but has to be distilled at high cost out of ordinary water (see Chapter 2) and therefore must be carefully confined;

   c. graphite, which has low absorption but is a solid which cannot be pumped around and hence cannot serve also as coolant;

   d. carbon dioxide, a gas with greater density than helium;

   e. beryllium, which suffers from all the drawbacks of graphite plus high cost and toxicity; and

   f. ordinary light water, which absorbs 600 times as many neutrons as heavy water but is cheap and plentiful.

3. A *coolant* which will carry off the heat from the fission reaction to a steam boiler and turbines or to any other heat engine for transforming heat energy to free energy. The coolant may also serve as moderator if it has the right properties. It must be a fluid, of course, in order to be pumped into and out of the reactor. If light or heavy water is chosen, it must be under high pressure in order to keep it in the liquid phase at temperatures high enough to effect economical heat transfer out of the reactor to the steam boiler. If a solid moderator is chosen, a fluid coolant will be necessary. If a gaseous moderator-coolant is chosen, a great deal of it must be circulated because its heat capacity is small.

4. *Controls* to adjust the rate of the fission reaction to the desired operating level, and to shut down the reactor if anything goes wrong.

5. A *breeding blanket* outside the reactor core to absorb neutrons that would otherwise be a hazard (and incidentally to use them to generate more fissionable fuel).

6. A *reflector and shield,* to return stray neutrons to the breeder and to absorb the harmful radiations. The shield is usually in three parts: a thermal shield (perhaps of cast iron) to absorb most of the radiation; a main shield of

barium concrete in layers 4 to 8 feet thick; and a biological shield of wood fiber to absorb the kind of γ rays and x-rays that would harm people.

There are thousands of engineering questions connected with the design of a reactor, and many other questions connected with the refueling operation, the processing of spent fuel to get rid of the fission products that "poison" the reaction, the disposal of the fission products waste, and the disposal of waste heat from the turbines. A book devoted primarily to chemistry cannot be expected to treat these questions exhaustively, but they should be kept in mind while considering some typical designs.

Types of Power Reactors. Because the U.S. was first in the field of designing nuclear power reactors, many think that we know most about it and our designs must be best. This is not necessarily so, and is part of the problem in the U.S. The designs used in American power plants are all offshoots of the reactors developed for submarine and aircraft carrier propulsion in the late 1940's and 1950's, and are handicapped both by the intricate construction characteristic of weapons (where economy is secondary) and by further association with the military at a time when all things connected with war are so clearly abhorred. In other words, those who are trying to sell the American public the present breed of nuclear power plants are carrying the atomic bomb and two unpopular, demoralizing wars on their backs. The emotions run correspondingly high.

Figure 9–2 shows the construction of the principal type of reactor used commercially in the U.S. Notice the large and very thick containment vessel and the intricate machined clad-alloy fuel assemblies. Enriched uranium is essential here, despite the very high cost of isotope separation, because the neutron absorption of ordinary light water is so high that a chain reaction cannot be sustained with natural uranium no matter how much of it is assembled in the reactor. This is a fundamental point.

In any reactor 99.9% of the original mass of the fuel consumed remains behind as fission products, while only 0.1% is converted to energy. All these fission products absorb neutrons, and some of them have enormous appetites for the precious neutrons, with thousands or even millions of times the neutron absorption of light water. These products have to be removed periodically or they will kill the chain reaction. To refuel the reactor of Figure 9–2 it is necessary to shut down the reactor for weeks, allow the fuel assembly to "cool off" (that is, decay from its initial extremely high level of radioactivity), pull out the spent fuel assembly, and replace it with new fuel units. Meanwhile some other source has to supply the continued demand for electric power.

Other countries have not had the resources and the motivations of the U.S. and have gone about the design of nuclear power reactors in their own ways to suit their own needs. Figure 9–3 shows the radically different design of a Canadian reactor used in six operating power plants and in ten more under construction. These reactors use *natural* uranium, avoiding the expensive isotope separation. To operate with natural uranium they must use heavy water as a moderator and coolant. Heavy water of the desired purity is obtained by fractional distillation, using waste heat from a reactor. Notice that each fuel element resides in its own pressure tube of zirconium alloy which need be only 4 mm thick because its diameter is so small. The containment vessel consequently is only 1.25 inches thick, compared to an average of 8 inches thick in the reactor of Figure 9–2. As a further consequence of its design, the Canadian reactor can be refueled continuously *while still in operation*, because the first fuel element to be poisoned by fission products is simply pushed out of its pressure-tube container and replaced, and then the next fuel element is replaced, and so on while

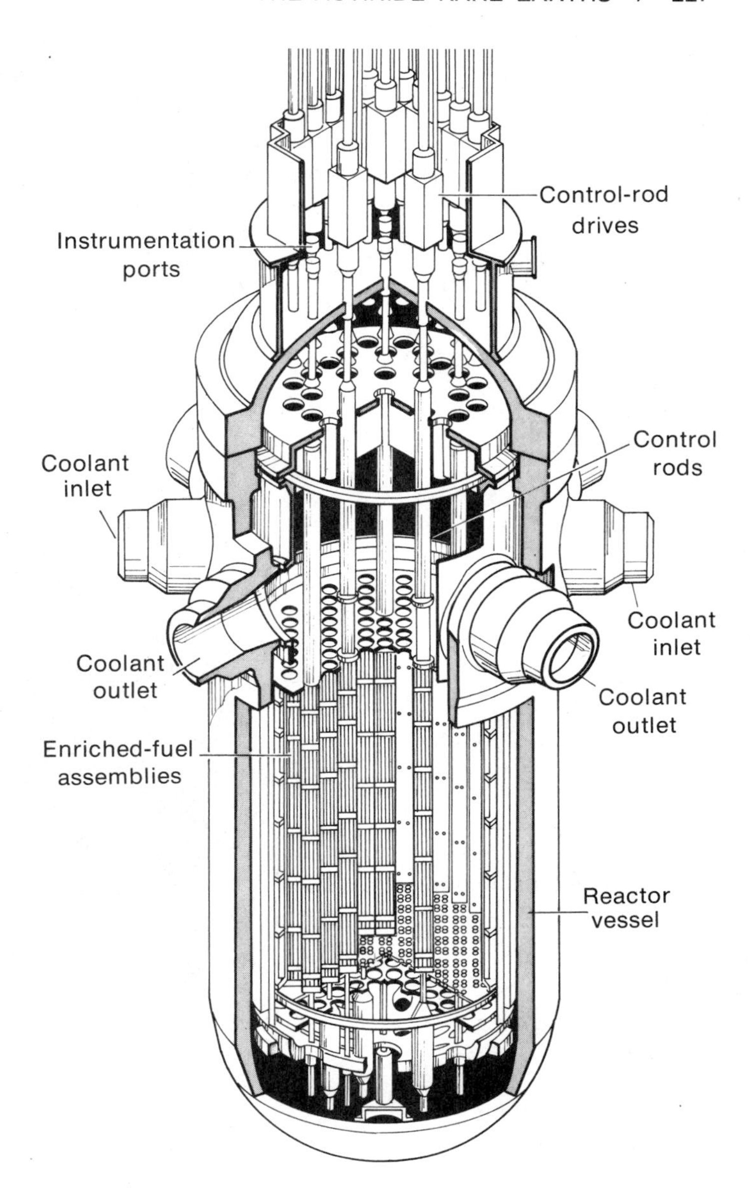

Figure 9–2. Pressurized light-water reactor, developed to power U. S. nuclear submarines, is one of the two systems used in all U. S. nuclear power plants now operating (about 55). Because light water captures neutrons about 600 times more readily than heavy water does, a light-water reactor will not operate unless the fuel is enriched to contain 1 to 4 per cent uranium-235 instead of the 0.72 per cent naturally present. A pressurized light-water reactor capable of generating 1100 MWe requires a pressure vessel about 15 feet in diameter, 45 feet tall, and 6 to 11 inches thick. When the reactor is charged with 196,000 pounds of uranium oxide containing an average of 3.2 per cent ^{235}U, it will operate for 10 to 12 months before fresh fuel is needed. Light-water coolant is heated to 320°C, and it circulates at 2250 pounds per square inch. A separate steam generator produces steam for the turbines at a temperature of 285°C and a pressure of 1000 pounds per square inch. (Adapted from H. C. McIntyre, "Natural-Uranium Heavy-Water Reactors," *Scientific American,* **233**:17, 1975.)

the rest keep working. This keeps the reactor in service, and limits the level of radioactivity to be handled outside of the reactor at any one time.

In Britain, where there were no producing oil wells until 1975 and coal mining has been a critical limitation, nuclear power plants were necessary at an early date. There a still different design evolved, making use of graphite as moderator for natural uranium (as in the very first self-sustaining experimental reactor at Chicago in 1942) and cooling the solid assembly with a stream of CO_2 under pressure. The very hot gas is pumped into a steam boiler, where it loses its heat by boiling the water, and is then returned to the reactor. Pure carbon and its dioxide have a low enough absorption of neutrons (only 1% of that of ordinary hydrogen) to allow use of natural uranium. A spherical steel pressure vessel is used, and provision is made to reoxidize any CO_2 that is reduced by the hot graphite.

Any power plant, like any other heat engine, must discard low-grade heat in order to extract free energy from higher-grade heat. This is a thermodynamic

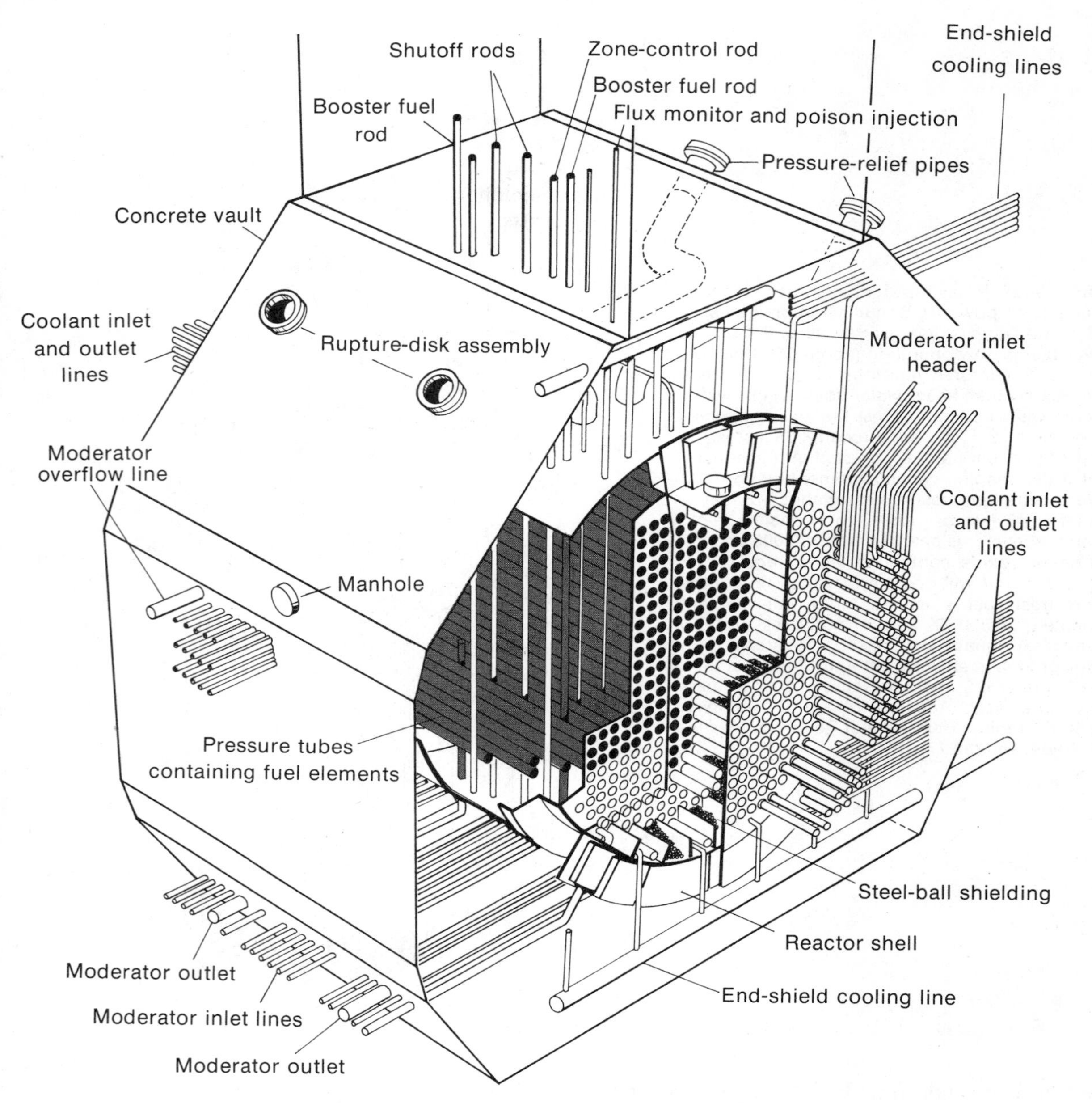

Figure 9–3. CANDU nuclear reactor differs from U. S. nuclear power reactors not only in using natural uranium and heavy water but also in that the coolant for the CANDU reactor flows through several hundred individual pressure tubes containing the fuel. In American systems the fuel elements are contained in a single massive pressure vessel through which the coolant flows. As this figure shows, the pressure tubes in the CANDU reactor vessel are arranged horizontally. The outer shell for a reactor capable of generating 750 MWe is 28 feet in diameter, 19.5 feet long, and 1.25 inches thick. The tubes are made of a zirconium alloy 4 mm (one-sixth inch) thick. The heavy-water coolant, at a pressure of 1450 pounds per square inch, leaves the reactor at a temperature of about 310°C and passes to a steam generator (not shown) where ordinary light water is converted into steam at a temperature of about 250°C and a pressure of 570 pounds per square inch. The steam drives the turbogenerator that generates electricity. The many pipes that convey the coolant into and away from the reactor ultimately are gathered into two large headers that are connected to the steam generator. The CANDU reactor is designed so that the fuel bundles in the individual pressure tubes can be replaced while the reactor is running at full power. (Adapted from H. C. McIntyre, "Natural-Uranium Heavy-Water Reactors," *Scientific American,* **233**:17, 1975.)

fact of life, in accord with the Second Law. It matters not whether fossil fuels or nuclear fuels are used; the plant must discard about 60% of the heat obtained from the fuel while it puts out the remaining 40% as useful work. In Britain the waste heat is transferred to the air by large conical cooling towers which are a feature of the landscape near any large city. Here in the U.S. it is customary to transfer the waste heat to rivers or lakes by using their water to cool the turbine condensers. There is considerable objection to this because the warmth encourages growth of algae and weeds. Cooling with seawater is possible along the coasts, but inland cooling towers are needed.

ARGUMENTS AGAINST NUCLEAR POWER. There are many new, insidious, and overwhelming dangers associated with a greatly enlarged network of nuclear power stations. The most elaborate control systems can fail or be countermanded, allowing lethal quantities of radioactive material to escape into the air, water, and earth in case of accident. While the boiling-water reactors may not explode, they certainly will suffer massive heat damage if their cooling systems fail, and may well breach their containers and spread disaster. The next generation of reactors, the fast-neutron breeders, will be much more difficult to control and much more liable to disastrous failure or explosion. They also contain an intolerable amount of plutonium. The intensely radioactive fission waste poses an impossible disposal problem, for exceedingly dangerous waste material has to be isolated from possible human contact or exposure for 100,000 years—far longer than our social and governmental institutions have endured in the past. The lack of experience in such a monumental task guarantees that costly and tragic mistakes will occur before we learn any effective methods of disposal.

All radiation is destructive to living organisms, and human beings are particularly susceptible to irreparable injury. Even the smallest increase in the environmental background of radiation will cause more suffering and more deaths from cancer, and will cause genetic damage. The only way to ensure no future trouble from this source is to abolish nuclear power.

The plutonium produced by fission reactors poses a separate serious threat, quite apart from its toxicity. If stolen or diverted, it could be used to make clandestine atomic bombs that could destroy a large city. In a world given over increasingly to anarchy and political terrorism, the possibility of nuclear weapons in the wrong hands is intolerable. Even the possibility of nuclear blackmail must be ruled out by generating no more plutonium and by using up what there is.

ARGUMENTS FOR NUCLEAR POWER. At the time of writing there is no long-term energy policy or program in the U.S.A. Everyone seems to agree that it would be better to get all the energy needed for industry, transportation, and domestic use from the sun (which certainly sends us enough), rather than from fossil fuels, because oil must be imported at exorbitant prices and coal must be obtained by strip-mining, and burning coal or oil puts cancer-producing pollutants in the air. However, no one knows how to harness solar energy to the task. In the meantime, it is argued, nuclear fission energy can fill the gap while research on solar energy, fusion energy, and other sources is pressed to the limit.* The safety record of uranium-burning reactors has been very good so far, with hundreds in routine operation. Although some segments of the public still equate nuclear power with nuclear weapons, and some unscrupulous writers exploit that fear, the fact is that no water-boiling reactor is an atomic bomb, nor can it ever act as one. If all controls fail, the water vaporizes, and when the

*See "The Necessity of Nuclear Power," by Hans A. Bethe, *Scientific American* **234**:28 (1976).

chain reaction is deprived of its moderator it stops. Water-moderated uranium fission reactors operated in a natural formation for hundreds of thousands of years at Oklo in Africa* and never blew up; when water was scarce, the reaction simply stopped. As for the plutonium produced, that remained in place and did not migrate more than 1 mm per half-life of 24,000 years!

Radiation is always injurious, and any radiation at all from reactor wastes is unwelcome. At its worst, it could endanger millions of lives if it were carried by air or water into a large city. Those officials who are entrusted with guarding the public from dangerous radiations would have to be as tireless and proficient as those who safeguard the public from communicable diseases and poisonous foodstuffs. But to say that this cannot be done is to sell short the human spirit that has had to cope with staggering dangers for hundreds of thousands of years. For example, water is an ever-present danger, and thousands drown each year; young children are especially prone to accidents involving water. Yet we have found how to cope with the danger and to navigate the seven seas. Similarly, infectious diseases are an insidious threat, and have claimed millions of lives in the past. Like radiation, the infectious organisms cannot be seen or felt at the time of exposure, and can be devastating later. Yet mankind has learned how to isolate the causative organisms, how to detect and obliterate them, and how to treat those who had the misfortune to be exposed. A similar but more difficult task awaits those who would establish a mature nuclear power utility. Fortunately, the only essential experiment which is impossible for them to carry out (whether or not fission products and plutonium can be stored safely underground for millions of years) has already been carried out by Mother Nature, and the answer is "yes." (See footnote.)

THE TRANSURANIUM ELEMENTS

Plutonium, element 94, was first made in a cyclotron in 1940, but since 1942 it has been a product of nuclear fission reactions in ^{235}U. The plentiful isotope of uranium, ^{238}U, absorbs slow neutrons while it is in a reactor, but the capture of these thermal neutrons does not release enough binding energy to cause fission. Instead, a heavier unstable isotope of uranium results from the absorption, and this decays by two successive beta emissions to an isotope of plutonium, ^{239}Pu:

$$^{1}_{0}n + ^{238}_{92}U \rightarrow ^{239}_{92}U \xrightarrow[\text{23 minutes}]{-\beta} ^{239}_{93}Np \xrightarrow[\text{2.3 days}]{-\beta} ^{239}_{94}Pu$$

When the fuel assemblies from uranium reactors are processed to remove the fission products, they are dissolved in nitric acid and then the solution is subjected to fractionation to recover the uranium for reuse. During the separation procedure the plutonium can be recovered.

The isolation of plutonium depends on the fact that in plutonium compounds the +6 oxidation state is less stable than it is in uranium compounds, while the +3 state is more stable than in uranium. To the solution of $UO_2(NO_3)_2$ and $PuO_2(NO_3)_2$ from the dissolution of the fuel assemblies, SO_2 is added to reduce plutonium to Pu^{+4} and Pu^{+3}, while uranium stays as the UO_2^{+2} ion. The solution is then extracted with tributyl phosphate dissolved in kerosene. Uranyl nitrate goes into the kerosene layer but plutonium nitrate stays in the

---

*See especially "A Natural Fission Reactor," by George A. Cowan, *Scientific American* **235**:36 (1976). Two billion years ago in the Gabon Republic in West Africa a vein of uranium ore "went critical" and operated intermittently for 150,000 years by natural control. The fission products are still there.

aqueous layer and can be recovered by decomposing the nitrate to oxide and converting this to fluoride with HF.

Metallic plutonium is best obtained by reducing PuF_3 with calcium, as was done with uranium:

$$2PuF_3(s) + 3Ca(s) \rightarrow 2\ Pu(l) + 3CaF_2(l)$$

It is a silvery-white reactive metal which crystallizes in six distinct phases, depending on the temperature. It oxidizes by the heat of its own radioactivity. Acids and halogens attack it readily, leading to compounds in the +3, +4, +5, and +6 oxidation states. These all have characteristic colors: Pu^{+3} is violet, Pu^{+4} is tan, PuO_2^+ is reddish, and PuO_2^{+2} is pink. The oxidation potentials of Pu^{+3}, Pu^{+4}, and PuO_2^{+2} are so close together that their ions can coexist in aqueous solution, a rather unusual situation. The compounds $Pu(OH)_4$ and $Pu(IO_3)_3$ are insoluble, and PuF_6 is volatile and covalent (mp 50.8° C).

The fission of ^{239}Pu by slow neutrons proceeds in much the same way as that of ^{235}U, and there is no doubt that power reactors and fast breeder reactors can be fueled with plutonium. The principal objection to it for this or any other use is its extreme toxicity. It is a bone-seeking element and a strong alpha emitter, with a half-life of 24,000 years. Even a single dose of 1 microgram ingested or inhaled can cause serious injury or death because the body cannot be rid of it. As for constant exposure, if a person were to pick up a 1-cm cube of metallic plutonium, put it on a balance pan, and then take it off, the amount that would have rubbed off on his fingers would exceed the allowable accumulation for that day. With this in mind, there is naturally a preference for ^{233}U produced from thorium as a fuel for fission reactors, if the supply of uranium grows short.

After plutonium, the remaining transuranium elements have been shown to constitute a 5*f* series, rather than a fourth sequence of transition metals, as was first thought. The first few members can be obtained by neutron or alpha particle bombardment of plutonium, but after that the preparations become far more difficult. Heavy ions such as those of carbon have been accelerated and used for bombardments, and intense neutron bursts such as were encountered during the first hydrogen-bomb tests have been found to produce some of the new elements. Due to the remarkable development of a theory of alpha emission systematics, the energies of emitted alpha particles can now be predicted before a new element is prepared for the first time. Consequently, researchers now know what they are looking for, ahead of time, and can confirm the transitory existence of even a few atoms of a new element just by recognizing its alpha particles when they appear. As a further result, the chemical properties of the new 5*f* elements can be studied by tracer techniques using far less than the nanogram quantities formerly required; now a hundred or so atoms suffice for some definitive experiments. What has been learned can be summarized in a few notes about each element; some physical properties are listed in Table 9–1.

Americium results from the intense neutron irradiation of plutonium in nuclear reactors, and can be concentrated by solvent extraction techniques. It is a distinctly amphoteric element, with oxidation potentials much lower in basic solution than in acid. Final separation is achieved by precipitating $(NH_4)_4AmF_8$. In strongly acid solution Am(V) disproportionates to Am(III) and Am(VI) by the reaction

$$3AmO_2^+ + 4H^+ \rightarrow 2AmO_2^{+2} + Am^{+3} + H_2O$$

and Am(VI) becomes gradually reduced by its own intense alpha radiation.

Curium, named after the Curies, was first made in a cyclotron by alpha bombardment of ^{239}Pu, but after americium became available in milligram amounts it was possible to produce curium by neutron irradiation of americium. A visible amount of the metal has been prepared by reduction of CmF_3 with barium. It is a silvery reactive metal which corrodes rapidly in air, probably because its own intense radioactivity heats the sample. Curium corresponds to gadolinium in the lanthanide rare-earth series, having half-filled 5*f* orbitals. Only the +3 oxidation state persists in solution, although CmF_4 can be made by dry fluorination. The absorption spectrum and magnetic properties of CmF_3 closely resemble those of GdF_3.

Berkelium, named after the city where it was discovered, has one more electron than curium and loses it readily to revert to the $5f^7$ electronic structure. It forms a yellow oxide, BkO_2.

Californium also follows rare-earth precedent in being named after its place of discovery. It was identified (along with berkelium) in debris from the first hydrogen bomb, where it was formed within an intense neutron flux that existed for only a few microseconds. Under these conditions the uranium or plutonium used to set off the H bomb undergoes a rapid succession of neutron captures, and isotopes of the heavy 5*f* elements form before the intermediates have a chance to decompose. They are separated by elegant ion-exchange techniques. It appears that californium, einsteinium, fermium, and mendelevium all adopt +3 as their only oxidation state in solution.

Very little is known about the chemistry of the last two members of the 5*f* series, nobelium and lawrencium. Einsteinium was named after Albert Einstein, of course, and fermium was named after Enrico Fermi, who first proved that nuclear fission could be self-sustaining. Lawrencium was named for E. O. Lawrence, who invented and developed the cyclotron used to prepare so many of the new elements.

There is a possibility that "islands of stability" will appear among the elements that follow the 5*f* series, so there is strong interest in carrying the synthesis of new heavy elements further. Even the short-lived transplutonium elements now on hand do have some limited utility: their high level of activity makes them good power sources for spacecraft instruments and heart pacemakers.

The physical properties of actinium and the fourteen 5*f* elements, insofar as they are known, are listed in Table 9–1.

TABLE 9–1 PHYSICAL PROPERTIES OF THE ACTINIDE RARE-EARTH ELEMENTS

Atomic Number	Symbol	Atomic Weight*	Melting Point, °C	Boiling Point, °C	Density g/cm³	State at Room Temperature
89	Ac	(227)	1050	3200	–	radioactive gray metal
90	Th	232.04	1750	4790	11.7	radioactive gray metal
91	Pa	(231)	1600	4200	15.4	radioactive gray metal
92	U	238.03	1132	3818	18.9	radioactive gray metal
93	Np	(237)	637	3902	20.5	radioactive gray metal
94	Pu	(242)	640	3235	19.7	radioactive gray metal
95	Am	(243)	995	2607	13.7	radioactive gray metal
96	Cm	(247)	1340	–	–	radioactive gray metal
97	Bk	(249)	–	–	–	?
98	Cf	(249)	–	–	–	?
99	Es	(254)	–	–	–	?
100	Fm	(253)	–	–	–	?
101	Md	(256)	–	–	–	?
102	No	(253)	–	–	–	?
103	Lr	(257)	–	–	–	?

*Atomic weights shown in parentheses are those of the most stable isotope.

SUMMARY

The uranium or actinide elements fill out a 5*f* series very much like the 4*f* rare earths, as shown by parallels in their oxidation states and their absorption spectra. All the isotopes of all the elements in this series are radioactive, with half-lives that shorten drastically as the series proceeds. All the elements beyond uranium are synthetic in the sense that they were first made by nuclear reactions conducted in accelerators or fission piles. (Neptunium and plutonium have since been found in trace amounts in uranium minerals, now that we know what to look for.) The preparation of one synthetic element from another, with yields of only 10^{-8} or 10^{-9} in each step because of the low probability of nuclear reactions, is a very difficult and expensive process; it would be self-defeating if it were not for the concurrent rapid development of instrumental methods that allow recognition of only a few atoms of the newest elements. Fortunately the decay characteristics of the 5*f* elements were found to follow a systematic pattern, so the emissions of the newest elements could be predicted and identified.

The +3 oxidation state which is characteristic of all the rare-earth elements is observed among all the 5*f* elements, too. However, the most stable oxidation state for thorium is +4, and for uranium it is +6. From there on, the most stable state declines to +4 for plutonium and +3 for americium. From californium on, the only oxidation state observed is +3. All the elements have small atomic radii in relation to their atomic weights, with consequent strong tendencies to covalency and coordination. The hexafluorides UF_6, NpF_6, and PuF_6 are all very volatile, and their stabilities decline in that order.

All other aspects of the 5*f* elements pale in comparison with the importance of uranium, thorium, and plutonium in nuclear fission and its applications. As the only source of nuclear energy that can be set up and controlled at will, fission energy offers an immediate but highly controversial alternative to the dwindling fossil fuels as sources of industrial and domestic energy. For this reason the fission reaction was explored in some detail, noting the many elements produced and the proportion of each. The requirements of any nuclear reactor for research or for power production were then explored, and some examples of reactor design were included to illustrate widely different types of construction. With the factual background for appreciating the problems involved, the arguments used by opponents and proponents of nuclear power were summarized.

Thorium is four times as abundant as uranium, and offers a chance to extend the supply of fissionable uranium by "breeding" ^{233}U from it. Plutonium is unique in its rate of production as a synthetic element, in the possibility of its use as a reactor fuel, and in the fantastic toxicity of the metal and all its compounds. Americium, curium, and californium are products of neutron exposure in reactors and have been explored not only for their chemical interest but also as compact sources of energy for isolated instruments. Attempts to make more new synthetic elements continue, in the hope of finding new areas of nuclear stability and better understanding of atomic structure.

GLOSSARY

critical: in the area of nuclear fission, an assembly of fissionable material and moderator "goes critical" at the exact point where enough neutrons are retained to sustain the chain reaction.

critical mass: the amount of fissionable material required to sustain a nuclear fission reaction under the conditions set up by the structure of the reactor.

criticality: The state of being at the critical condition in nuclear fission. Criticality requires a balance between the rates of fission, neutron capture, and neutron leakage.

isotropic: having the same properties in all directions. Annealed glass and unstrained cubic crystals are examples of isotropic materials. Every material which is not isotropic is anisotropic and will exhibit different physical properties in different directions.

microgram: 10^{-6} gram.

nanogram: 10^{-9} gram.

thermal neutrons: neutrons which are moving at velocities no greater than those imposed by thermal agitation at the specified temperature; neutrons with the lowest possible average kinetic energy at that temperature.

tracer method: the technique of following the reactions and fate of a radioactive isotope by following its radioactive emissions with appropriate instruments, rather than by observing it visually or gravimetrically.

QUESTIONS

1. If no information were given about UF_6, what could you predict about its physical properties and reactivity from the general principles outlined previously in this book?

2. Accepting the premise that uranium is part of a developing $5f$ series of elements, could you predict that UF_6 would or would not be a fluorinating agent? Why? Write sample equations to illustrate your answer.

3. Taking dimensions given in Figures 9–2 and 9–3, calculate the weights of steel needed for the containment vessels of the two reactor designs. Assume that the vessels are of pure iron, although they are of steel. With steel at 38c per kg, what are the relative costs?

4. The composition of the mineral pitchblende is sometimes given as U_3O_8. What is the oxidation state of uranium in pitchblende, according to this formula? Can there in fact be such an oxidation state? What does this tell us about the probable condition of uranium in this mineral? Are there any other analogous situations involving well-known minerals containing common metals?

5. If you saw something in an antique shop made of yellow-green glass, slightly fluorescent, how could you tell whether it was uranium glass without harming the object in the slightest?

6. If the supply of uranium ore were to become short, what would be done to supplement ^{235}U as a reactor fuel? Give equations and details, and summarize the advantages and disadvantages of any methods you propose.

7. In what ways do the $5f$ elements differ from the $4f$ elements? In what ways are they very similar?

8. List the necessary components for any fission reactor intended for research or power production, and draw a labeled sketch to illustrate the relation of these components.

9. Summarize the arguments for and against the large-scale generation of electric power through the use of nuclear energy, and outline your stand on the issue. Cite any references you have read on the matter.

THE NOBLE GASES

10

The zero group of elements, the noble gases helium, neon, argon, krypton, and xenon, were not envisioned in Mendeleev's time, so he made no provision for them in the Periodic Table. They came to light in a very literal sense with the development of spectroscopy at the end of the 19th century. Helium was the first to be discovered; its characteristic strong yellow lines were noted in the spectrum of the surface flares of the sun by Janssen in 1968 and later by Lockyer and Frankland. Then in 1882 Palmieri found the same lines in gases trapped in the lava from Vesuvius, and in 1895 Ramsay found helium in uranium minerals and was able to isolate it. It was found to be absolutely indifferent to chemical combination and to consist of monatomic molecules.

Once it was realized that helium was an earthly element that did not fit into the Periodic Table, a search began for other inert gases. In 1894 Rayleigh found that nitrogen obtained from the atmosphere was 0.5% heavier than nitrogen obtained by the decomposition of pure ammonium nitrite, and concluded that there must be a heavy impurity in atmospheric nitrogen. He found spectral lines of a new unreactive element which he isolated by distillation and named argon, meaning "the lazy one." It constitutes about 1% of the atmosphere by volume (1.3% by weight) and so is quite easy to obtain by distillation of liquid air. Further careful fractionation of air by Ramsay and Travers revealed the presence of three more inert gases: neon, "the new one" (0.00182% of air); krypton, "the hidden one" (0.00114% of air); and xenon, "the stranger" (0.0000087% of air). Ramsay classified them all with helium and the newly discovered inert but highly radioactive gas called radon (Pierre and Marie Curie, 1898), calling them the Zero Group of the Periodic Table. There they have had a profound influence on the development of the theory of atomic structure.

The gases of the zero group are all monatomic, odorless, tasteless, noninflammable, and nonoxidizing, with low boiling points and high ionization energies (see Table 10–1). Helium has been discussed in Chapter 2, and the others will be considered here in order.

Neon has an unusual combination of a stable filled-shell nucleus and a stable filled-orbital complement of electrons. It is correspondingly abundant on the cosmic scale, being as plentiful as silicon and 20 times as plentiful as sodium. However, with a molecular weight of only 20.17 and nearly zero attraction for other gas molecules, it has gradually been lost from our atmosphere. Enough is recovered during the production of nitrogen and oxygen by the distillation of liquid air to supply the only major use of neon, in gaseous-discharge lamps. Electrical excitation of neon at a pressure of 10 mm in a glass tube causes emission of its many orange, yellow, and pink spectral lines, readily observed in advertising signs and neon signal lamps with the aid of a simple hand spectroscope. The emission of neon light is efficient and economical in terms of lumens per watt or per dollar, and only its color keeps it from wider use. Neon is sometimes mixed with other noble gases to obtain special color effects.

Argon is only 1/77 as abundant as neon on the cosmic scale, but its higher molecular weight (39.95) has enabled the earth's gravitational field to hold on to some of it. It is separated in commercial quantities from air and used as a protective unreactive atmosphere for the arc welding of aluminum, as an inert gas to protect sensitive reagents in the laboratory, as the operating gas in lasers, and (most of all) as an unreactive low-pressure filler for incandescent lamps. Its function in lamps is to reduce the evaporation of tungsten from the very hot filament, as explained in Chapter 7.

Many attempts have been made to bring argon into chemical combination, always without success. It is trapped and held by some solids, notably ice, phenol (C_6H_5OH), and boron trifluoride, but apparently it exists in these solid "compounds" only as uncombined argon molecules that fit into the spaces left in the rather open crystal structures of H_2O, C_6H_5OH, and BF_3. The hydrates of argon, krypton, and xenon are the best known of these "compounds," with the following melting points and dissociation pressures:

	$Ar(H_2O)_x$	$Kr(H_2O)_x$	$Xe(H_2O)_x$
Melting point, °C	8	13	24
Dissociation pressure at 0°C, atm	98	15.5	1.3

These are prepared by freezing ice under high pressure of the noble gas, whereupon some of the gas is retained in the solid (but with the high escape pressures shown). The hydrates exist only in the solid phase. There is a general name for such solids in which one substance is trapped in the interstices of another substance as it freezes: they are called *clathrates*, from the Latin word which means to enclose by a lattice or by prison bars. Clathrates are obviously not compounds in the chemical sense, but they do have a limiting composition which is fixed by the number of appropriate crystal vacancies.

Krypton is a very rare element and has no important uses. It has been proposed as a superior gas for filling incandescent lamps because its higher molecular weight (83.8) would lead to less convective heat loss from the filament than with argon, while still retarding evaporation of the tungsten, by interfering with the free flight of tungsten atoms from the surface. Unfortunately there is not enough krypton for this purpose; what we do have would be too expensive to use. An isotope of krypton is one of the most plentiful species produced in the fission of uranium, but is radioactive and decays to cesium.

After it was discovered that xenon could actually be brought into chemical combination (see next section), the same techniques were applied to krypton with partial success. When krypton is mixed with fluorine and an electric discharge is passed through the mixture at low pressure, a white solid of the composition KrF_2 is formed. This sublimes at 0° C and decomposes to the elements at room temperature. It can be kept at low temperature in dry glass or in Teflon containers, but reacts vigorously with moisture. An addition or coordination compound of the composition $KrF_2 \cdot 2SbF_5$ has also been prepared, and is more stable than KrF_2 alone.

Xenon was the first of the Group 0 gases found to be capable of forming a series of true chemical compounds. The event destroyed a long-standing article of faith in chemistry concerning the inviolability of completed octets of electrons, and demoted the Group 0 elements from inert gases to noble gases.

This remarkable achievement began with the experiments of Neil Bartlett with PtF_6 in the early 1960's. Platinum is reluctant to assume an oxidation state of +6 and, when forced to do so by the action of elementary fluorine under vigorous conditions, the resulting PtF_6 is a very strong oxidizing agent and potent fluorinating agent. Bartlett noticed that it reacted with glass to give an orange-brown solid which he thought was a hydrolysis product, but careful experiments with pure oxygen showed that the orange substance was actually a compound O_2PtF_6, composed of O_2^+ ions and PtF_6^- ions. This meant that PtF_6 was an oxidizing agent sufficiently strong to tear an electron out of the O_2 molecule chemically, something previously not thought possible.

Since the first ionization potential of molecular oxygen is 313.9 kcal/per mole, and that of xenon is 280 kcal, Bartlett reasoned that PtF_6 might oxidize xenon. It did. He obtained yellow crystalline $XePtF_6$, and so set off an intensive search for more compounds. It was soon found that xenon combines directly with fluorine when the two gases are heated to 400° C under moderate pressure in a nickel capsule, forming XeF_2, XeF_4, and XeF_6, depending on the proportions. Now two dozen or more compounds are known with xenon in oxidation states of +2, +4, +6, and +8. The most important compounds can be classified according to the oxidation state of the xenon.

COMPOUNDS OF Xe(II): XeF_2 is a colorless crystalline compound which melts at 140° C and decomposes water:

$$2XeF_2 + 2H_2O \rightarrow 2Xe + 4HF + O_2$$

It forms a solid yellow addition compound, $XeF_2 \cdot 2SbF_5$, which melts at 63° C.

COMPOUNDS OF Xe(IV): XeF_4 is also a colorless crystalline solid, but more stable than XeF_2. It melts at 114° C and its heat of formation is −68 kcal/mole. There is also an oxyfluoride, $XeOF_2$, but it is unstable and poorly characterized.

COMPOUNDS OF Xe(VI): XeF_6 is obtained when a large excess of fluorine (20:1) is used at a pressure of 50 atm. It is also a colorless crystalline solid, and it melts at 47.7° C. Its heat of formation is −96 kcal/mole. It reacts with fluorides of the alkali metals to form salts such as $CsXeF_7$ and Cs_2XeF_8. The sodium salts of this type decompose at 100° C, but the cesium salts are stable to 400° C. Partial hydrolysis of XeF_6 gives an *oxide* of xenon:

$$XeF_6 + H_2O \rightarrow XeOF_4 + 2HF$$

$$XeOF_4 + 2H_2O \rightarrow XeO_3 + 4HF$$

This oxide is an endothermic compound ($\Delta H_f = +96$ kcal/mole) which explodes easily but can be used to make other compounds in water solution.

COMPOUNDS OF Xe(VIII): When strong bases are added to a water solution of XeO_3 the xenon disproportionates to the element and *perxenate* ions:

$$XeO_3 + OH^- \rightarrow HXeO_4^-$$

$$2HXeO_4^- + 2OH^- \rightarrow XeO_6^{-4} + Xe + O_2 + 2H_2O$$

Since two gases come off, the solution contains only salts of perxenic acid, H_4XeO_6. Insoluble stable perxenate salts, such as Na_4XeO_6 and Ba_2XeO_6, can be precipitated from this solution as hydrates. All are powerful oxidizing agents. A little potassium perxenate sprinkled into a dilute solution of manganese sulfate oxidizes the Mn^{+2} ions to purple permanganate at once:

$$8Mn^{+2} + 5XeO_6^{-4} + 4OH^- \rightarrow 8MnO_4^- + 5Xe + 2H_2O$$

From the considerable amount now known about the aqueous chemistry of xenon compounds, it seems that the reduction potential for $Xe^{+8} + 8e^- \rightarrow Xe$ is 4.8 volts in acid solution and 2.9 volts in alkaline solution. The fact that xenon has any aqueous chemistry at all is a big surprise, considering the "nobility" of the other Group 0 gases.

Radon is the gaseous product of the radioactive decay of all three natural heavy-metal decay series. The uranium series gives ^{222}Rn by way of ^{226}Ra, the thorium series gives ^{220}Rn by way of ^{224}Ra, which used to be called thorium X, and the actinium series gives ^{219}Rn by way of ^{223}Ra:

$$^{238}_{92}U \rightarrow ^{222}_{86}Rn + 4\alpha + 2\beta^-$$

$$^{232}_{90}Th \rightarrow ^{220}_{86}Rn + 3\alpha + 2\beta^-$$

$$^{235}_{92}U \rightarrow ^{219}_{86}Rn + 4\alpha + 2\beta^-$$

Of these isotopes, ^{222}Rn has the longest half-life, 3.82 days. With so short a half-life the radiation level associated with radon is extremely intense—it discolors glass, decomposes most compounds, and destroys living tissue. Under these circumstances it is difficult to try to prepare compounds and study them, but from what has been done so far it is believed that radon forms a fluoride, probably RnF_4.

Because of its intense emission of 5.49 meV alpha rays, radon is sometimes used to destroy cancer tissue. If a malignant tumor is well-defined and in an appropriate place, a slender glass tube in which a minute amount of radon has been sealed can be implanted surgically in the center of the tumor for a calculated time and then removed. The radiolysis products are gradually absorbed by the body.

SUMMARY

The elements of Group 0 are all monatomic gases with high ionization potentials and very low reactivity. Their boiling points are close to their melting points, and are all low. They are all rather rare, but helium, neon, and argon are recovered in commercial quantities for their own special uses. Helium is obtained from natural gas by distillation and purification over charcoal adsorbent; neon and argon are obtained during the distillation of liquid air. Neon is used in signs and signal lamps, and argon is used in incandescent lamps to retard evaporation of tungsten. All three light gases are used as protective atmospheres during welding and chemical operations.

Krypton and xenon have no practical uses, but are interesting because they form chemical compounds. In 1962 Bartlett showed that PtF_6 could oxidize xenon, thus opening the door to the preparation of the xenon fluorides XeF_2, XeF_4, and XeF_6. From these, various oxides and oxyfluorides can be made by hydrolysis. The oxides are explosive, but in water solution lead to stable products. In basic solution XeO_3 disproportionates to xenon and the perxenate ion, XeO_6^{-4}. Barium and sodium perxenates are stable in the dry state and are powerful oxidizing agents.

Radon results from the radioactive disintegration of ^{232}Th, ^{235}U, and ^{238}U. All its isotopes are intensely radioactive, and some use is made of their radiations in cancer therapy. Radon combines with fluorine, but the study of its products is hampered by the high level of radiation.

The physical properties of the zero-group noble gases are listed in Table 10–1. And, with that, our survey of the chemical elements and their compounds comes to an end. As promised, some of the elements have turned out to be intriguing, some are indispensable, most are helpful, and a few just seem to stand by idly. Some are even antagonistic, but not many. I hope that you have enjoyed the tour, and I hope that it has improved your understanding of the world in which we live.

GLOSSARY

clathrate: a solid substance in which molecules of one component (usually a gas) are enclosed and trapped within the crystal lattice of a second component. The first component cannot escape entirely unless the forces which hold together its surroundings are overcome, usually by melting the second component.

dissociation pressure: the pressure generated by a decomposition product evolved from a substance held at a fixed temperature in a closed container.

laser: an instrument for generating or amplifying coherent light; an acronym for *l*ight *a*mplification by *s*timulated *e*mission of *r*adiation.

TABLE 10–1 PHYSICAL PROPERTIES OF THE ELEMENTS OF GROUP ZERO

Atomic No.	*Symbol*	*Atomic Wt.*	*Melting Point, °C*	*Boiling Point, °C*	*Gas Density**	*Ionization Energy, kcal/mole*
2	He	4.0026	−272.3	−268.6	0.177	567
10	Ne	20.183	−248.7	−245.9	0.900	497
18	Ar	39.948	−189.2	−185.7	1.784	363
36	Kr	83.80	−156.6	−152.3	3.733	323
54	Xe	131.30	−111.9	−107.1	5.887	280
86	Rn	(222)*	−71	−61.8	9.73	248

*Gas density is given in grams per liter, not g/cm³. Atomic weight shown in parentheses here is that of the most stable isotope.

lumen: the unit of luminous flux, being the light emitted within a unit solid angle (1 steradian) by a point source having an intensity of one standard candle. The output of light sources is given in lumens.

radiolysis: the decomposition of a substance brought about by exposure to radiation of the required intensity. Example: the radiolytic decomposition of water into its elements by the action of the radiations within a nuclear fission reactor.

QUESTIONS

1. The ionization energies of all the available elements have been known for many years, and are listed in older textbooks of advanced inorganic chemistry. Hence it has been known for a long time that xenon has about the same ionization energy as oxygen. It has also been known since 1930 that fluorides of oxygen can be made. Why should it have taken until 1962 to find that xenon and fluorine unite to form solid compounds? List all the reasons you can think of, taking into account any interim developments in chemistry pointed out in these last two chapters.

2. Helium and argon are far more expensive than pure nitrogen, yet are used to supply a protective atmosphere during the arc welding of aluminum. Why not use nitrogen for this purpose?

3. Are the noble gases of Group 0 ideal gases? Give facts to support your answer. If you believe not all of them are ideal gases, point out which ones are, or which ones come closest to being ideal.

4. Underwater explorers who stay at depths of 100 feet or more for days or weeks breathe a mixture of four fifths helium and one fifth oxygen instead of air. This is done to avoid the dissolving of nitrogen in the blood under pressure, for it is release of dissolved nitrogen that causes painful or even fatal embolisms (bubbles in the blood vessels) upon decompression. Why is helium any better? Support your answer with figures from a chemical handbook.

5. Men who live underwater in an atmosphere of four fifths helium and one fifth oxygen complain of two things: they feel cold, and their voices sound high-pitched and odd, even difficult to understand. Is there any physical or chemical basis for these complaints, or are they psychological? Justify your answer.

6. Argon is cheaper and more plentiful than helium. Would not argon be a better choice than helium for the application of Questions 4 and 5? Explain why or why not. If it could be used, would it be subject to the same two complaints given in Question 5?

7. Write an unsigned report or criticism of this book, pointing out the chief areas for improvement.

APPENDIX: THE NAMING OF INORGANIC COMPOUNDS

There need be no mystery in the naming of ordinary inorganic compounds if a few general rules are understood. The rules outlined below are those adopted by the International Union of Pure and Applied Chemistry and approved by a committee of the American Chemical Society. This nomenclature system attempts to provide rules so that the name of a compound can always be derived from its formula, and a formula can always be written when the name is known.

I. Elements

The symbols for the chemical elements are given in Table 1–1, on the Periodic Table of Chapter 1, and also inside the back cover of this book.

II. Compounds of Two Elements

A. *Formulas* are written with the symbol of the more positive (metallic) element first, and suitable subscripts to show the atomic proportions: $CuCl$, $CuCl_2$, CF_4, H_2O.

B. *Names* are written in the same order, with Greek prefixes to indicate the proportions and with the ending *-ide:*

NO	nitrogen oxide
N_2O	dinitrogen monoxide
NO_2	nitrogen dioxide
NCl_3	nitrogen trichloride
N_2O_4	dinitrogen tetroxide
PCl_5	phosphorus pentachloride
SF_6	sulfur hexafluoride
FeS_2	iron disulfide

Other prefixes: 7 hepta, 8 octa, 9 ennea, 10 deca, 11 undeca, 12 dodeca. *Mono* is usually omitted unless it should be emphasized.

C. In speech, either the formula or name may be used. When there is little possibility of error, or when the elements have one fixed combining capacity (a single oxidation state), the prefixes may be omitted:

$NaCl$	sodium chloride
CaO	calcium oxide
$CaCl_2$	calcium chloride
AlF_3	aluminum fluoride

D. For elements with variable combining capacity (with more than one oxidation state), the oxidation state is shown by the Greek prefixes in the names, or by Roman numerals placed after the name:

$FeCl_2$, FeO, and $Fe^{+2}\ X^{-2}$ are iron (II) compounds.
$FeCl_3$, Fe_2O_3, and $Fe^{+3}\ X^{-3}$ are iron (III) compounds,
where X is any appropriate element.

E. Common or trivial names are retained for some common substances like water and ammonia, but all other compounds are named by the general method.

III. Compounds of Three or More Elements

A. *Formulas* and *names* follow the same order as above with the symbol of the most positive (most metallic) element placed first. Complex ions having special names are treated like elements:

NH_4Cl	ammonium chloride
$CaCO_3$	calcium carbonate
$NaHCO_3$	sodium hydrogen carbonate (not sodium bicarbonate)
$LiAlSi_2O_6$	lithium aluminum disilicate
$Fe(OCN)_3$	iron (III) cyanate
$Fe(SCN)_3$	iron (III) thiocyanate

B. Acids containing oxygen are given particular names:

H_2SO_3	sulfurous acid
H_2SO_4	sulfuric acid
$H_2S_2O_3$	thiosulfuric acid (*thio* comes from the Greek θειοω, meaning sulfur)
H_2SO_5	peroxy(mono)sulfuric acid
$H_2S_2O_8$	peroxydisulfuric acid
HNO_2	nitrous acid
HNO_3	nitric acid

IV. Coordination compounds with complex ions as the negative ions use the names of the complex ions as modifiers, and sometimes use the Latin names of metals for euphony:

$K_2[PtCl_6]$	potassium hexachloroplatinate
$K_4[Fe(CN)_6]$	potassium hexacyanoferrate(II)
$K_3[Fe(CN)_6]$	potassium hexacyanoferrate(III) or tripotassium hexacyanoferrate

For hydrates and ammoniates:

$CaCl_2(H_2O)_6$	calcium chloride hexahydrate
$AlCl_3(NH_3)$	aluminum chloride ammoniate
$[Cr(NH_3)_6]Cl_3$	hexamminechromium (III) chloride

V. For naming other classes of compounds not mentioned in this outline, see "Rules for Naming Inorganic Compounds," *Journal of the American Chemical Society,* **63**:889 (1941); or *Nomenclature of Inorganic Chemistry,* a report of the International Union of Pure and Applied Chemistry, London: Butterworths (1959).

AUTHOR INDEX

SUBJECT INDEX

Page numbers in *italics* indicate illustrations; page numbers followed by (t) indicate tables; (f) indicates footnote.

Please insert this complete Table of Atomic Weights
on the inside of the back cover.

TABLE OF ATOMIC WEIGHTS (Based on Carbon-12 = 12 amu)

	Symbol	Atomic No.	Atomic Weight
Actinium	Ac	89	(227)
Aluminum	Al	13	26.98154
Americium	Am	95	(243)
Antimony	Sb	51	121.75
Argon	Ar	18	39.948
Arsenic	As	33	74.9216
Astatine	At	85	(210)
Barium	Ba	56	137.34
Berkelium	Bk	97	(247)
Beryllium	Be	4	9.01218
Bismuth	Bi	83	208.9804
Boron	B	5	10.81
Bromine	Br	35	79.904
Cadmium	Cd	48	112.40
Calcium	Ca	20	40.08
Californium	Cf	98	(251)
Carbon	C	6	12.011
Cerium	Ce	58	140.12
Cesium	Cs	55	132.9055
Chlorine	Cl	17	35.453
Chromium	Cr	24	51.996
Cobalt	Co	27	58.9332
Copper	Cu	29	63.546
Curium	Cm	96	(247)
Dysprosium	Dy	66	162.50
Einsteinium	Es	99	(254)
Erbium	Er	68	167.26
Europium	Eu	63	151.96
Fermium	Fm	100	(253)
Fluorine	F	9	18.99840
Francium	Fr	87	(223)
Gadolinium	Gd	64	157.25
Gallium	Ga	31	69.72
Germanium	Ge	32	72.59
Gold	Au	79	196.9665
Hafnium	Hf	72	178.49
Helium	He	2	4.00260
Holmium	Ho	67	164.9340
Hydrogen	H	1	1.0079
Indium	In	49	114.82
Iodine	I	53	126.9045
Iridium	Ir	77	192.22
Iron	Fe	26	55.847
Krypton	Kr	36	83.80
Lanthanum	La	57	138.9055
Lawrencium	Lr	103	(257)
Lead	Pb	82	207.2
Lithium	Li	3	6.941
Lutetium	Lu	71	174.97
Magnesium	Mg	12	24.305
Manganese	Mn	25	54.9380
Mendelevium	Md	101	(256)
Mercury	Hg	80	200.59
Molybdenum	Mo	42	95.94
Neodymium	Nd	60	144.24
Neon	Ne	10	20.179
Neptunium	Np	93	237.0482
Nickel	Ni	28	58.71
Niobium	Nb	41	92.9064
Nitrogen	N	7	14.0067
Nobelium	No	102	(254)
Osmium	Os	76	190.2
Oxygen	O	8	15.9994
Palladium	Pd	46	106.4
Phosphorus	P	15	30.97376
Platinum	Pt	78	195.09
Plutonium	Pu	94	(242)
Polonium	Po	84	(210)
Potassium	K	19	39.098
Praseodymium	Pr	59	140.9077
Promethium	Pm	61	(147)
Protactinium	Pa	91	231.0359
Radium	Ra	88	226.0254
Radon	Rn	86	(222)
Rhenium	Re	75	186.207
Rhodium	Rh	45	102.9055
Rubidium	Rb	37	85.4678
Ruthenium	Ru	44	101.07
Samarium	Sm	62	150.4
Scandium	Sc	21	44.9559
Selenium	Se	34	78.96
Silicon	Si	14	28.086
Silver	Ag	47	107.868
Sodium	Na	11	22.98977
Strontium	Sr	38	87.62
Sulfur	S	16	32.06
Tantalum	Ta	73	180.947
Technetium	Tc	43	(99)
Tellurium	Te	52	127.60
Terbium	Tb	65	158.9254
Thallium	Tl	81	204.37
Thorium	Th	90	232.0381
Thulium	Tm	69	168.9342
Tin	Sn	50	118.69
Titanium	Ti	22	47.90
Tungsten	W	74	183.85
Uranium	U	92	238.029
Vanadium	V	23	50.9414
Xenon	Xe	54	131.30
Ytterbium	Yb	70	173.04
Yttrium	Y	39	88.9059
Zinc	Zn	30	65.38
Zirconium	Zr	40	91.22

*Values in parentheses are estimates and denote, in most cases, the most stable isotopes.

Note: There is general agreement that elements 104, 105, and 106 have been prepared but there is disagreement concerning who prepared them first and what the names should be. See *Chemical and Engineering News,* Sept. 16, 1974, p. 4.